Endangered and Threatened Wildlife of Michigan

David C. Evers, Editor

CONTRIBUTORS
Dennis Albert
John B. Burch
David C. Evers
David N. Ewert
James H. Harding
Larry Master
Mogens C. Nielsen
Gerald Smith
Leni Wilsmann

Ann Arbor

THE UNIVERSITY OF MICHIGAN PRESS

For Laura and Nicole

Library of Congress Cataloging-in-Publication Data

Endangered and threatened wildlife of Michigan / David C. Evers,
   editor ; contributors, Dennis Albert . . . [et al.].
      p.   cm.
   Includes bibliographical references (p.   ) and index.
   ISBN 0-472-10367-9 (alk. paper)
   1. Endangered species—Michigan. 2. Zoology—Michigan.
I. Evers, David C., 1962–   . II. Albert, Dennis A.
QL84.22.M5E63   1994
591.52'9'09774—dc20   94-9032
                                                          CIP

Technical illustrations by Laura Evers, Mary Evers, Amelia Hansen, and Gerry Wykes.

Acknowledgments

In December, 1983, Judy Soule (at the time zoologist with the Michigan Natural Features Inventory) helped me initiate a short publication with the Michigan Department of Natural Resources' Nongame Wildlife Fund. The goal was to develop brief accounts of Michigan's endangered and threatened animals. Thanks to cheerful encouragement and inspiration from Soule and the rest of the staff, the framework for this cause was established. Several subsequent grants from the Nongame Wildlife Fund kept this long-term project going.

Over the next ten years, this book became larger and more encompassing. More and more species were added as surveys further defined their actual status. The interdisciplinary scope of this work required the expert technical assistance of amateur and professional people. The accuracy of information for each species account was of foremost importance and, because of the dedicated efforts of over 80 people, the content has been thoroughly and tediously examined. Many of the people responsible for this behind-the-scenes research deserve high praise. A special thanks goes to Robert Hess and Thomas Weise for ensuring the final push for publication. Tom reviewed the final manuscript in its entirety.

Early drafts of the material on mammals were reviewed by Rollin Baker, Ralph Bailey, Chris Carmichael, James Cope, Dan Cristol, Robert Downing, Rich Earle, Jim Hammill, John Hendrickson, Allen Kurta, John Lerg, Jim Ludwig, Larry Master, Gail McPeek, David Mech, Jim Paruk, Rolf Peterson, John Stuht, Dick Thiel, Merlin Tuttle, and Mike Zuidema. The material on birds was first reviewed by Raymond Adams, Jr., Tom Allen, William Anderson, Don Beaver, Glenn Belyea, Michael Benuc, Tom Cade, Dan Cristol, Michael DeCapita, James Dinsmore, Bert Ebbers, Susan Haig, Jim Hammill, John Hendrickson, Bill Irvine, Marge Kolar, Anne Lambert, John Lerg, Jim Ludwig, Janea Little, Al Maley, Larry Master, Harold Mayfield, Judy McIntyre, Gail McPeek, T. J. Miller, Jim Paruk, Robert Payne, Greg Petersen, Ed Pike, Sergej Postupalsky, Harold Prince, Bob Pacific, Mary Rabe, Bill Robinson, Rick Sawicki, Bill Scharf, Bob Seppala, William Southern, Sylvia Taylor, John Urbain, Lawrence Walkinshaw, and Terry Wiens. The material on reptiles and amphibians was initially reviewed by Jim Harding, Max Hensley, and Mark Sellers, and the material on fish was reviewed by Tom Doyle, Carl Latta, Gerry Smith, and Tom Todd.

Extensive updates of the earlier drafts necessitated one final inspection. The following people generously contributed their time and expertise, and I am grateful for their efforts: Lowell Getz (least shrew and prairie vole), Merlin Tuttle (Indiana bat), Rolf Peterson (gray wolf), Bill Berg (cougar and lynx), Rich Earle (marten), Judy McIntyre (common loon), Cortez Austin (least bittern), Joe Johnson (trumpeter swan), Sergej Postupalsky (osprey, bald eagle, and red-shouldered hawk), Alan Poole (osprey), Dave Drummond (merlin), Pat Redig (peregrine falcon), Ted Bookhout (yellow rail), Brooke Meanley (king rail), Susan Haig (piping plover), Francesca Cuthbert (Caspian tern), Bill Scharf (common tern), Bruce Colvin (barn owl), Richard Clark (short-eared owl), Tom Carpenter (long-eared owl), Bonnie Brooks (loggerhead shrike), Paul Sykes, Jr. (yellow-throated warbler), Jerry Weinrich (Kirtland's warbler), Val Nolan (prairie warbler), John Zimmerman (lark sparrow), Jim Harding (reptiles and amphibians), and Dan Rice and Gerry Smith (fish).

Introductions discussing Michigan's landscape and each group of animals were written by authorities in the respective fields. Their acceptance of the task of providing a thorough, yet condensed overview, attests to their strong conservation ethics. I am extremely grateful for their contributions to this project. These contributing authors include Dennis Albert (Landscape), Larry Master (Mammals), David Ewert (Birds), Jim Harding (Reptiles and Amphibians), Gerry Smith (Fish), Mo Nielsen (Insects), and John Burch (Mollusks).

Leni Wilsmann and John Burch wrote the species accounts for the insects and mollusks, respectively. Their expertise with these groups precluded outside efforts.

Baseline information on the historical and recent distribution of each species was primarily provided by the Michigan Natural Features Inventory (now part of the Michigan Natural Heritage Program). Its role was instrumental in the process of researching and confirming museum and other records. This is the first comprehensive work to publish their original data. Museum records on the herps and especially the fish, were painstakingly compiled by Larry Master. The map for the lake herring was drawn after a statewide questionnaire to all fishery district offices. Mo Nielsen and Wayne Miller unselfishly provided their personal county records of lepidopteran specimens. The Michigan Natural Features Inventory, with the help of Jim Best, filled in gaps in the insect maps and provided all the data for the five nonlepidopteran insects. Rollin Baker's publication "Michigan Mammals" provided historical data on the state's mammal ranges; correspondence with Department of Natural Resource biologists provided current maps. Nearly all the information for the bird maps came directly from the Michigan Breeding Bird Atlas, thanks to Gail McPeek and Ray Adams, Jr. Maps were not made for the mollusks; however, John Burch provided original range information for each species account.

The line-art was skillfully accomplished by four artists; although I have relayed my deep appreciation many times, I once again would like to convey how obliged I am to them for their contributions. Laura, my wife, pursued this project through its entirety, drawing at every free moment. Somehow she unfailingly stayed with the project's goals and sketched the mammals, most birds, and fish. The excellent sketches of reptiles and amphibians by Amelia Hansen speak for her talent. I remember when Gerry Wykes called me one day and voluntarily offered to draw. At the time, I could never have imagined how important his high-quality work would be to this cause. He contributed sketches for the mussels, snails, and nonlepidopteran insects. I am lucky to have a sister-in-law with a fine arts degree: Mary Evers contributed her artistic skills with her drawings of the birds and all the lepidoptera. Gjisbert Frankenhuzen provided critical comments on most of the sketches over the years and encouraged Laura in the early stages.

Most sketches were developed from a photo or specimen used as a reference. Museum loans from the University of Michigan and Michigan State University were immeasurably useful, particularly some uncommon insects that Mark O'Brien was able to provide. Wayne Miller and Mo Nielsen kindly loaned several lepidopteran specimens to be used as sketch templates. Dan Rice helped by supplying many photos of fish in their natural habitat. Nearly all the mollusks were drawn from specimens provided by the University of Michigan.

Phone and mail correspondance and copying costs were substantial and were mostly covered by the Avian Research Program of the Kalamazoo Nature Center. This is no surprise, because the Center always has been instrumental in state conservation efforts. The Center's vision was established by the executive director, H. Lewis Batts, Jr., and continues today with the new director, Willard Rose. Gail McPeek graciously reviewed much of the book and provided many helpful and instrumental comments. Her unselfish style is greatly appreciated and shows her professionalism and caring attitude toward this and all the conservation causes in which she participates. Sue Smith and Mark Betz served as word processors. SSG Laserworks graciously provided computer logistical support. I would also like to thank the Nature Center, its staff, and particularly Ray Adams for seeing this project through to completion. Without their emotional and logistical support it would not have been possible to complete this book.

Under the direction of the University of Michigan Press, Matthew Douglas reviewed the entire finished manuscript and made sure each of the variously authored introductions flowed together. The rest of the University of Michigan Press staff worked hard to insure the completion of this high-quality publication.

Of course, this book would not have been written without the tremendous support of my family and friends. I thank them all, especially Laura, for caring and understanding my obsession. We have lived with this project since we have been married; it is a tremendous relief to finally have it completed.

Finally, I am not sure of my source of motivation and inspiration, but I now realize who helped form my deep environmental convictions when I was growing up in Roscommon. Thank you, Don Fenton, for teaching me the names of all those shrubs!

Contents

Introduction

David C. Evers

Global warming, the greenhouse effect, tropical forest destruction, nuclear war, desertification, ozone depletion, mass extinction, acidification: all these are catastrophic environmental events that potentially affect us worldwide. On a smaller and shorter space-time scale, degradation of natural plant communities, pollution, soil erosion and river siltation, chemical contamination, and threats from human activity have a more personal effect on our natural surroundings.

These environmental threats, no matter how local or difficult to identify, modify the living earth that supports us. One measure of the earth's health is the survival of its flora and fauna. If landscapes, ecosystems, and plant communities are intact and functioning within their natural dynamic structure, all earth's inhabitants benefit.

The investigation of biological diversity is one method of monitoring the earth's health. Each species of plant and animal and their interrelationships are important to their ecosystem. Removal of one or several species may be ecologically imperceptible; however, it could start a chain of events that ends with unnatural chaos and causes major ecological perturbations.

The monitoring of endangered and threatened species serves to measure changing biological diversity. Species identified as endangered and threatened are typically those most sensitive to environmental disturbances caused by human activity. For our purpose, we need to establish a timeline to determine which species are of concern, because, although extinction is a natural process, the present accelerated rate of this process is not. Extinction caused by human activity probably began with the arrival of Paleo-Indian hunters following the end of the last glacial period, 10,000 to 11,000 years ago. Although there is some evidence for this impact, European settlement (beginning about 350 years ago) has unquestionably changed our natural environment. This is the point in time when we should begin the reconstruction of our historical faunal diversity and compare it with our present diversity.

Much of Michigan's former flora and fauna was vastly different from today's. The Lower Peninsula's extensive upland prairies and oak savannas were inhabited by grunting herds of bison and a host of field and forest edge species such as the lark sparrow, loggerhead shrike, and regal fritillary butterfly. The oak-hickory forests were filled with massive flocks of passenger pigeons, and the boreal forests were home to caribou and wandering solitary wolverines. Trumpeter swans probably nested in the extensive emergent wetlands, and certain inland rivers were teeming with Arctic grayling. In the Great Lakes, ciscoes dominated the faunal assemblage and lake sturgeon were abundant.

These and other fauna were integral components of Michigan until Europeans settled the state in large numbers. The plowing under of grassy openings, draining of wetlands, logging of forests, and exploiting of the Great Lakes fisheries completely altered Michigan's presettlement landscape and waters. Opportunistic species invaded the newly created habitats, while biologically less flexible species disappeared or were reduced in number.

The result of these environmental disturbances brought about by European settlement is the current, official state listing of 97 endangered, threatened, extirpated, and extinct species of vertebrates and invertebrates. An *endangered species* is one that is threatened with extinction throughout all or a significant portion of its Michigan range. A *threatened species* is one that is likely to become endangered within the foreseeable future throughout all or a significant portion of its Michigan range. An *extirpated species* is one that is lost from a region or locale, but still exists in other parts of its range or in captivity, whereas an *extinct species* is one that no longer exists anywhere. We now have a working timeline and a list of endangered and threatened species with which we can guide the process of restoration.

The following species accounts describe and discuss each of these 97 animals and provides information on their current and historical distribution, habitat requirements, ecology and life history, past and current limiting factors, and the conservation measures that can be applied for their recovery. Each of the endangered and threatened animals are accompanied by a line drawing and a county dot map (except mollusk accounts, which have no maps) showing historical and recent distribution. Although not officially recognized by the state list, three other species are included in this book: one is an extinct subspecies of mammal (eastern elk), one is a federally endangered bird that may migrate through Michigan (Eskimo curlew), and one is a federally endangered bird that may have once nested in Michigan and currently is being considered for reintroduction into the central Upper Peninsula (whooping crane).

A species-by-species conservation effort is crucial for the survival of many endangered and threatened animals. However, the preservation of intact plant communities and aquatic ecosystems is the best long-term ecological solution for ensuring the survival of these listed species. In Michigan, plant communities are divided into 16 types; 9 are upland and wetland forest, 3 are nonforested uplands, and 4 are nonforested wetlands. These plant communities and the factors that determine their composition are described in the first chapter, "Michigan's Landscape."

Since European settlement, a total of 68 mammals are known from Michigan; 4 species are endangered, 3 are threatened, and 3 are extirpated. No species are extinct; however, a subspecies of the elk that formerly occurred in Michigan is gone forever. Nearly 15% of Michigan's mammals are officially listed. Direct human persecution is responsible for the precarious status of the gray wolf, mountain lion, and lynx, and the extirpation of the caribou, wolverine, bison, and marten (which is now reestablished due to successful restoration efforts). Similar attempts coupled with conservation measures are needed for recovery of the other species.

The number of birds known to have nested in Michigan is not known; an approximate number is 238 (about 400 species are confirmed in state records). Birds are well studied; for example, intensive statewide surveys from 1983–88 by the Michigan Breeding Bird Atlas confirmed only 3 new breeding species. Yet, despite their popularity, many birds are rare or gone. Seven species are endangered, 12 are threatened, 4 are extirpated, and 1, the passenger pigeon, is extinct. Two other federally listed species are included, even though they are not officially state listed. Recent increases in the abundance of the whooping crane and Eskimo curlew permit their inclusion here to alert Michigan observers. All told, 10% of Michigan's native nesting birds are officially listed.

Most of the birds are rare because of habitat degradation; for example, less than 1% of the original prairie and oak savanna plant communities remain. In turn, this severely limits the population of birds that depend on those habitats. Market hunting has had historical effects on several species, such as the piping plover and common tern, and has contributed heavily to the extirpation of the trumpeter swan and the passenger pigeon's extinction. Today, intense land-use patterns in the southern Lower Peninsula, continuing wetland loss, chemical environmental contamination, and the overuse of our natural resources for recreational purposes are threatening the survival of the 19 listed survivors.

Our impact on reptiles and amphibians is not clear, mainly because there are few historical baseline studies to compare with today's research. There are 28 species of reptiles and 23 species of amphibians with known established populations in Michigan. Two snakes and one salamander are endangered and one snake and one salamander are threatened. This means that 11% of our native reptiles and 9% of the amphibians are listed. Several other herpetofauna populations also are in danger from the same threats affecting the listed species: encroachment by human-development on selected habitats, chemical contamination (possibly acidification), and indiscriminate collecting.

Michigan's fish fauna totals 159 species; 136 are native, 21 are introduced, and 2 entered the Great Lakes through artificial canals. Of this total, 6 are endangered, 13 threatened, 2 extirpated, and 3 species and 1 subspecies are extinct. Therefore, nearly 18% of Michigan's native fish fauna are listed. The Great Lakes and southeastern Lower Peninsula fish faunas have experienced the greatest reduction in their diversity. Commercial fishery pressures and the introduction of exotic fish have caused 6 of the 8 ciscoes to disappear from much of their Great Lakes range. Three ciscoes are now extinct. Siltation, caused by deforestation and degradation of riparian zones, has severely limited the survival of several minnow, shiner, and darter species.

New emphasis on our insect diversity has defined a list of 20 rare and declining species: 7 are endangered and 13 are threatened. Most research has been done on the lepidoptera, but further investigations will undoubtedly add more species from other orders. Habitat loss and alteration are the primary culprits for the rare status of many species. Although the exact number of insects is unknown in Michigan (ranges between 15,000 and 20,000), the diversity of some orders has been studied. For example, approximately 1,300 moths and 158 butterflies (135 known to breed, 8% are listed) are confirmed in Michigan.

Michigan's freshwater mussel fauna (family Unionidae) is the most endangered group of organisms in the state. Of the 46 species, 10 or 22% are listed. Urban sprawl around the Detroit River, western Lake Erie, and their tributaries coincides with the center of this family's local distribution. The 4 listed snails, 2 endangered and 2 threatened, generally are restricted to a single lake or small area and represent 2% of the listed species in this group.

Box Score Listings

Animal Class	Total Number of Native Species	Total Number of Listed Species	Percentage of Listed Species
Mammals	68	10	15%
Birds	238*	23	10%
Reptiles	28	3	11%
Amphibians	23	2	9%
Fish	136	24	18%
Insects	15,000–20,000	20	<1%
(Lepidoptera)	(135*)	(11)	(8%)
Snails and slugs	180	4	2%
Freshwater mussels	46	10	22%

*confirmed breeding species

Extinction should be a major concern to all of us. Some authorities estimate that one-fifth of all living species will disappear by the turn of the century. The effect of this mass extinction cannot be forecast; however, it will likely have catastrophic consequences on all the world's ecosystems as well as the survival of human society. For the first time in the earth's history there is a creature that consciously and deliberately causes the extinction of other species. Our aggressive and self-centered actions are not compatible with our small world, a place with finite resources. We all need to lean back, contemplate, and strive to understand and accept our interrelationship with all the earth's animals and plants. Diversity is the sustenance of life, and we are but one link in its interrelated and all-encompassing web.

Michigan's Landscape

Dennis Albert
Michigan Natural Heritage Program

Approximately 70% of the borders of Michigan's Upper and Lower peninsulas are on the Great Lakes (fig. 1). The area of the entire state is 97,107 square miles (251,507 square kilometers), of which 56,959 square miles (147,524 square kilometers) are land (Rand McNally 1982). Most of the remaining 40,148 square miles (103,983 square kilometers) consist of Great Lakes. The Lower Peninsula comprises about 72% of the land area and the Upper Peninsula the remaining 28%.

Understanding the distribution and abundance of the present fauna and flora of Michigan requires an understanding of their relationships to both the existing natural framework of the state and also the human management history of the state, beginning with the native Americans and continuing to the present.

Human Land Use History

Native American Land Use

Michigan was occupied by several native American tribes prior to settlement by Europeans, and their subsistence patterns varied markedly according to environmental conditions. Due to climate limitations, intensive agriculture was limited to the southern half of the Lower Peninsula and localized areas of northern lower Michigan, along Lake Michigan, and the Inland Water Route (Tanner 1987; Trigger 1978). The droughty high-plains area of the northern Lower Peninsula was an area used almost exclusively for hunting. Some agriculture was practiced in the rest of the northern Lower Peninsula and along the northern shore of Lake Michigan in the Upper Peninsula, but in combination with hunting and fishing. Hunting and cultivation of wild rice (*Zizania aquatica*) on small lakes and marshes were the primary form of subsistence in most of the Upper Peninsula. Fisheries were an important natural resource in the northern Lower Peninsula and in the eastern Upper Peninsula, with the rapids of the St. Marys River being an exceptional fishing site (Tanner 1987).

Changes in the flora and fauna could be expected where native American land use was most intensive; widespread dominance of red oak (*Quercus rubra*) along many inland lakes may be an example of such agricultural influence (Albert and Minc 1986). As a result of wildfires and native American fire management, vast expanses of landscape in the southern half of the Lower Peninsula were maintained

5

Fig. 1. Map of Michigan, showing Great Lakes and counties

as oak savanna. The savannas converted to closed-canopy oak forest when fire was suppressed by European settlers.

European Land Use

The earliest European settlers, the French, established trading posts at Michili-mackinac (now Mackinaw City) as early as 1640 (Tanner 1987) and developed an extensive fur trade that greatly depleted the beaver (*Castor canadensis*) popula-tions. At least two mammal species, the fisher (*Martes pennanti*) and marten (*Martes americana*) were extirpated from the state as a result of the intensity of the fur trade.

In 1787, the Northwest Territory, which included Michigan, was created by the Continental Congress (Jamison 1945). Settlement of Michigan began as soon as its lands were surveyed; township surveys began in southeastern Michigan in 1815 and continued into the 1850s in the western Upper Peninsula. As township surveys were finished, the land became available for purchase. The State Land Office was established in 1843 to dispose of the 12.0 million acres of land under state jurisdiction, and by 1890 all but 0.5 million acres were privately owned.

Logging, the first major land use following the fur trade, greatly altered the flora and fauna of the state. At first, the lumber industry concentrated on har-vesting the vast pine forests along major rivers in the Lower Peninsula. Among the heavily logged areas was the Saginaw basin, where timbering began in the early 1830s and continued into the 1880s (Bacig and Thompson 1982). Beginning about 1870, the introduction of narrow-gauge logging railroads allowed pineries isolated from the river to be harvested. Between 1873 and the early 1900s, most of the eastern white pine (*Pinus strobus*) and red pine (*Pinus resinosa*) stands of both the northern Lower Peninsula and the Upper Peninsula were cut. Following logging, slash fires swept repeatedly through the landscape, leaving extensive areas of the state without forest cover. Fires often destroyed pine seedlings, largely eliminating pines from many of the state's second-growth forests.

Following the harvesting of white and red pine, later cuttings removed other conifer species, including eastern hemlock (*Tsuga canadensis*), northern white ce-dar (*Thuja occidentalis*), white spruce (*Picea glauca*), and balsam fir (*Abies bal-samea*) and eventually hardwoods including sugar maple (*Acer saccharum*), bass-wood (*Tilia americana*), and white ash (*Fraxinus americana*). By the late 1930s almost all of Michigan's forests were cut; many of the remaining stands of old-growth timber have since been incorporated within the state's system of state parks and natural areas, the U.S. Forest Service's wilderness areas or research natural areas, or private nature preserves.

The impact of this intensive logging upon the state's fauna was not well documented. It was probably responsible for eliminating wolves from the state and greatly reducing the populations of many other large mammals. Logging also ef-fected wildlife through slash fires, river logjams, and increased siltation in streams. The population of at least one large mammal, the white-tailed deer (*Odocoileus virginianus*), increased dramatically with the cutting of Michigan's forest, to the extent that it negatively impacted reproduction of both eastern hemlock and northern white-cedar.

Farming by early settlers from Europe during the early 1800s also dramatically impacted forest wildlife by destroying the forests of the southern Lower Peninsula. Fine-textured soils (loams) were settled and cleared early. Most of these lands had supported dense forests, most of which were eliminated by burning. Although much of the swamp land was not cultivated initially, poorly drained sites with clay and loam soils were almost completely converted to agriculture as early as 1900 through the construction of extensive county drain systems. Many poorly drained organic soils over sand were not farmed, because drainage of these soils resulted in severe wind erosion.

Agricultural settlement began immediately after logging in the northern part of the Lower Peninsula, even on the most drought-prone sandy lands, but most of these lands were soon abandoned and soon reverted to forest. Erosion, low productivity, and extreme climatic conditions limited utilization of organic soils for agriculture in the north.

Where agriculture has been successful, the native flora and fauna have largely been replaced by alien species, both cultivars and weeds. Many remaining areas of undisturbed or recovering natural habitat may be too small to support a full complement of native fauna. Agriculture indirectly contributed to the problems of the endangered Kirtland's warbler (*Dendroica kirtlandii*); its nests are parasitized by the brown-headed cowbird (*Molothrus ater*), a bird not native to Michigan before European settlement. The brown-headed cowbird extended its range from the western prairies into Michigan when fields were plowed and forests were cleared.

Prairies and savannas were among the first lands farmed by European settlers, and, for this reason, surviving prairie and savanna remnants are few and typically quite small. The best remaining examples of these are often narrow bands persisting along railroad rights-of-way, most too small to support the fauna associated with these communities.

Agriculture and settlement have also greatly impacted our wetlands, lakes, and streams. Siltation of streams and lakes has eliminated several fish (Trautman 1981) and mollusks. Further degradation has resulted from the steady runoff of fertilizers and pesticides into our streams.

In the early 1900s, heavy industrialization caused large-scale air and water pollution. This problem escalated until the eutrophication and so-called death of Lake Erie in the 1960s from pollution. A threat to both wildlife and humans, is the introduction into our environment of invisible, odorless, toxic chemicals as by-products of Michigan's diversified manufacturing.

Residential and industrial expansion continues to destroy wildlife habitat. Wetlands, some of the last remaining wildlands in the southern part of the state, are being filled at an alarming rate. Alteration of remaining wetlands by road construction and increased runoff has reduced their usefulness for wildlife.

Public Lands

Public lands provide the majority of remaining habitat for Michigan's native flora and fauna. In southern Michigan, many tracts of public lands were acquired by tax forfeit of poor farm lands; these lands presently comprise a large part of our State Game Areas and Recreation Areas. Michigan has 293,361 acres (118,721 hectares) of state game area or state wildlife area, and 65,890 acres (26,665 hectares) of state recreation area (16 areas).

State forests were similarly formed from lands abandoned in the northern part of the state following logging. Currently, Michigan has 3,864,951 acres (1,564,124 hectares) of state forest lands. Michigan also contains nearly 3 million acres (1.2 million hectares) of National Forest land; 928,221 acres (375,646 hectares) in the Ottawa National Forest, 879,600 acres (355,969 hectares) in the Hiawatha National Forest, and 950,000 acres (384,460 hectares) in the Huron-Manistee National Forest. State parks, totaling approximately 251,000 acres (101,578 hectares) in 89 parks, account for the remainder of the large tracts of wildlands. In the southern half of the Lower Peninsula, state game areas, state parks, and state recreation areas account for almost all of the remaining large forested or wetland tracts available to wildlife.

Abiotic Framework

Regional Landscape Ecosystems

Biogeographers have long recognized several factors influencing the distribution of plants and animals. Among the most important of these factors are climate, bedrock geology, landform, and soil composition. The distribution of many animals is linked directly to either a specific species of plant or to a plant community, thus linking animal distribution to these same factors.

Using the factors mentioned above, a Regional Landscape Ecosystem Classification of Michigan (Albert et al. 1986) was developed to provide a regional framework for studying and managing the biota of the state. Individual mapping units within this statewide classification delineate relatively homogeneous areas of climate, bedrock, geology, or physiography; within each mapping unit, there is usually a distinctive assemblage of plant communities that distinguish it from adjacent mapping units. The distribution of Michigan's fauna relative to these mapping units has not yet been closely investigated, but since the distributions of several plant communities and some plant species correlate well with the mapping units, it is assumed that similar relationships will hold for the fauna. Kirtland's warbler and Mitchell's satyr (*Neonympha mitchellii*) are examples of species with distributions restricted to a single mapping unit or a pair of closely related mapping units.

Climate

Climate, an important factor influencing the state's biota, is largely determined by several factors, including latitude, the effect of the Great Lakes, and elevation (Denton 1985; Albert et al. 1986). Latitude is probably the most important factor determining statewide temperature patterns; the southern portion of the Lower Peninsula (fig. 2, Region I) has the highest average annual temperature and the longest growing season, followed by the northern half of the Lower Peninsula (fig. 2, Region II). The lowest average annual temperature (coldest) and the shortest growing season occur farther to the north in the Upper Peninsula (fig. 3, Regions III and IV). The average length of the growing season in the southern Lower Peninsula is 154 days, in the northern Lower Peninsula 126 days, and in the Upper Peninsula 109 days (based on Albert et al. 1986). The average annual temperature in the southern Lower Peninsula is 8.9°C, in the northern Lower Peninsula 7.0°C, and in the Upper Peninsula 4.8°C. Possibly more important is the average temperature during the growing season: 18.6°C in the southern Lower Peninsula, 16.8°C in the northern Lower Peninsula, and 15.0°C in the Upper Peninsula.

This latitudinal climate difference can easily be seen in the tree distribution pattern within the state. In the southern Lower Peninsula there are several tree species with primarily southern distribution, such as northern hackberry (*Celtis occidentalis*), sassafras (*Sassafras albidum*), pin oak (*Quercus palustris*), tulip tree (*Liriodendron tulipifera*), and black gum (*Nyssa sylvatica*). These species are uncommon or absent in the northern half of the Lower Peninsula, where conifers, such as balsam fir, white spruce, and jack pine (*Pinus banksiana*), with a more northern distribution become increasingly common.

The effect of the Great Lakes upon climate, commonly called "lake effect," is also of considerable importance to the biota. Four of the Great Lakes, Lake Michigan, Lake Huron, Lake Superior, and Lake Erie, form the majority of Michigan's boundary. Lake effect occurs along all of the shorelines of the Great Lakes, increasing the length of the growing season and influencing average temperature, extreme temperatures, and the amount and timing of precipitation. Lake effect is most easily observed along the northern Lake Michigan shoreline, where a longer growing season allows orchards to be planted as far north as Leelanau and Grand Traverse counties. Lake effect can also be seen on the Keweenaw Peninsula in Lake Superior, where the growing season is the longest in the Upper Peninsula (134 days). Also on the Keweenaw Peninsula (fig. 3, Districts 20.1 and 20.2), Lake Superior is responsible for the lowest average temperature in the state during the growing season (14.4°C) and the lowest percentage of precipitation during the growing season (44%).

The western half of the Upper Peninsula (fig. 3, Region IV) is less impacted by the Great Lakes than the rest of the state, because winds are generally from the Great Plains to the southwest. As a result, the climate is considered more "continental," with the most extreme minimum winter temperatures (-29 to -34°C inland) and the shortest growing season (87–107 days inland) in the state. The elevation rises very rapidly from Lake Superior's shoreline, greatly reducing the impact

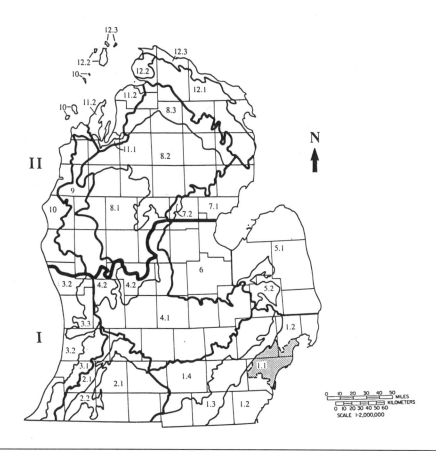

No.	District	Subdistrict	Site Condition	Area Sq Mi	(km²)
Region I: Southern Lower Michigan					
1.1	Washtenaw	Detroit	Heat island		
1.2		Maumee	Lake plain	2300	(5960)
1.3		Ann Arbor	Fine and medium-textured moraine	1635	(4235)
1.4		Jackson	Interlobate; coarse-textured end moraine, outwash, and ice-contact typography	2060	(5335)
2.1	Kalamazoo	Battle Creek	Outwash and ground moraine	2770	(7175)
2.2		Cassopolis	Coarse-textured and end moraine and ice-contact terrain	720	(1865)
3.1	Allegan	Berrien Springs	End and ground moraine	760	(1970)
3.2	Benton Harbor		Lake plain	1355	(3510)
3.3	Jamestown		Fine-textured end and ground moraine	490	(1270)
4.1	Ionia	Lansing	Medium-textured ground moraine	4810	(12460)
4.2		Greenville	Coarse-textured end and ground moraine	760	(1970)
5.1	Huron	Sandusky	Lake plain	3210	(8319)
5.2		Lum	Medium and coarse-textured end-moraine ridges and outwash	480	(1245)
6.0	Saginaw		Lake plain	2390	(6190)
Region II: Northern Lower Michigan					
7.1	Arenac	Standish	Lake plain	1295	(3355)
7.2		Wiggins Lake	Fine-textured end and ground moraine	110	(285)
8.1	Highplains	Cadillac	Coarse-textured end and ground moraine	2860	(7405)
8.2		Grayling	Outwash	4085	(10580)
8.3		Vanderbilt	Steep end- and ground- moraine ridges	1505	(3900)
9.0	Newaygo		Outwash	1920	(4975)
10.0	Manistee		End moraine and sand lake plain	1480	(3835)
11.1	Leelanau	Williamsburg	Coarse-textured end-moraine ridges	100	(260)
11.2		Traverse City	Coarse-textured drumlin fields on ground moraine	750	(1940)
12.1	Presque Isle	Onaway	Drumlin fields on coarse-textured ground moraine	1845	(4780)
12.2		Stutsmanville	Steep and ridges	270	(700)
12.3		Cheboygan	Lake plain	835	(2165)

Fig. 2. Regional landscape ecosystems of lower Michigan, Regions I and II. (From Albert et al. 1986.)

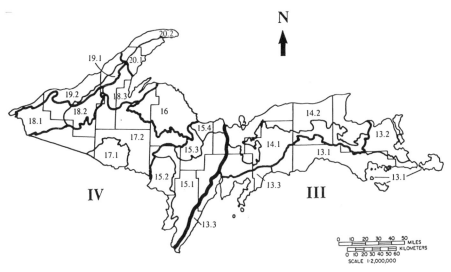

No.	District	Subdistrict	Site Condition	Area Sq Mi	(km²)
Region III: Eastern Upper Michigan					
13.1	Mackinac	St. Ignace	Limestone bedrock and sand lake plain	1580	(4090)
13.2		Rudyard	Clay lake plain	600	(1555)
13.3		Escanaba	Limestone bedrock and sand lake plain	780	(2020)
14.1	Luce	Seney	Poorly drained sand lake plain	1515	(3925)
14.2		Grand Marais	Sandy end moraine, shoreline, and outwash plains	1905	(4935)
Region IV: Western Upper Michigan					
15.1	Dickinson	Hermansville	Drumlins and ground moraine	1855	(4805)
15.2		Norway	Granitic bedrock and end moraine	595	(1540)
15.3		Gwinn	Poorly drained sandy outwash	265	(685)
15.4		Deerton	Sandstone bedrock and high, sandy ridges	225	(580)
16.0	Michigamme		Granitic bedrock	1160	(3005)
17.1	Iron	Iron River	Drumlinized ground moraine	465	(1205)
17.2		Crystal Falls	Kettle-kame topography, outwash, and sandy ground moraine	2390	(6190)
18.1	Bergland	Bessemer	Large, high, coarse-textured ridges and metamorphic bedrock knobs	745	(1930)
18.2		Ewen	Dissected clay lake plain	450	(1165)
18.3		Baraga	Broad ridges of coarse-textured rocky till	575	(1490)
19.1	Ontonagon	Rockland	Narrow, steep bedrock ridge	135	(350)
19.2		White Pine	Clay lake plain	655	(1695)
20.1	Keweenaw	Gay	Coarse-textured broad ridges and swamps	275	(710)
20.2		Calumet	High igneous and sedimentary bedrock ridges and knobs	285	(740)
20.3		Isle Royale	Island of igneous bedrock ridges and swamps	230	(595)

Fig. 3. Regional landscape ecosystems of upper Michigan, Regions III and IV. (From Albert et al. 1986.)

of lake effect. Because Lake Superior is the coldest of the Great Lakes, its ameliorating effect on climate is less than that of the other Great Lakes.

Elevation is generally recognized as an important factor influencing both temperature and precipitation, but in Michigan, elevations range only from 580 ft. to 1,980 ft. (180 m to 605 m); making elevation less of a climatic factor than latitude or lake effect. However, some elevation effect can be observed, especially in the highest portions of the state, the Highplains (fig. 2, District 8) and portions of the western Upper Peninsula (fig. 3, Districts 16, 17, and 18 of Region IV), where the growing seasons are among the shortest in the state (115 days in District 8, 89 days in District 16, 87 days in District 17, and 106 days in District 18).

Bedrock Geology

Michigan's bedrock can be divided into two broad categories based on age: those of the Precambrian era and those of the Paleozoic era (Dorr and Eschman 1970). Precambrian bedrock, exposed in the western part of Upper Michigan (fig. 3, Region IV), is considered to be a part of the Canadian Shield; extensive deformation events (downwarping, folding, and mountain building) occurred during Precambrian time. During the Paleozoic era, shallow marine waters occupied an intracratonic basin that occupied most of what is now the Great Lakes region. Paleozoic marine and nearshore deposits underlie eastern Upper Michigan and all of Lower Michigan (figs. 2 and 3, Regions I, II, and III). Bedrock is exposed only over a relatively small part of Michigan's land surface. The majority of Michigan's surface is mantled with thick deposits of glacial drift (Dorr and Eschman 1970).

Precambrian igneous and metamorphic bedrock is exposed over relatively large areas in western Upper Michigan. Paleozoic limestone, dolomite, and sandstone (sedimentary bedrock types) are exposed in smaller areas along or near the shores of the Great Lakes and along larger streams farther inland, for example, along the Grand and Escanaba rivers.

In the western Upper Peninsula extensive exposures of basaltic bedrock occur on Isle Royale, the Keweenaw Peninsula, and the Porcupine Mountains (fig. 3, Districts 19.2, 20.2, and 20.3). The flora of two of these areas, Isle Royale and the Keweenaw Peninsula, is rich in disjunct species from the Pacific northwest (Cordilleran region), presumably because of the chemical characteristics of the basalt, a bedrock type that is also abundant in the west. Granitic bedrock is exposed in portions of the Michigamme Highlands (fig. 3, District 16), west of Marquette, and also near Norway and Iron Mountain (fig. 3, District 15.2). Highly metamorphosed sandstones characterize the exposed bedrock knobs of the Huron Mountains, northwest of Marquette.

More recent sedimentary rock of Paleozoic age is exposed in the eastern Upper Peninsula (fig. 3, Region III). Resistant sandstones form the small bedrock knobs near Munising and also the cliffs of Pictured Rocks east of Munising (fig. 3, District 14.2). Resistant sandstone is also exposed in several waterfalls along the northern edge of the eastern Upper Peninsula.

The resistant limestones and dolomites of the Niagara Escarpment form cliffs along Lake Michigan's northern shoreline on the Garden and Stonington peninsulas (fig. 3, District 13.3). Flat exposures of limestone, dolomite, and occasionally shale are relatively common within a few miles of the Lake Michigan shoreline from Escanaba in the west to Drummond Island in the east. Where the limestone contains numerous fissures, it is typically forested by northern hardwood forest; where the limestone contains few fissures and drainage conditions are poorer, either cedar swamps or herbaceous wetlands occur.

It is easy to underestimate the impact of bedrock geology upon the biota of Michigan, since bedrock is exposed on less than 5% of the state's surface. But bedrock is important in determining the size and orientation of landforms as well as drainage conditions. Surface glacial deposits are also derived from local bedrock sources in many parts of the state, thus the mineralogy of the bedrock affects important soil characteristics.

In the Upper Peninsula, underlying bedrock largely determines the topography. Highly resistant igneous and metamorphic Precambrian bedrock creates the large, high elevation features of the western Upper Peninsula, whereas easily eroded Paleozoic limestones, dolomites, and sandstones form the low-elevation plain of the eastern Upper Peninsula (Dorr and Eschman 1970). The topographic differences of the eastern and western Upper Peninsula result in major climatic and soil drainage differences: the high-elevation ridges in the west have colder minimum temperatures than the east and are primarily well drained; in contrast, the eastern plain is poorly drained for the most part and supports large tracts of wetland.

Soil differences in the Upper Peninsula can also be traced to the bedrock; soils derived from the silica-rich igneous and metamorphic bedrock are more acidic than the soils derived from the calcium carbonate-rich limestones and dolomites. Silt-sized mineral fragments result from the glacial abrasion of the igneous and metamorphic bedrock; windblown deposits (loess) of this silt are locally common in the western Upper Peninsula. In contrast, clays resulting from glacial abrasion of limestone and dolomite are more common in the eastern Upper Peninsula and in the Lower Peninsula, where they are deposited as lacustrine (lake) clay.

In the Lower Peninsula, the influence of the bedrock geology is not so easily observed, as thick glacial deposits blanket much of the bedrock surface. Bedrock is exposed along the northern shorelines of Lakes Michigan and Huron, but only locally inland and along the shoreline of southern Lake Huron or western Lake Erie. The entire Lower Peninsula is underlain by sedimentary bedrock of Paleozoic age, primarily sandstone, limestone, and shale. Glacial deposits reflect the parent bedrock, which largely determines soil texture, pH, and other chemical characteristics. In the northern part of the Lower Peninsula, thick sand deposits were derived from sandstone; these deposits consist primarily of acidic silica sand. In the southern part of the Lower Peninsula, a much greater portion of the glacial drift was derived from Paleozoic limestone, resulting in calcareous loamy soils.

Glacial History and Glacial Landforms

Glaciers of Wisconsinan age covered all of Michigan as recently as 15,000 years ago (Farrand and Eschman 1974). By approximately 10,000 years ago, the ice had retreated north, leaving all of Michigan free of ice. The thick ice of the continental glaciers reshaped the landscape, eroding and abrading soils and bedrock as the glaciers advanced, and then redeposited the drift elsewhere on the landscape (Dorr and Eschman 1970).

Following the retreat of the Wisconsin continental ice sheets, most of Michigan's land surface was covered with glacial deposits (drift). The character of glacial deposits is determined, in part, by the process of deposition and by the source of the parent material (i.e., the material from which the glacial deposits were derived).

Common glacial features include moraines, drumlins, eskers, kames, lake plains, and outwash plains (fig. 4). Lakes are also common within a glacial landscape; those formed by glacial scouring of the underlying substrate are shallow and linear, whereas those formed by the melting of ice blocks (called kettle lakes) are nearly round, steep sided, and deep. The major glacial landforms are briefly defined and discussed.

End moraines (terminal moraines) are ridgelike accumulations of glacial drift built along the margin of an active glacier. End moraines are sometimes distinctive hilly features several miles wide and several hundred feet high, but they can also be relatively low, with gently sloping ridges. End moraines can generally be traced for several miles across the adjacent flatter landscape of outwash, ground moraine, or lake plain. Small kettle lakes or depressions commonly occur within end-moraine features; these kettle lakes were originally ice blocks surrounded by glacial drift. The soil texture of end moraines ranges from sand to clay, depending on the source of the soil or bedrock that the glacier was eroding and then depositing as end moraine. Even sandy end-moraine soils generally contain small amounts of silt or very fine sand, making them more mesic than sands of outwash or lake plains. End moraines were typically farmed, unless the ridges were too steep or rocky for agricultural equipment, in which case the ridges were used as pastures for grazing livestock or allowed to remain forested.

Ground moraine is gently sloping, rolling topography formed when drift was deposited under an active glacier. Long linear lakes are sometimes common on ground moraine; the long axis of these lakes is parallel to the direction of the glacier's movement. The soil texture of ground moraine is dependent upon the source material for the glacial drift and can range from sand to clay. Where climatic conditions are suitable, ground moraine is typically farmed, regardless of soil texture. Farming on ground moraine often requires either ditching or tiling to improve the drainage conditions.

Drumlins are long, linear ridges oriented parallel to each other, formed beneath glacial ice. They form on ground moraine and are often relatively close to large bodies of water, such as the Great Lakes. Their soil texture can vary greatly, just as that of the associated ground moraine. The drumlins of Menominee County and those of Charlevoix, Emmet, Antrim, Grand Traverse, and Leelanau counties are often very rocky, containing rocks derived from nearby limestone bedrock.

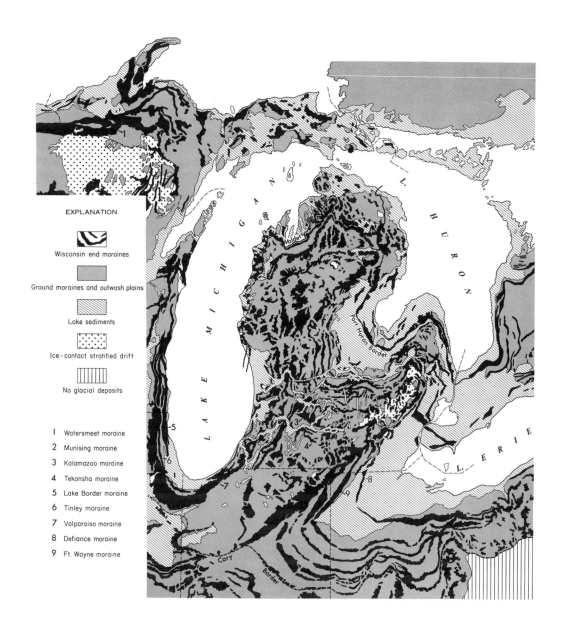

Fig. 4. Glacial map of Michigan. (From Dorr and Eschman 1970.)

Lake plains formed either in the basins of proglacial lakes (lakes formed in front of the glacier) or beneath earlier basins of the Great Lakes, that had higher lake levels than the present Great Lakes basins. Approximately one-third of Michigan's surface is on glacial lake plain. A broad band of lake plain, in some places greater than 40 miles (64 km) wide, extends along portions of all of Michigan's Great Lakes.

The topography of lake plains is relatively flat, generally with slopes of less than 12 feet (3.6 m) per mile. Locally, postglacial stream erosion has created highly eroded, steep topography on the clay-rich soils of the lake plain, such as in Ontonagon County along Lake Superior. Soils of the lake plain can be either coarse textured (sand) or fine textured (clay or silt). Silt and clay were carried by glacial meltwater into the relatively quiet waters of the proglacial lakes, where they gradually settled out, forming laminar deposits (varves) several feet thick. Sand was also carried by the meltwater streams, but it was deposited in relatively narrow channels near the ice front, where the stream current was strongest. Some of these sands (along with sands carried by postglacial streams) were reworked by lake waves to form beach ridges. Beach ridges are still visible on the lake plains associated with shorelines of proglacial lakes or earlier levels of the Great Lakes.

The clay- and silt-rich soils of the lake plain are extremely fertile; as a result, most of their acreage has been converted to agriculture. This has required extensive drainage projects. Only the most poorly drained areas of clay lake plain near the present Great Lakes shoreline retain their original vegetation, either swamp forest, marsh, or wet prairie. In contrast, much of the sand lake plain has not been farmed, either because the fertility of the sand dunes is low or because the sands become unstable when forest vegetation is removed. Some of the organic soils over sand were relatively fertile, but drainage of those soils resulted in severe wind erosion.

The topography of the sand lake plain is quite variable, especially within embayments of earlier glacial Great Lakes. Along the shoreline of glacial lake Algonquin in the eastern Upper Peninsula there are large embayments containing dozens of transverse dune ridges (Futyma 1981 and 1982). These ridges are 30–60 feet (9–18 m) high, supporting open forests of jack pine, white pine, and red pine; broad expanses of peatland surround the dune ridges. Patterned peatlands also occupy the margins of Lake Algonquin, where sand outwash was deposited into the shallow waters of the lake.

Most of the large sand dunes along the Lake Michigan or Lake Superior shorelines are not of glacial origin; they formed during the Lake Nippising highwater period, approximately 4,000 years ago. These dunes, many of which are perched upon glacial till, can be as high as 600 feet (180 m). The dunes are vegetated with grasses and shrubs during early successional stages. Later, mesic forest composed of trees such as American beech (*Fagus grandifolia*), sugar maple, basswood, eastern hemlock, and red oak dominate large areas of stabilized dunes. Parabolic blowouts, which are unstable areas of bare, shifting sand, occur near the shoreline.

Extensive series of parallel beach ridges occur along former embayments of the postglacial Great Lakes shoreline, the result of both postglacial uplift of the land surface (isostatic rebound) and the gradual lowering of Great Lakes water levels.

The higher ridges support open, mixed forests of white pine, red pine, jack pine, balsam fir, or bigtooth aspen (*Populus grandidentata*); the lower ridges support northern white cedar or other swamp conifers. The wettest interdunal depressions (swales) contain emergent or submergent aquatic plants, whereas the driest support swamp or upland forest.

Outwash plains are formed by glacial meltwaters carrying a suspended load of sediment, including clay, silt, sand, and rock-sized particles. Outwash plains are usually flat and consist of thick deposits of sand and gravel. Several other landforms are common on outwash plains. *Kettle lakes*, formed when ice blocks were surrounded by outwash sands, are often locally abundant; areas where kettle lakes and depressions are concentrated are called *pitted outwash*. *Eskers* formed when rocks and gravel were deposited at the bottoms of glacial crevasses; when glacial ice retreated, the steep-sided, esker ridges remained. Similarly, *kames* were formed when rocks and gravel were deposited in a hole in the glacier's surface, resulting in the formation of steep-sided conical ridges.

Most large outwash plains consist of thick sand deposits, resulting in excessively well drained soils. Fire occurred regularly on these droughty sites. Jack pine and northern pin oak (*Quercus ellipsoidalis*) are typical dominant trees on droughty outwash; the extensive jack pine plains of the High Plains are the habitat of one of Michigan's most critically endangered species, the Kirtland's warbler. Other well known outwash plains are the Raco and Dahner Plains of the eastern Upper Peninsula and the Yellow Dog, Sagola, and Baraga Plains of the western Upper Peninsula.

Outwash deposits occupy most of our large stream valleys. Outwash deposits along streams vary in thickness; the thicker deposits may form droughty plains 1 to 2 miles wide along the river. Thinner deposits may be poorly drained due to characteristics of underlying bedrock or glacial deposits. Such poorly drained outwash deposits support either marsh or swamp vegetation. For example, the outwash adjacent to the Muskegon River supports extensive hardwood and conifer swamp along almost the entire length of the river, over 100 miles (166 km).

Lakes and Streams

Lakes and streams cover a large part of Michigan's surface. Colby and Humphrys (1962) estimates that Michigan has 35,000 lakes and ponds, of which approximately 8,600 are named. Streams are equally common. Several threatened or endangered fish and invertebrates occupy these lakes, ponds, and streams; several other endangered or threatened birds nest along their shores and feed within these bodies of water.

Great Lakes. There are approximately 3,222 miles (5,155 km) of Great Lakes shoreline in Michigan (Voss 1972). This shoreline is quite variable in topography and vegetation. In areas exposed to the full force of wave action, bare bedrock, sand, or clay are exposed, with little or no colonizing vegetation. In protected embayments, broad marshes may occupy the shallow waters and marsh, shrub swamp, or forested swamp may extend far inland. The protected embayments support a diverse fauna. Vegetation of the shorelines can vary considerably, depending on the nature of the substrate (MNFI 1987, 1988, and 1989). The acidic substrates of the western Upper Peninsula support boglike coastal wetlands with few species in common with those of the calcareous substrates of the eastern Upper Peninsula.

Although some protection of Great Lakes shorelines is provided by federal and state ownership, large areas of natural coastline have been destroyed for homes, marinas, and recreational sites. Large areas of marsh have been eliminated in the southern part of the Lower Peninsula. Similar encroaching development is also occurring along the remainder of the Great Lakes shoreline.

Inland Lakes. Smaller inland lakes can be subdivided into several types: (1) deep, rock-shored lakes formed in bedrock, (2) kettle lakes, and (3) long, linear lakes on ground moraine. The rock-shored lakes are often deep and almost devoid of aquatic macrophytes. Examples are found in the Porcupine and Huron mountains of the western Upper Peninsula.

Kettle lakes can be either deep or shallow; generally they have marshes, fens, or bogs along their margins. In southern Michigan, many of these lakes have had their shorelines mined for marl or muck. Many kettle lakes are now surrounded by homes and have sand beaches instead of the original margin of aquatic vegetation.

The linear lakes on ground moraine typically support emergent marsh or swamp forest at their margins. Floating or submergent vegetation may cover the entire bottom of relatively shallow examples of this lake type.

Streams. Streams within the state vary greatly both in substrate and channel characteristics, largely as a result of bedrock geology or glacial geology. These differences have not been well studied, but are probably highly significant for the fauna occupying the streams. Trautman (1981) discussed the relationship of fish distribution to stream characteristics in both glaciated and unglaciated terrain in Ohio. All of the glacial landforms discussed by Trautman as occurring in Ohio, as well as others, also occur in Michigan.

Vegetation

Although the southern part of the Lower Peninsula has been vegetated for approximately 15,000 years and the Upper Peninsula for 10,000 years, the ranges of tree species are not static and continually expand and contract as climatic conditions change. For this reason, plant communities are described only generally; the species composition at any specific site will differ somewhat from nearby sites, due to both historic land use and abiotic site differences.

Plant Communities

Plant communities can be generally divided into Upland Forest, Wetland Forest, Upland Nonforested, and Wetland Nonforested. Each of these divisions can be further divided into several plant communities.

Upland and Wetland Forests

The major forested communities of the state are classified and mapped differently by several authors, including Barnes and Wagner (1981), Braun (1950), Kuechler (1964), and the Michigan Natural Features Inventory (1986). For this introduction, the map of potential forest communities by Barnes and Wagner (1981) from *Michigan Trees* will be used (fig. 5), because it provides the greatest available accuracy and detail for the entire state. I have modified this map where more detailed information is available. For a thorough discussion of the forest communities, the reader should consult *Michigan Trees*, which also contains a list of common species found within each forest community. More recent, detailed maps for portions of the state are included in the discussions of individual forest communities.

Oak Savanna. The oak savannas found in the southwestern part of Lower Michigan (fig. 5) occurred on the sandy ridges at the margins of prairies and were dominated primarily by bur oak (*Quercus macrocarpa*). Almost all of these savannas are now farmed. Much of the area shown as Oak-Hickory Forest (fig. 5), was actually savanna dominated by white oak (*Quercus alba*) and black oak (*Quercus velutina*). Chapman (1984) shows the approximate distribution of savannas and prairies based on the original land survey notes and soils and physiographic maps (fig. 6). These savannas were called oak openings or oak barrens by the original government land surveyors (Chapman 1984, Albert 1990). The oak openings were probably the result of frequent fires, many started by native Americans (Chapman 1984). The extent of oak savannas in southern Michigan is demonstrated by the map of *Presettlement Vegetation of Kalamazoo County, Michigan* (Hodler et al. 1981).

Oak-Hickory Forest. At present, oak-hickory forest occupies a broad range of well-drained soils, from sand to loam. When the original land surveys were conducted in southern lower Michigan, beginning in 1815, much of the present oak-hickory forest was savanna. Savanna was best developed on excessively to well-drained sandy soils, whereas closed-canopy, oak-hickory forest primarily occupied loamy soils. Fire suppression resulted in the conversion of savanna to forest (Chapman 1984). At the time of the original land surveys, white oak was the most common species of oak on dry to dry-mesic sites, but logging and grazing opened up the savannas sufficiently for black oak to become more common in many places. Hickories (*Carya* spp.) generally make up only a small proportion of the forest canopy.

Fire suppression has often changed open savannas into closed-canopy forests and has also caused significant changes in the ground cover and fauna. Originally, the ground cover contained many plants common within prairies, including cream wild indigo (*Baptisia leucophaea*), prairie false indigo (*B. lactea*), and lupine (*Lupinus perennis*). Reduced populations of lupine have greatly reduced populations of karner blue butterfly (*Lycaeides melissa samuelis*). Other species impacted by the conversion of savanna to forest include lark sparrow (*Chondestes grammacus*), loggerhead shrike (*Lanius ludovicianus*), barn owl (*Tyto alba*), prairie vole (*Microtus ochrogaster*), and least shrew (*Cryptotis parva*).

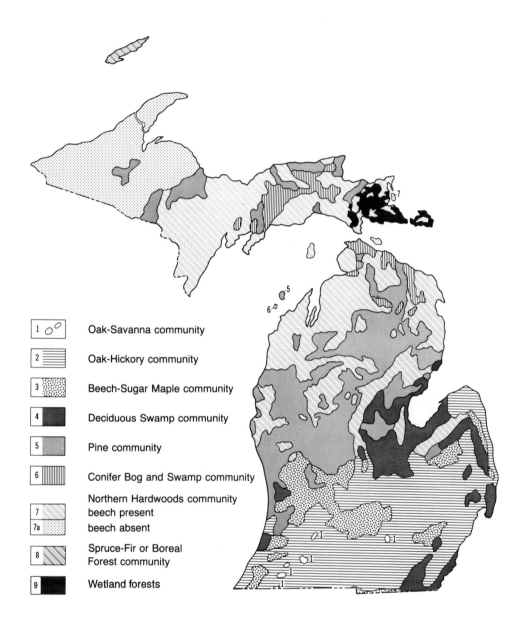

Legend:

1	Oak-Savanna community
2	Oak-Hickory community
3	Beech-Sugar Maple community
4	Deciduous Swamp community
5	Pine community
6	Conifer Bog and Swamp community
7	Northern Hardwoods community beech present
7a	beech absent
8	Spruce-Fir or Boreal Forest community
9	Wetland forests

Fig. 5. Potential forest communities of Michigan. (From Barnes and Wagner 1981.)

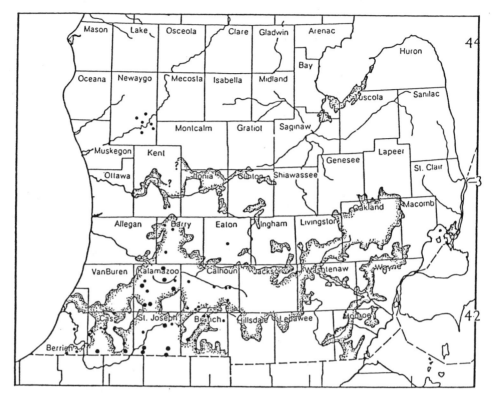

Fig. 6. Location of blacksoil (●) and historical upland prairies (•) and limits of lake plain prairie/savanna and oak savanna at the time of settlement. (After Brewer et al. 1984, Ferrand and Bell 1982, Veatch 1959, and Trygg 1964.)

Beech–Sugar Maple Forest. This forest community, dominated by American beech and sugar maple, is similar to the Northern Hardwoods Community, but includes several species (the walnuts [*Juglans nigra* and *J. cinerea*], bitternut hickory [*Carya cordiformis*], and tuliptree, among others) that are not hardy in northern Michigan (Barnes and Wagner 1981). This community occupies the more mesic loams and clays of the ground moraine and lake plain within southern lower Michigan. Most of these fertile lands are presently farmland and the community diversity is greatly reduced.

Deciduous Swamp Forest. The largest areas of deciduous swamp in southern lower Michigan occupy the postglacial lake basins adjacent to Lake Erie and Lake Huron (Barnes and Wagner 1981). Extensive areas of swamp are also present on flat, ground-moraine topography. Only the most poorly drained swamp lands remain on the ground moraine and lake plain; most of the swamps on loam or clay have been drained, cleared, and farmed. Many of our present swamps were marshes prior to drainage. Small swamps are also present in ice-block depressions and kettles within steep end moraines or pitted outwash.

Several tree species occupy our deciduous swamps, including red maple (*Acer rubrum*), silver maple (*Acer saccharinum*), American elm (*Ulmus americana*),

Northern Hardwoods Forest. Northern hardwood forests occupy large areas of moist (mesic) upland in the northern half of the Lower Peninsula and in the Upper Peninsula. Common dominant trees include sugar maple, American beech, eastern hemlock, yellow birch, hop-hornbeam, and basswood. The forest canopy is typically quite dense, restricting the herbs and shrubs to those tolerant of low light conditions. Northern hardwoods provide important habitat for such forest interior birds as the warblers and thrushes; older stands are important to the marten and the red-shouldered hawk. In the southern half of the Lower Peninsula, a similar forest type, beech–sugar maple forest occurs on moist sites. (Photograph by Gary Reese.)

Oak Savanna. At the time of Michigan's settlement, open oak savannas or, as they were often called, oak openings were common on most of the sandy soils and some of the loamy soils of the southern half of the Lower Peninsula. Almost no savannas remain in Michigan; lack of fire has resulted in the conversion of the savannas to oak forests and the loss of many plant and animal species, such as the Karner blue and frosted elfin butterflies and their host plant, lupine. (Photograph by Kim A. Chapman.)

Oak and Pine Barrens. The droughtiest sand plains supported open barrens dominated by widely spaced, stunted trees. In the southern half of the state, black oaks were common on the barrens. Farther north, jack pine and northern pin oaks were the dominants. Although fire is important for maintaining the open conditions of the barrens, the droughty soils alone are sufficient to prevent rapid closing in of the forest. Pennsylvania sedge and several of the prairie grasses, including big and little bluestem, form a dense turf that also prevents rapid colonization of trees within the barrens. Jack pine barrens provide habitat for the federally endangered Kirtland's warbler; farther to the south, oak and pine barrens support several rare butterflies, including the Karner blue, ottoe skipper, phlox moth, and frosted elfin. (Photograph by Gary Reese.)

Pine Forest. Open forests of red pine, often mixed with jack pine, occupy the sand plains and dune ridges of northern Michigan. Some of the largest remaining pine stands are found near the Great Lakes shorelines of the eastern Upper Peninsula. White pine stands occupied moister sites than red or jack pine; postlogging fires destroyed most seedlings and small understory white pine, resulting in a tremendous reduction of white pine forest acreage. (Photograph by Don Les.)

Spruce-Fir or Boreal Forest. Upland forests of white spruce and balsam fir are restricted to Isle Royale and a few other islands in Lake Superior. White spruce and balsam fir, along with several other conifers, form extensive swamp forests on the flat, poorly drained topography of the eastern Upper Peninsula. Two state-listed endangered mammals, the lynx and the gray wolf, may be found here. (Photograph by Kim A. Chapman.)

Prairie. Michigan's prairies, originally concentrated in southwestern lower Michigan, harbored a number of insects and plants that are now threatened in Michigan. The most extensive prairies, located in Cass, St. Joseph, Kalamazoo, and Calhoun counties, were among the first lands farmed by European settlers. Less extensive wet and wet-mesic prairies persist along some railroad rights-of-way and also along or near the Great Lakes shorelines in the southern half of lower Michigan. The silphium borer moth and Great Plains spittlebug are restricted to prairies in Michigan. (Photograph by Susan R. Crispin.)

Sand Dunes. Michigan has the most freshwater sand dunes in the world. Most of Michigan's large dunes are located along the Lake Michigan shoreline, but there are also dunes along the Lake Huron and Lake Superior shorelines. The largest dunes, which date from the Lake Nipissing high-water period of approximately 4,000 years ago, can be 200 to 300 feet high. The dunes support an endemic plant, Pitcher's thistle, and an endemic insect, Lake Huron locust. This habitat may also harbor the majority of the remaining prairie warbler population in Michigan. (Photograph by Susan R. Crispin.)

Alvar. Alvar is a grassland type occurring on exposed limestone and dolomite bedrock. The most extensive occurrences of alvar are on Drummond Island, Michigan, and nearby Cockburn and Manitoulin islands in Ontario. A narrow strip also lines a portion of the Escanaba River shoreline. Seasonal flooding followed by drought prevents successful establishment by trees. Alvar has a distinctive flora of prairie and boreal plants and likely supports rare prairie insects, as well. (Photograph by Susan R. Crispin.)

Deciduous Swamp or Floodplain Forest. Deciduous swamps are found along the floodplains of most streams and also in depressions within the landscape, providing habitat for a diversity of plants and animals. Floodplains provide important migration corridors in southern Michigan, especially for forest interior birds. Dukes' skipper, copperbelly water snake, Indiana bat, and yellow-throated warbler are just some of the many endangered and threatened animals that utilize swamp and floodplain forests during at least part of their life cycle. (Photograph by Gary Reese.)

Conifer Swamp. Extensive swamps of northern white cedar grow on the calcareous soils and bedrock of northern lower Michigan and eastern upper Michigan. Other conifers, including tamarack, black spruce, white spruce, balsam fir, hemlock, and white pine, form extensive swamps over a broad range of soil, drainage, and disturbance conditions. Osprey are typically found nesting in dead tamaracks or spruce or in snags in conifer swamps throughout the northern two-thirds of the state. (Photograph by Susan R. Crispin.)

Kettle Bog. Kettle bogs form on ponds and lakes throughout the state. The bog mat begins to form along the lake margin, gradually expanding to cover the entire lake surface. The open mat of sedges and peat mosses may eventually be colonized by black spruce and tamarack, as seen here. Only a small number of plants and animals are capable of surviving in the extremely acid conditions of the bog. The common loon nests on these relatively undisturbed lakes. (Photograph by Gary Reese.)

Muskeg or Paludified Peatland. Extensive shallow peatlands occupy the flat glacial lake plains of the eastern Upper Peninsula. Sedges or peat mosses share dominance with stunted tamarack, black spruce, or jack pine; 100-year-old trees may only be 10 to 20 feet tall and 1 to 2 inches in diameter. Yellow rails are known only from this type of habitat in Michigan. (Photograph by William Brodowicz.)

Northern Fen. Near the Straits of Mackinac, marly shorelines and seepages support a distinctive flora of calciphiles. Woody plants are periodically killed by water level fluctuations, especially along the Great Lakes shoreline and wind storms regularly blow down trees. Several birds that nest nearby routinely forage in northern fens, including the long-eared owl, short-eared owl, and merlin. (Photograph by Dennis A. Albert.)

Prairie Fen. In southern lower Michigan, a distinctive wetland type, prairie fen, forms in and adjacent to calcareous seeps. Numerous prairie plants persist in these wetlands, probably as a result of calcareous groundwater and fires from the adjacent sandy uplands. Tamaracks and poison sumac are common along the fen margins. Mitchell's satyr, a state endangered butterfly, is found exclusively in prairie fens. (Photograph by Gary Reese.)

Marsh. Extensive marshes occupy the shallow waters of Great Lakes embayments and estuaries, as well as the shorelines of thousands of smaller inland lakes and streams. The largest marshes occur along the St. Marys River, Saginaw Bay, and Lake St. Clair, covering hundreds to thousands of acres. Although the plant diversity of the marshes is often low, the fauna is diverse. The marshes provide critical habitat for the short-eared owls, king rail, and eastern fox snake, as well as more common migratory waterfowl. (Photograph by Dennis A. Albert.)

Intermittent Ponds. Intermittent ponds are common on the sandy glacial lake beds outwash plains. Water level fluctuations can be extreme and rapid. As a result of these fluctuations, many of these ponds, especially those in southwestern lower Michigan, support a diverse flora of disjunct plants from the Atlantic and Gulf coastal plains. A number of invertebrates characteristic of the coastal plain are expected to occur within these ponds. (Photograph by Susan R. Crispin.)

Aquatic Habitats. Within Michigan there are approximately 36,000 miles of inland rivers and streams and 35,000 lakes and ponds, providing aquatic habitat for a broad diversity of plants and animals. Among the threatened and endangered species requiring high-quality aquatic habitat are several clams, snails, and fish. (Photograph by Gary Reese.)

basswood, red ash (*Fraxinus pennsylvanica*), black ash (*Fraxinus nigra*), swamp white oak (*Quercus bicolor*), and pin oak. The tolerance of each of these species to the poor drainage conditions characteristic of swamps varies considerably. Several conifers, including tamarack (*Larix laricina*), eastern white pine, eastern hemlock, and northern white cedar, also occupy the swamps of southern Michigan. Tamarack is common adjacent to the calcareous seepages flowing from sandy end-moraine ridges, whereas eastern hemlock, eastern white pine, and, locally, northern white cedar are restricted to the lake plains along Lake Michigan and Lake Huron.

Pine Forest. Pine forests occupied large areas in central lower Michigan and in upper Michigan. Pine forest habitat is characterized by sandy, droughty, acid, nutrient-poor soils, and high fire frequency (Barnes and Wagner 1981). Jack pine occupied the most extreme, droughty sites, especially extensive outwash plains. Northern pin oak was a common associate. Red pine occupied slightly more mesic sites, either on outwash deposits, sandy end moraine, or sandy beach ridges and small transverse dunes. White pine was most common on sand lake plain, occupying a broad range of sites, from poorly drained embayments to excessively drained sand dunes. White pine frequently grew with white oak or red oak. Following intensive logging of eastern white pine and red pine between 1850 and 1900, widespread fires resulted in the replacement of many of our pine forests by red maple and the light-seeded, wind-dispersed aspens and white birch (*Betula papyrifera*).

Conifer Bog and Swamp Forest. In northern lower Michigan and in upper Michigan, conifers dominate most of the wetlands. However, hardwoods can still be found along stream margins and well-aerated wetlands in stream headwater areas. The most extensive conifer swamps are located on the glacial-lake basins of eastern Upper Michigan. Conifer swamps occur in several different site conditions, including shallow organic soils over limestone bedrock, lacustrine sands, lacustrine clays, and outwash sands.

Northern Hardwoods Forest. This community is widespread in the northern United States; it occurs from Minnesota to Maine, southward to North Carolina and Tennessee in the Appalachian Mountains (Barnes and Wagner 1981). Common species occupying this forest type include sugar maple, red maple, American beech hop hornbeam (*Ostrya virginiana*), basswood, yellow birch (*Betula alleghaniensis*), and eastern hemlock. Barnes and Wagner (1981) break this community into two types in Michigan, one with beech and one without beech. The western half of the Upper Peninsula does not contain beech, probably because of the extremely low winter temperatures in this part of the state (Denton and Barnes 1987). This community occupies a zone with cool, moist climatic conditions and can be found on soils with a broad range of moisture and nutrient conditions. Fire is not common within the community, but occurs frequently enough to maintain the presence of white pine and red oak. When fires do occur, they are most common along lake margins and areas adjacent to sandy outwash or lake plains.

Spruce-Fir Forest or Boreal Forest. Within Michigan, this community is best represented on Isle Royale, but it is also scattered throughout upper Michigan. It is best developed in Canada, especially along the north shore of Lake Superior (Barnes and Wagner 1981). The habitat is characterized by cold climates, poor drainage conditions, and shallow, rocky, acid soils. Fire is an important factor maintaining this forest community. Common dominant species of these forests are white spruce, balsam fir, black spruce (*Picea mariana*), and jack pine, but trembling aspen (*Populus tremuloides*) and paper birch are also common.

River Floodplain and Bottomland Hardwood Forest. This forest community occurs in narrow bands along all of the larger streams within the state, although its distribution is not shown in figure 5. The habitat is subject to periodic flooding and siltation; its climate is warmer and more humid in the summer and cooler in the spring than that of the surrounding upland (Barnes and Wagner 1981). Species found throughout its range include American elm and red ash. In the southern part of the Lower Peninsula, river floodplains contain several species with a predominantly southern geographic distribution, including redbud (*Cercis canadensis*), honey locust (*Gleditsia triacanthos*), Kentucky coffeetree (*Gymnocladus dioicus*), sycamore (*Platanus occidentalis*), northern hackberry, red mulberry (*Morus rubra*), and others (Barnes and Wagner 1981). Evidence of the local amelioration of climate along floodplains can even be seen in upper Michigan, where butternut and silver maple can be found along both the Sturgeon River (Delta County) and Menominee River (Menominee County).

Nonforested Upland Plant Communities

Sand Dunes. Dunes line the Lake Michigan shoreline from the Indiana border to the Straits of Mackinac. Several of the islands in Lake Michigan, including North and South Manitou, North and South Fox, Beaver, and High islands, have large dunes along their western shoreline. The Grand Sable dunes are located on Lake Superior's shoreline just west of Grand Marais. All of these large dunes are relatively recent, dating from Lake Nipissing times, approximately 4,000 years ago. Large areas of these dunes are vegetated only by herbs, beach grasses (*Amophila breviligulata* and *Calamovilfa longifolia*), and shrubs, including willows (*Salix cordata* and *S. myricoides*) and beach cherry (*Prunus pumila*). Portions of the dunes have been stabilized by the grasses and shrubs, and now support forests. The open dunes contain several characteristic insects, some of which are found only on the dunes. One of these is the Lake Huron locust (*Trimerotropis huroniana*), found only on blowouts.

Mesic Prairie. Mesic prairies were originally restricted to a relatively small portion of southwestern Michigan (see fig. 6). This area has been called the "prairie peninsula" (Veatch 1927; Kenoyer 1940) and was considered to be a northeastern extension of the prairies from Indiana and Illinois. Native American management with fire appears to have been an important factor maintaining the open condition of these mesic prairies (Chapman 1984). The prairies were among the first areas settled and plowed by European settlers; as a result, only small areas of prairie persist, and these are generally found on wet sites. Some of the common herbaceous plants found on the mesic prairie are big bluestem (*Andropogon gerardii*), switch grass (*Panicum virgatum*), slough grass (*Spartina pectinatus*), cream wild indigo, black-eyed susan (*Rudbeckia hirta*), culver's root (*Veronicastrum virginicum*), and hairy phlox (*Phlox pilosa*).

Small areas of dry prairie are also locally common in ice-block depressions of the outwash plains of Muskegon and Newaygo counties and on sand lake plain in Allegan County. These prairies are possibly the result of either microclimatic conditions (frost pockets) or more extreme fires in the depressions. Little bluestem (*Schizachyrium scoparium*) dominates the depressions, along with Pennsylvania sedge (*Carex pensylvanica*), poverty grass (*Danthonia spicata*), and other dry-site grasses. Prairie smoke (*Geum triflorum*), blazing star (*Liatris aspera*), and prickly pear (*Opuntia humifusa*) are common herbs.

Alvar. An unusual community, alvar is a thin-soiled limestone that supports a flora with both prairie and boreal components. In Michigan, alvar is restricted to the banks of the Escanaba River and the northern portion of Drummond Island, but, in adjacent Ontario, alvar covers extensive areas on Manitoulin Island and the Bruce Peninsula. The persistence of the prairie flora on alvar is probably caused by both periodic flooding during the growing season and by periodic summer droughts that combine to limit the invasion of large shrubs and trees. Periodic fire may also inhibit succession. Distinctive flora of the alvar includes prairie dropseed (*Sporobolis heterolepis*), prairie smoke, flattened spike rush (*Eleocharis compressa*), and two sedges, *Carex scirpoides* and *C. richardsonis*. Portions of the alvar are wet during most or all of the growing season and are better treated as wetland.

Nonforested Wetlands

Marsh. Michigan marsh lands consist of sedge-dominated and grass-dominated wetlands along the saturated or flooded shorelines of lakes and streams. Some of the largest marshes are found along the shorelines of large, protected embayments of the Great Lakes. These marshes are characterized by distinctive vegetation zones (MNFI 1987, 1988, and 1989). Near the upland margins is a wet meadow zone, where sedges (*Carex stricta, C. lacustris,* and *C. aquatilis*) and grasses, primarily bluejoint (*Calamagrostis canadensis*), grow on shallow organic soils. In the shallow water, an emergent marsh zone, dominated by bulrushes (*Scirpus acutus, S. americanus,* and *S. validus*), spike rush (*Eleocharis smallii*), pickerel weed (*Pontederia cordata*), arrowhead (*Sagittaria latifolia*), bur-reed (*Sparganium eurycarpum*), pondweeds (*Potamogeton* spp.), spatterdock (*Nuphar variegata* and *N. advena*), water lily (*Nymphaea cordata*), and several other floating and submergent plants. The emergent marsh beds can extend into water 4 to 5 feet (1.2–1.5 m) deep. In some of the more protected marshes, submergent marsh (containing floating or submergent plants) extends into much deeper water.

Marshes are also found along the shoreline of most small lakes and along many large creeks and rivers. On some of the large rivers flowing into the Great Lakes, where water levels near the river mouths are determined by those of the Great Lake, marshes are quite extensive for several miles inland. These stream-side marshes contain a distinctive flora that includes species not typically found along Great Lakes shorelines, such as water arum (*Peltandra virginica*) and wild rice (*Zizania aquatica* var. *aquatica*).

Marshes are important habitats to waterfowl, fish, and aquatic invertebrates for breeding and feeding. The tremendous pressure caused by human perturbation and alteration of marshland to create marinas, swimming beaches, and other recreational facilities is one of the major natural resource problems facing Michigan.

Fens. Michigan fens are calcareous herbaceous wetlands, most of which are located either along calcareous shorelines of lakes or in seepage areas at the base of end-moraine ridges. Extensive fens are located along glacial and present-day Great Lakes shorelines in the Straits of Mackinac area, where limestone-rich sediments are common. The distinctive flora of these northern fens includes such species as bird's eye primrose (*Primula mistassinica*), butterwort (*Pinguicula vulgaris*), grass-of-parnassus (*Parnassia glauca*), shrubby cinquefoil (*Potentilla fruticosa*), and numerous other species, including boreal rarities such as black crowberry (*Empetrum nigrum*) and hyssop-leaved fleabane (*Erigeron hyssopifolius*).

In the southern half of lower Michigan, fens are concentrated in the so-called interlobate (fig. 2, Districts 1.4 and 2.1), an area of calcareous, gravelly end-moraine and kettle lakes formed between lobes of the last glacier. These fens, often called prairie fens, are typically located at the seepy margins near the edge of the end moraines. Peat accumulation within the prairie fens has resulted in the gradual enlargement of the wetland from the center outward; the peat in the center of the wetland can be greater than 10 feet (3 m) deep. Open marly pools, similar to those occurring in northern fens, are common features. Calciphiles (calcium-loving plants) such as spike-rush (*Eleocharis rostellata*) are common in the seepages; sedges (*Carex stricta, C. aquatilis,* and *C. lacustris*), and grasses, primarily bluejoint, dominate the center of the wetland. Other common dominants of prairie fens include *Cladium mariscoides,* shrubby cinquefoil, Indian grass (*Sorgastrum nutans*), willow (*Salix candida*), and pitcher-plant (*Sarracenia purpurea*). Michigan rarities sometimes present in prairie fens include prairie dropseed, Richardson's muhly (*Muhlenbergia richardsonis*), and white lady-slipper (*Cypripedium candidum*).

Bogs. These formations are acidic wetlands with low concentrations of nutrients and thick accumulations of peat. Precipitation provides most of the water and nutrients to bogs. Sedges form the bog mat, onto which pioneering sphagnum mosses grow. The low pH of bogs is partially the result of acidification caused by the growth of sphagnum mosses.

Two types of bog are common in Michigan: kettle-lake bogs and bogs resulting from paludification (swamp-forming activity). The kettle-lake bog occupies an ice-block depression. In the classical development of a kettle-lake bog, sedges and mosses initially form a floating mat at the margins of the lake. The mat gradually expands to cover the entire surface of the lake. The irregular end moraine of western upper Michigan contains a large number of these bog lakes. In contrast, paludification begins with sedges and mosses occupying the moist center of a depression. Organic material from these wetland plants impedes drainage conditions, and the wetland gradually increases laterally in size. The wetland can expand to fill an entire glacial embayment or lake bed. Paludification in many of Michigan's lowlands is the result of a climatic change to cooler, moister conditions. The eastern half of upper Michigan contains several large paludified wetlands.

Submergent and floating aquatic macrophytes cover the bottoms of many of our shallow lakes and streams. Although these aquatic plant beds serve as habitat for many aquatic animals, there are few detailed classifications or descriptions of them.

LITERATURE CITED

Albert, D. A. 1990. *A regional landscape ecosystem classification of Michigan stressing physiographic, geologic, and soil factors.* Ph.D. diss. Univ. Mich., Ann Arbor.

Albert, D. A., and L. D. Minc. 1986. *The natural ecology and cultural history of the Colonial Point red oak stands.* Univ. Mich. Biological Station, Tech. Rept. No. 14.

Albert, D. A., S. R. Denton, and B. V. Barnes. 1986. *Regional landscape ecosystems of Michigan.* Univ. Mich., Ann Arbor.

Bacig, T., and F. Thompson. 1982. *Tall timber: A pictorial history of logging in the upper Midwest.* Voyageur Press, Bloomington, Minn.

Barnes, B. V., and W. H. Wagner, Jr. 1981. *Michigan trees.* Univ. Mich. Press, Ann Arbor.

Braun, E. L. 1950. *Deciduous forests of eastern North America.* Free Press, New York.

Brewer, L. R., H. A. Raup, and T. W. Hodler. 1984. Presettlement vegetation of Southwestern Michigan. *Mich. Bot.* 23:153–56.

Chapman, K. A. 1984. *An ecological investigation of native grassland in southern lower Michigan.* M.S. thesis, W. Mich. Univ., Kalamazoo.

Colby, Joyce, and C. R. Humphrys. 1962. *Summary of acreage analysis charts from Lake Inventory Bulletins 1 to 83.* Mich. Agr. Exp. Sta., Dept. of Resource Development, Mich. State Univ., E. Lansing.

Denton, S. R. 1985. *Climatic patterns and tree distributions in Michigan.* Ph.D. diss., Univ. Mich., Ann Arbor.

Denton, S. R., and B. V. Barnes. 1987. Tree species distributions related to climatic patterns in Michigan. *Can. J. For. Res.* 17:613–29.

Dorr, J. A., Jr., and D. F. Eschman. 1970. *Geology of Michigan.* Univ. Mich. Press, Ann Arbor.

Farrand, W. R., and D. F. Eschman. 1974. Glaciation of the southern peninsula of Michigan. *Mich. Acad. Sci.* 7:31–56.

Futyma, R. F. 1981. The northern limits of glacial Lake Algonquin in upper Michigan. *Quat. Res.* 15:291–310.

Futyma, R. F. 1982. *Postglacial vegetation of eastern upper Michigan.* Ph.D. diss., Univ. Mich., Ann Arbor.

Hodler, T. W., R. Brewer, L. G. Brewer, and H. A. Raup. 1981. Presettlement vegetation of Kalamazoo County Dept. Geog., W. Mich. Univ., Kalamazoo.

Jamison, K. 1945. Base and principal meridian lines in Michigan. *Mich. Conserv.* 14 (3): 8–9.

Kenoyer, L. A. 1940. Forest distribution in southwestern Michigan as interpreted from the original land survey (1826–1832). *Pap. Mich. Acad. Sci., Arts, Ltrs.* 19:107–11.

Kuechler, A. W. 1964. The potential natural vegetation of the conterminous United States. *Am. Geog. Soc. Spec. Publ.* No. 36.

MNFI (Michigan Natural Features Inventory). 1986. Draft descriptions of Michigan natural community types. Unpubl. Rept.

MNFI. 1987. A survey of Great Lakes marshes in Michigan's Upper Peninsula. Unpubl. Rept.

MNFI. 1988. A survey of Great Lakes marshes in the southern half of Michigan's Lower Peninsula. Unpubl. Rept.

MNFI. 1989. A survey of Great Lakes marshes in the northern half of Michigan's Lower Peninsula. Unpubl. Rept.

Rand McNally. 1982. *Universal world atlas.* Rand McNally and Co., Chicago.

Tanner, H. H. 1987. *Atlas of the Great Lakes Indian history.* Univ. Oklahoma Press, Norman.

Trautman, M. B. 1981. *The fishes of Ohio.* Ohio State Univ. Press, Columbus.

Trigger, A. 1978. *Handbook of North American Indians, the Northeast.* Smithsonian Inst., Washington, D.C.

Trygg, J. W. 1964. Composite maps of U.S. land surveyors' original plats and field notes (map). Ely, Minn.: Trygg Land Office.

Veatch, J. O. 1927. The dry prairies of Michigan. *Pap. Mich. Acad. Sci., Arts, Ltrs.* 8:269–78.

Veatch, J. O. 1959. *Presettlement forest in Michigan.* Dept. of Resource Development, Mich. State Univ., E. Lansing.

Voss, E. G. 1972. *Michigan flora (part 1). Gymnosperms and monocots.* Cranbrook Inst. Science, Bloomfield Hills, Mich.

Mammals

Endangered and Threatened

Least shrew (*Cryptotis parva*)
Indiana bat (*Myotis sodalis*)
Gray wolf (*Canis lupus*)
Marten (*Martes americana*)
Cougar (*Felis concolor*)
Lynx (*Felis lynx*)
Prairie vole (*Microtus ochrogaster*)

Extirpated

Wolverine (*Gulo gulo*)
Woodland caribou (*Rangifer tarandus caribou*)
Bison (*Bison bison*)

Extinct

Eastern elk (*Cervus elaphus canadensis*)*

---

* This subspecies is not officially on the state list.

An Introduction to Mammals

Larry Master
The Nature Conservancy

Michigan is home to a diverse assemblage of mammal species. Some of these, such as the gray wolf and the lynx, are rare, retiring, and/or restricted to remote parts of the state. Others are small and difficult to observe (e.g., shrews, mice, lemmings, and bats). Mammal specimens, mostly collected prior to 1945 and deposited in collections around the country, form the basis for much of our knowledge of the historical distribution of Michigan's mammals. In the early decades of this century, zoologists and their students at Michigan universities and colleges conducted many studies that form the foundation of our knowledge of the habitat, distribution, and population dynamics of Michigan mammals (e.g., Dice 1920; Wood 1922; Murie 1934; Allen 1938 and 1939; Blair 1940; Burt 1940; Linduska 1950; Allen 1943; Manville 1949; Craighead and Craighead 1956; Getz 1961). The study of owl diets in Michigan, as revealed by regurgitated pellets containing small mammal remains, also has been a source of information about the distribution of Michigan's smaller mammals (e.g., Wilson 1938; Wallace 1948; Armstrong 1958).

In the past two decades, following the passage of Michigan's Endangered Species Act of 1974, numerous studies have been conducted to ascertain the current status, distribution, and ecological requirements of Michigan's mammals, particularly rare species. These studies, many of them supported by the Michigan Department of Natural Resources (e.g., the Natural Heritage Program and the Natural Features Inventory) and the U.S. Fish and Wildlife Service, include ongoing investigations into the ecology of the gray wolf on Isle Royale (e.g., Peterson 1977; Peterson and Page 1988), reintroductions of the marten into Michigan's Upper (Churchill et al. 1981) and Lower peninsulas (Irvine 1989), and surveys for small mammals possibly thought to be rare in the state (e.g., Master 1978; Kurta 1980; Wilkinson 1980; Shier 1981; Ryan 1982; Kurta et al. 1989).

Summaries of the distributions and natural histories of Michigan's mammals may be found in Burt's *The Mammals of Michigan* (1946); Burt's *Mammals of the Great Lakes Region* (1957); Baker's voluminous treatise, *Michigan Mammals* (1983); and Jones and Birney's *Handbook of Mammals of the North-Central United States* (1988). Sixty-eight species, representing 17 families in 7 orders, are known to have occurred in Michigan within historical times. This list includes 3 species that are extirpated from the state, 3 species that have been reestablished following extirpation, and 2 introduced species that are not native to Michigan or North America.

Mammal Distributions: Natural Factors

Several times during the past million years, Michigan's surface was covered with enormous ice sheets. As a result of these glacial advances, all of Michigan's present-day fauna arrived relatively recently, following the slow retreat of the last (Wisconsin) ice sheets between 10,000 and 20,000 years ago. The subsequent postglacial distribution of mammals within the state were particularly influenced by several factors.

Water barriers, in the form of narrow connections between the Great Lakes, have served as an effective barrier to the dispersal of many species of nonflying mammals (Baker 1983). For example, two species of mammals found in the Upper Peninsula, arctic shrew and least chipmunk, have not crossed the 5-mile wide Straits of Mackinac between the Upper and Lower peninsulas. Similarly, several species of mammals found in the Lower Peninsula, including eastern cottontail, thirteen-lined ground squirrel, and white-footed mouse, have not crossed the Straits of Mackinac to the eastern end of the Upper Peninsula. The St. Mary's River, which forms a narrow boundary less than 700 yards wide and is frozen over in winter, also serves as an effective barrier to the dispersal of some mammals. Three species found on the Ontario side (hairy-tailed mole [*Parascalops breweri*], heather vole [*Phenacomys intermedius*], and rock vole [*Microtus chrotorrhinus*]) have never been recorded in Michigan, and a fourth species, the smoky shrew, only recently was discovered on Sugar Island in Chippewa County (Master 1982). A similar barrier (the Detroit River) between extreme southern Ontario and southeastern Michigan has blocked the dispersal of at least seven mammal species.

The climate has affected mammalian fauna of the state directly as well as indirectly through its effects on vegetation. Eight of the mammal species found in the Upper Peninsula are restricted to the northern or more boreal portions of the Lower Peninsula. Six species occur only in the southern, more temperate portion of the Lower Peninsula. However, a number of other species with southern or more temperate affinities apparently have spread northward into the northern part of the Lower Peninsula in response to changes in the environment caused by humans.

Mammal Distributions: The Human Factor

The available evidence suggests that humans first arrived in Michigan about 12,000 years ago, as the glaciers retreated. The earliest human inhabitants of the area were likely nomadic hunters who pursued mammals and other animals for food and other products. The hunting activities of these early peoples may have contributed to the late Pleistocene extinctions of many large-bodied mammals (Jones and Birney 1988). Six of these now-extinct mammals are known to have inhabited Michigan between 4,000 and 11,000 years ago: giant beaver (*Castor ohioensis*), American mastodon (*Mammut americanus*), Jefferson mammoth (*Mammuthus jeffersoni*), peccary (*Platygonus compressus*), giant moose (*Cervalces scotti*), and woodland muskox (*Symbox cavifrons*) (Holman 1975; Baker 1983).

Since the arrival of European settlers about 350 years ago, Michigan's landscape has been altered severely. The cutting of mature forests, the plowing of the prairies and oak opening, the draining of wetlands, urbanization, and other human-related impacts on the landscape have affected the status and distribution of every species in the state in one way or another. Some mammals are more common now than they were historically. White-tailed deer, for example, have benefited from the increase in early-successional habitats. But many of the other larger mammals have suffered. In the past 200 years, seven species of medium-to-large mammals have been extirpated from the state as a result of human activities. These species and their approximate dates of extirpation (from Baker 1983) are as follows: bison (1800), elk (1877), wolverine (1880), caribou (1910), marten (1911), fisher (1936), and mountain lion or cougar (1937). Three of these species have been successfully reestablished, although one of these introductions (elk) involves a western subspecies (*Cervus elaphus nelsoni*) different from the eastern subspecies (*Cervus elaphus canadensis*) that was historically present but driven to extinction in the state.

Although some species are naturally rare, anthropogenic or human-caused factors can explain the current status of most mammal species on the state lists of endangered, threatened, and special concern mammals.

Overview of State-Listed Mammals

There are ten species of mammals listed as endangered, threatened, or special concern in the state. An additional three species are unquestionably extirpated from the state with little or no likelihood of natural reappearance; and one recently occurring subspecies is extinct, as discussed previously.

Six of the ten listed mammals (gray wolf, lynx, mountain lion, marten, wolverine, and woodland caribou) are medium-to-large mammals that were extirpated or nearly so from Michigan as a result of unregulated logging, shooting, and trapping. Early European settlers treated large carnivorous mammals as competitors or varmints to be shot on sight. A bounty existed on wolves until 1960, but wolves continued to be persecuted by humans unconcerned with the wolf's place in a balanced northern ecosystem. Mountain lions were similarly persecuted throughout Michigan and North America, and the eastern race (*Felis concolor couguar*) is perilously close to extinction.

Legal protection, the regrowth of the forests of Michigan's Upper Peninsula, and improved public awareness of the value of predators give hope for the future of some of these mammals. With recent restoration efforts, the marten has joined the fisher as an apparently secure, reestablished component of the state's fauna. As a result of protection activities in Minnesota and Wisconsin, the wolf has begun to reestablish itself on the Upper Peninsula mainland. With habitat conditions, the lynx and cougar may also be able to inc in the Upper Peninsula. Moose, as well as martens, now have into the Upper Peninsula. But because of their low numbers, remain vulnerable to accidental trapping and shooting and to biles. In addition to continued legal protection, long-term p species will require the protection of large areas with minimal human disturbance (e.g., Jensen et al. 1986).

Although some endangered species use Michigan lands only to breed, their habitats must be protected throughout the year. For example, the Indiana bat (*Myotis sodalis*) summers in Michigan but does not winter here. This species is federally and state listed as endangered because of documented declines of its wintering populations, which are now concentrated in a few caves. Some of their population decline can be attributed to human disturbances at the winter hibernacula, but disturbances to nursery colonies or to their summer habitat also may be a significant factor in their decline. Surveys need to be conducted throughout promising riparian and lowland forest habitat in southern Michigan to identify nursery colonies and their habitat so that these areas may be protected from inadvertent destruction or alteration (e.g., selective cutting or pesticide spraying) that would be harmful to the bats or to their food supply.

The remaining listed mammals in the state are small and inconspicuous: the least shrew and the prairie vole. Both state listed as threatened, they are more common south of Michigan and barely reach the southern part of the state. Despite the absence of recent records, both species will likely be found if systematic efforts are made to locate them. The prairie vole is probably constrained in Michigan by a natural shortage of native grassland or prairie habitat; the least shrew, at the northern edge of its range, possibly occurs somewhat erratically due to climatic factors. If some of the predictions of climate change due to global warming are accurate, southern species such as the least shrew may move northward and increase in numbers. Similarly, Franklin's ground squirrel (*Spermophilus franklinii*), known to occur in Indiana just south of Michigan, may be expected to move into the state.

The loss of native, grassy upland habitats as a result of agriculture and development likely has adversely affected both the prairie vole and the least shrew. The protection of habitats for such species at the peripheries of their ranges is important, both as a first line of defense against a gradual loss of a species' range and to conserve genetically divergent populations that might be expected to occur at the periphery of a relatively sedentary species' range.

The long-term protection of Michigan's endangered and threatened mammals and other species depends on informed citizens and the will of those citizens to protect the best of what remains of our natural world. There is still time for conservation organizations, public land planners and managers, and others to work together to effect lasting protection of suitable habitat for all of Michigan's mammals and other existing fauna and flora—beginning with habitat for known rare and endangered species and with the best examples of the natural communities and ecosystems.

LITERATURE CITED

Allen, D. L. 1938. Ecological studies on the vertebrate fauna of a 500-acre farm in Kalamazoo County, Michigan. *Ecol. Monogr.* 8:347–436.

Allen, D. L. 1939. Winter habits of Michigan skunks. *J. Wildl. Mgmt.* 3:212–28.

Allen, D. L. 1943. Michigan fox squirrel management. Game Div. Publ., Mich. Dept. Conserv., *Game Div. Publ.* No. 100.

Armstrong, W. H. 1958. Nesting and food habits of the long-eared owl in Michigan. *Mich. State Univ., Publ. Mus.* No. 1:63–96.

Baker, R. H. 1983. *Michigan mammals.* Mich. State Univ. Press, E. Lansing.

Banks, R. C., R. W. McDiarmid, and A. L. Gardner. 1987. Checklist of vertebrates of the United States, the U.S. Territories, and Canada. *U.S. Fish Wildl. Serv., Resour. Publ.* 166.

Blair, W. F. 1940. A study of prairie deer-mouse populations in southern Michigan. *Am. Midl. Nat.* 24:273–305.

Burt, W. H. 1940. Territorial behavior and populations of some small mammals in southern Michigan. *Univ. Mich. Mus. Zool. Misc. Publ.* No. 45.

Burt, W. H. 1946. *The mammals of Michigan.* Univ. Mich. Press, Ann Arbor.

Burt, W. H. 1957. *Mammals of the Great Lakes region.* Univ. Mich. Press, Ann Arbor.

Churchill, S. J., L. A. Herman, M. F. Herman, and J. P. Ludwig. 1981. Final report on the completion of the Michigan marten reintroduction program. *Ecol. Res. Serv., Inc.*, Iron River, Mich., Unpubl. Rept.

Craighead, J. J., and F. C. Craighead, Jr. 1956. *Hawks, owls, and wildlife.* Dover, New York.

Dice, L. R. 1920. The mammals of Warren Woods, Berrien County, Michigan. *Univ. Mich. Mus. Zool. Occas. Pap.* No. 86.

Getz, L. L. 1961. Factors influencing the local distribution of *Microtus* and *Synaptomys* in southern Michigan. *Ecology* 42:110–19.

Holman, J. A. 1975. Michigan fossil vertebrates. *Mich. State Univ. Mus. Publ., Education Bull.* No. 2.

Irvine, G. W. 1989. Evaluation of marten translocations on the Manistee national forest and the Pere Marquette State Forest. *Mich. Dept. Nat. Resour.*, Unpubl. Rept.

Jensen, W. F., T. K. Fuller, and W. L. Robinson. 1986. Wolf, *Canis lupus*, distribution on the Ontario-Michigan Border near Sault Ste. Marie. *Can. Field-Nat.* 100:363–66.

Jones, J. K., Jr., and E. C. Birney. 1988. *Handbook of mammals of the north-central United States.* Univ. of Minn. Press, Minneapolis.

Kurta, A. 1980. Status of the Indiana bat, *Myotis sodalis*, in Michigan. *Mich. Acad. Sci.* 13:31–36.

Kurta, A., T. Hubbard, and M. E. Stewart. 1989. Bat species diversity in central Michigan. *Jack-Pine Warbler* 67:80–87.

Linduska, J. P. 1950. Ecology and land-use relationships of small mammals on a Michigan farm. *Mich. Dept. Conserv., Fed. Aid Project 2-R.*

Manville, R. H. 1949. A study of small mammal populations in northern Michigan. *Univ. Mich. Mus. Zool. Misc. Publ.* No. 73.

Master, L. L. 1978. A survey of the current distribution, abundance, and habitat requirements of threatened and potentially threatened species of small mammals in Michigan. *Univ. Mich. Mus. Zool.*, Unpubl. Rept.

Master, L. L. 1982. The smokey shrew: A new mammal for Michigan. *Jack-Pine Warbler* 60:28–29.

Murie, A. 1934. The moose of Isle Royale. *Univ. Mich. Mus. Zool. Misc. Publ.* No. 25.

Peterson, R. O. 1977. Wolf ecology and prey relationships on Isle Royale. *U.S. Natl. Park Serv., Sci. Monogr. Serv.* No. 11.

Peterson, R. O., and R. E. Page. 1988. The rise and fall of Isle Royale wolves, 1975–1986. *J. Mamm.* 69:89–99.

Ryan, J. M. 1982. Distribution and habitat of the pigmy shrew, *Sorex (Microsorex) hoyi*, in Michigan. *Jack-Pine Warbler* 60:85–86.

Shier, J. L. 1981. *Habitats of threatened small mammals on the Rose Lake Wildlife Research Area, Clinton County, Michigan—early forties and late seventies.* M.S. thesis, Mich. State Univ., E. Lansing.

Wallace, G. J. 1948. The barn owl in Michigan: Its distribution, natural history and food habits. *Mich. State Coll. Agr. Exp. Sta. Tech. Bull.* No. 208.

Wilkinson, A. M. 1980. *Status and distribution of threatened and rare mammals in certain wetland ecosystems in Upper Michigan.* M.S. thesis, Mich. Tech., Houghton.

Wilson, K. A. 1938. Owl studies at Ann Arbor, Michigan. *Auk* 55:187–97.

Wood, N. A. 1922. The mammals of Washtenaw County, Michigan. *Univ. Mich. Mus. Zool. Occ. Pap.* No. 123.

Least Shrew

· *Cryptotis parva* (Say)

· **State Threatened**

Map 1. Confirmed individual least shrews. ▲ = before 1961 (Baker 1983). No records known since 1961.

Status Rationale: Common in much of its North American range, the least shrew had a scattered and limited range in the southern Lower Peninsula. The reasons for this shrew's apparent decline since the 1950s are unclear, since it has general habitat requirements and is not directly persecuted. Increased efforts in locating populations are needed before considering it extirpated in Michigan.

Field Identification: Among Michigan's seven shrews, the least shrew is one of the smallest. The elongated head, pointed nose, barely distinguishable eyes and ears, and grayish brown fur are characteristic of this and many other shrews. Adults have a body length of $2^1/2$ to $3^1/4$ inches (6.4 to 8.3 cm), and a tail of 1/2 to 3/4 inches (1.3 to 1.9 cm). The short tail separates the least shrew from Michigan's longer tailed shrews. In the southern Lower Peninsula, only the masked shrew (*Sorex cinereus*) compares in overall body size, but it has a tail length of $1^1/4$ to 2 inches (3.2 to 5.1 cm). The nominate subspecies occurs in Michigan.

Total Range: The least shrew ranges throughout much of the eastern United States (northeast to New York), west through the southern Great Plains, and south through Central America to Panama. Nine subspecies are recognized throughout this range (Whitaker 1974). In the Great Lakes region, this typically common species is found in abundance as far north as extreme southeastern Minnesota and southwestern Wisconsin. Regional declines have been noted in the southern Great Lakes basin and the Northeast.

State History and Distribution: Michigan historically served as the northernmost limit of this shrew's range. Occurring in a wide variety of field and forest habitats, it has been recorded in 12 Lower Peninsula counties (Baker 1983), primarily in the southernmost three-county tier. The first confirmation of its presence was in 1874 (Wood and Dice 1923). The last individual found in the state was in Livingston County in the autumn of 1960. Since then, a few researchers have investigated historical sites but have not found evidence of its presence (Master 1978; Shier 1981).

Habitat: The least shrew generally inhabits upland grasslands, but frequently occurs in abandoned fields (forblands), hedgerows, wet meadows, orchards, and forest edges. Ungrazed tallgrass areas may be preferred over actively grazed pastures (Grant et al. 1982).

Many of Michigan's least shrews have been trapped in grassy areas with rank, herbaceous vegetation. The globular nests are placed under logs or other debris providing cover (Broadbrooks 1952; Kilham 1954). Least shrews use existing surface tunnels constructed by voles (*Microtus* spp.); underground burrows are rarely used. In winter, they commonly seek cover under hay bales remaining in fields.

Ecology and Life History: Shrews are normally solitary; the least shrew is an exception, however, particularly in winter. Several dozen may be found in one communal nest (McCarley 1959; Davis and Joeris 1945). This social behavior also has been shown in captive conditions and during the breeding season, when several adults may care for one litter (Conaway 1958). Hoffmeister and Mohr (1957) estimated 10 to 15 individuals per acre (0.4 ha). The upper limit of its home range size is 3 acres (1.2 ha) (Choate and Fleharty 1973). In Michigan, the least shrew may breed as early as March and continue through autumn, producing 3 to 4 litters a year. After a gestation period of 3 weeks, litters of 3 to 6 young are produced. The young are weaned in 3 weeks.

Like other insectivores, the least shrew has an extremely high metabolic rate. It demands as much food as its total body weight every 24 hours (Barret 1969). Invertebrates, many considered insect pests, are its primary food (Whitaker and Mumford 1972). Shrews are opportunistic foragers, however, and will prey on most small, living organisms. Shrews are most active at night, and for the least shrew, its seasonal activity peaks are most pronounced in spring and autumn (Briese and Smith 1974). When prey availability is high, the least shrew will hoard food items (Hamilton 1934 and 1944; Davis and Joeris 1945), particularly females (Formanowicz et al. 1989).

Conservation/Management: The least shrew's peripheral status in Michigan explains its historically low populations. This species only occurred locally even though habitat selection was apparently ubiquitous. Populations may have become isolated and eventually lost due to natural phenomena, genetic limitations, and/or intensive land use by humans. Whatever the case, the least shrew still may exist in Michigan, since few systematic and widespread trapping surveys have been conducted.

Since effective habitat management is difficult for this apparently opportunistic species, a first step is to simply document its presence. The use of pitfall traps and drift fences are the most effective techniques to capture shrews (Briese and Smith 1974; Williams and Braun 1983). Least shrews typically do not build burrows and prefer to remain under objects during the day; by remembering this behavioral trait, searches and surveys can prove more effective. In winter, Barbour and Davis (1974) suggest that an efficient method for locating least shrews is to search the undersides of hay bales left in fields. Analyzing owl pellets is another technique that may provide clues on the existence of least shrews in an area (Wallace 1948; Armstrong 1958).

LITERATURE CITED

Armstrong, W. L. 1958. Nesting and food habits of the long-eared owl in Michigan. *Mich. State Univ., Publ. Mus. Biol. Ser.* No. 1:61–96.

Baker, R. H. 1983. *Michigan mammals.* Mich. State Univ. Press, E. Lansing.

Barbour, R. W., and W. H. Davis. 1974. *Mammals of Kentucky.* Univ. Kentucky Press, Lexington.

Barret, G. W. 1969. Bioenergetics of a captive least shrew, *Cryptotis parva. J. Mamm.* 50:629–30.

Briese, L. A., and M. H. Smith. 1974. Seasonal abundance and movement of nine species of small mammals. *J. Mamm.* 55:615–29.

Broadbrooks, H. E. 1952. Nest and behavior of short-tailed shrew, *Cryptotis parva. J. Mamm.* 33:241–43.

Choate, J. R., and E. D. Fleharty. 1973. Habitat preference and spatial relations of shrews in a mixed grassland in Kansas. *Southwest Nat.* 18:110–12.

Conaway, C. H. 1958. Maintenance, reproduction and growth of the least shrew in captivity. *J. Mamm.* 39:507–12.

Davis, W. B., and L. Joeris. 1945. Notes on the life history of the little short-tailed shrew. *J. Mamm.* 26:136–39.

Formanowicz, D. R., Jr., P. J. Bradley, and E. O. Brodie, Jr. 1989. Food hoarding by the least shrew (*Cryptotis parva*): Intersexual and prey type effects. *Am. Midl. Nat.* 122:26–33.

Grant, W. E., E. C. Birney, N. R. French, and D. M. Swift. 1982. Structure and productivity of grassland small mammal communities related to grazing-induced changes in vegetative cover. *J. Mamm.* 33:241–43.

Hamilton, W. J., Jr. 1934. Habits of *Cryptotis parva* in New York. *J. Mamm.* 15:154–55.

Hamilton, W. J., Jr. 1944. The biology of the little short-tailed shrew, *Cryptotis parva. J. Mamm.* 25:1–7.

Hoffmeister, D. F., and C. O. Mohr. 1957. *Fieldbook of Illinois Mammals.* Ill. Nat. Hist. Surv. No. 4.

Kilham, L. 1954. Cow-pasture nests of *Cryptotis parva parva. J. Mamm.* 35:252.

McCarley, W. H. 1959. An unusually large nest of *Cryptotis parva. J. Mamm.* 40:243.

Master, L. L. 1978. A survey of the current distribution, abundance, and habitat requirements of threatened and potentially threatened species of small mammals in Michigan. *Univ. Mich. Mus. Zool.,* Unpubl. Rept.

Shier, J. L. 1981. *Habitats of threatened small mammals on the Rose Lake Wildlife Research Areas, Clinton County, Michigan—early forties and late seventies.* M.S. thesis, Mich. State Univ., E. Lansing.

Wallace, G. J. 1948. The barn owl in Michigan: Its distribution, natural history and food habits. *Mich. State Coll. Agr. Exp. Sta., Tech. Bull.* No. 208.

Whitaker, J. O., Jr. 1974. Mammalian Species: *Cryptotis parva. Am. Soc. Mamm.,* Mamm. Ser. No. 43.

Whitaker, J. O., Jr., and R. E. Mumford. 1972. Food and ectoparasites of Indiana shrews. *J. Mamm.* 53:329–35.

Williams, D. F., and S. E. Braun. 1983. Comparison of pitfall and conventional traps for sampling small mammal populations. *J. Wildl. Mgmt.* 47:841–45.

Wood, N. A., and L. R. Dice. 1923. Records of the distribution of Michigan mammals. *Mich. Acad. Sci.* 3:425–69.

Indiana Bat

· **Myotis sodalis** Miller and Allen
· **Federally and State Endangered**

Map 2. Confirmed individual Indiana bats. ▲ = before 1978;
● = 1978 to 1992 (Kurta 1980).

Status Rationale: The Indiana bat population has declined more than 50% since 1960. Since 85% of the population overwinters in only seven caves, all in the lower Midwest, it is vulnerable to catastrophic extinction. This species continues to decline over much of its breeding range even though the major hibernacula are now protected, indicating problems during migration and/or on the breeding grounds.

Field Identification: The Indiana bat is one of the smaller bats in Michigan (average body length of 3¹/₄ inches [8.3 cm]) with grayish brown fur, dark wing membranes, a pinkish cast to its underside, and short, rounded ears.

An Indiana bat needs to be in-hand to distinguish it from Michigan's other eight species (Kurta 1982; Baker 1983), particularly to distinguish it from the two similarly sized and colored *Myotis* species. Indiana bats can be readily identified by the following detailed characteristics: (1) the *tragus* (a fleshy projection in the ear) is at least half the length of the ear, (2) a distinct elevated ridge is present on the *calcar* (a structure extending from the heel to support the back margin of the tail), and (3) the hairs on the hind toes are less than the length of the toenail (Hall 1962; Barbour and Davis 1969). The other two Michigan *Myotis* species, the northern long-eared bat (*M. septentrionalis*) and the little brown bat (*M. lucifugus*) lack the calcar. In addition, the northern long-eared bat's ears, when laid forward, extend beyond the tip of the nose (those of the Indiana bat do not), and the little brown bat has hairs that are distinctly longer than the toenail.

41

Total Range: The Indiana bat summers throughout much of the eastern half of the United States. However, actual breeding of this migratory species (determined by observations of lactating females or young in summer) has been documented only in Illinois, Indiana, Iowa, Kentucky, Michigan, Missouri, New York, Ohio, and Tennessee. Other eastern states have summer records but are without confirmed breeding evidence. In winter, the majority of the population (85%) congregate in only seven caves (three in Missouri, two in Kentucky, and two in Indiana) for their six-month hibernation (Brady et al. 1983). Few caves provide the microclimatic conditions required by this species. Caves with large Indiana bat populations are now protected with specially designed cave gates. Additional limestone caves in these three states, as well as Tennessee, Arkansas, and others as far north as the New England area, harbor smaller groups. Indiana bat populations declined between 1960 and 1975 by 28% (Humphrey 1978). The total 1987 population was around 260,000 individuals (Clawson 1987), representing more than a 50% reduction from its 1960 population. Western populations are more unstable and declining faster than eastern ones.

State History and Distribution: Prior to an intensive survey beginning in 1978 (Kurta 1980), there were only nine known records of this species (all recorded from May to October) scattered in the southernmost three tiers of Michigan's Lower Peninsula counties. Between 1978 and 1980, 17 additional individuals were captured in mist nets in the following counties: Eaton, Hillsdale, Livingston, and St. Joseph. Each of the known 26 records in Michigan are either adult females or juveniles of both sexes. A lactating female caught in 1978 (Kurta 1980) was the first unquestionable evidence that this bat reproduces in the state. An exceptional area that may contain a self-sustaining population is the riparian habitat along the Thornapple River in Eaton County. A nursery colony was located in this area in 1991.

Clark et al. (1987) suggested this species has a small home range and competition for prey with other bats may limit its distribution. An intensive 1986 study on the distribution of bats in the central Lower Peninsula (an area from Muskegon-Saginaw county north to Manistee-Iosco county) did not locate the Indiana bat (Kurta et al. 1989). Humphrey et al. (1977) found that this species was difficult to capture in mist nets, even when it was known to be foraging over the netting area. Indiana bats migrate south of Michigan and, from limited band returns, appear to hibernate in the caves of Kentucky (Davis, in Kurta 1980). This species is not known to remain in Michigan for the winter.

Habitat: This species forms colonies and forages in riparian and mature floodplain habitats (Humphrey et al. 1977). Nursery roost sites are usually located under loose bark or in hollows of trees near riparian habitat. Indiana bats typically avoid houses or other artificial structures.

Foraging typically occurs over slow-moving, wooded streams and rivers as well as in the canopy of mature trees (United States Fish and Wildlife Service 1987), although it will use a wide variety of riverine habitat (Clark et al. 1987). Movements may also extend into the outer edge of the floodplain and to nearby solitary trees. A summer colony's foraging area usually encompasses a stretch of stream over a half-mile (0.8 km) in length (Humphrey et al. 1977; LaVal and LaVal 1980). Upland areas isolated from floodplains and nonwooded streams are generally avoided (Humphrey et al. 1977).

The summer habitat of the male Indiana bat is poorly understood. Summering males in Missouri were found mainly to be foraging among trees rather than over water. During winter, Indiana bats hibernate exclusively in caves and mines south of the Michigan border, where cave temperatures stabilize at 50°F (10.0°C) or less. Typical hibernaculum temperatures are between 39 and 46°F (3.9 to 7.8°C) (Clawson 1987).

Ecology and Life History: Courtship and mating begins in late summer near a hibernaculum, and sometimes lasts into the winter. Sperm is stored throughout the winter, and pregnancy does not occur until the female leaves the cave in late April. Upon leaving the winter hibernaculum, where approximately half its life is spent (Mumford and Cope 1958; Hall 1962), the Indiana bat disperses widely to suitable maternity areas.

Females begin to leave the hibernacula in late March or early April, giving birth to a single offspring in June or early July. The females form nursery or maternity colonies, which may be as large as 100 individuals (including both the females and young). Optimal sites may be used in successive years. Warm temperatures in early summer are crucial to the growth and success of each year's progeny (Humphrey et al. 1977). Under normal conditions, the young are able to fly within one month after birth. By late August, most individuals have arrived at the hibernaculum (Thomson 1982; Brady et al. 1983). Indiana bats, particularly males, typically return to their winter cave in successive years (Hall 1962). Survival is relatively high for Indiana bats in undisturbed environments, with average lifespans of ten years in females and six years in males (Humphrey and Cope 1977).

Indiana bats are most active during crepuscular and nighttime hours. Echolocation is used to avoid obstacles and locate prey. Several studies indicate that Indiana bats may be dietary specialists. LaVal and LaVal (1980) found that moths comprised 60% to 95% of their diet in Missouri. Belwood (in LaVal and LaVal 1980) identified 70% of the lactating female's diet to be comprised of moths; Brack and LaVal (1985) found similar diet selectiveness, often in excess of proportional prey availability. Bats are extremely beneficial in the control of insects, and individual bats have been recorded to eat nearly 3,000 insects in one night.

Conservation/Management: The Indiana bat has been federally protected since 1973. Since that time, its numbers have continued to decline to a currently estimated population of 260,000. This large number of bats would appear secure from extinction, but catastrophes in the hibernacula could severely threaten its existence. A critical factor contributing to its population decline has been the extent of disturbance (natural or human related) to the few hibernacula. Floods, ceiling collapses, mortality during severe winters, and human disturbances (e.g., vandalism, caving, and indiscriminate collecting) have severely disrupted local populations. A primary limiting factor in the summer range is selective deforestation of riparian habitats, which usually occurs from the cutting of large, dead trees for firewood. Stream channelization, bank modification, and agricultural development along stream banks also contribute to habitat destruction.

To save the Indiana bat from extinction, a recovery plan was developed by the United States Fish and Wildlife Service (Brady et al. 1983). In Michigan, this bat's riparian habitat can be maintained by protecting mature, wooded areas, leaving large, dead trees standing, and maintaining wide vegetation buffer strips. Further research is needed on this species' summer habitat requirements, the effects of riparian habitat destruction, the effects of water pollution and siltation, and the possible chemical contamination due to the widespread use of pesticides.

The first maternity colony of this species was reported in 1971 in Indiana (Cope et al. 1974), an indication of its summer elusiveness. Intensive surveys to locate nursery colonies, following Kurta's successful methods (Kurta 1980), are crucial to the Indiana bat's future in Michigan.

Comments: Like any wild mammal, a bat should not be handled. Bats that are handled will bite in self-defense, and sick individuals are more likely to be caught. Less than one percent of bats contract rabies, and even these rarely become aggressive (Tuttle and Kern 1981). If bats are not handled, there is very little danger.

LITERATURE CITED

Baker, R. B. 1983. *Michigan mammals.* Mich. State Univ. Press, E. Lansing.

Barbour, R. W., and W. H. Davis. 1969. *Bats of America.* Univ. Kentucky Press, Lexington.

Brack, V., and R. K. LaVal. 1985. Food habits of the Indiana bat in Missouri. *J. Mamm.* 66:308–15.

Brady, J., R. L. Clawson, R. K. LaVal, T. Kunz, M. D. Tuttle, and D. Wilson. 1983. Recovery plan for the Indiana bat. *U.S. Fish Wildl. Serv.* Rockville, MD.

Clark, B. K., J. B. Bowles, and B. S. Clark. 1987. Summer status of the endangered Indiana bat in Iowa. *Am. Midl. Nat.* 118:32–39.

Clawson, R. L. 1987. Indiana bats: Down for the count. *Endangered Species Tech. Bull.* 12(9): 9–11.

Cope, J. B., A. R. Richter, and R. S. Mills. 1974. A summer concentration of the Indiana bat, *Myotis sodalis,* in Wayne County, Indiana. *Proc. Indiana Acad. Sci.* 83:482–84.

Hall, J. S. 1962. A life history and taxonomic study of the Indiana bat, *Myotis sodalis. Reading Publ. Mus. Art Gallery Sci. Publ.* No. 12.

Humphrey, S. R. 1978. Status, winter habitat and management of the endangered Indiana bat, *Myotis sodalis. Florida Sci.* 41:65–76.

Humphrey, S. R., and J. B. Cope. 1977. Survival rates of the endangered Indiana bat, *Myotis sodalis. J. Mamm.* 58:32–36.

Humphrey, S. R., A. R. Richter, and J. B. Cope. 1977. Summer habitat and ecology of the endangered Indiana bat. *J. Mamm.* 58:334–46.

Kurta, A. 1980. Status of the Indiana bat, *Myotis sodalis,* in Michigan. *Mich. Acad. Sci.* 13:31–36.

Kurta, A. 1982. A review of Michigan bats: Seasonal and geographic distribution. *Mich. Acad. Sci.* 15:295–312.

Kurta, A., T. Hubbard, and M. E. Stewart. 1989. Bat species diversity in central Michigan. *Jack-Pine Warbler* 67:80–87.

LaVal, R. K., and M. L. LaVal. 1980. Ecological studies and management of Missouri bats with emphasis on cave-dwelling species. *Terrestrial Ser., Missouri Dept. Conserv.* No. 8.

Mumford, R. E., and J. B. Cope. 1958. Summer records of *Myotis sodalis* in Indiana. *J. Mamm.* 39:586–87.

Thomson, C. F. 1982. *Myotis sodalis.* Mamm. Species No. 163.

Tuttle, M. D., and S. J. Kern. 1981. Bats and public health. Milwaukee Publ. Mus. Press. No. 48.

United States Fish and Wildlife Service. 1987. *Endangered Species Tech. Bull.* 12(2): 2–3.

Gray Wolf

· *Canis lupus* Linnaeus
· **Federally and State Endangered**

Map 3. Valid reported observations of gray wolves. Historically statewide (Baker 1983); ▲ = 1992, individuals (Michigan DNR); ● = 1992 packs (Michigan DNR).

Status Rationale: The gray wolf currently inhabits only a fraction of its original North American range. Except for Isle Royale National Park's wolf packs, Michigan probably had been without a breeding wolf population since the late 1950s. The wolf has yet to reestablish itself, even though dispersal into Michigan's Upper Peninsula from Wisconsin and Ontario occurs regularly. This trend is changing as wolves are forming activity centers in various Upper Peninsula locales. A pack in the central Upper Peninsula produced pups in 1991 and 1992.

Field Identification: The gray wolf is the largest wild member of the dog family. In the Great Lakes region, individuals are usually a grizzled gray with a darker shoulder mantle. The male's dimensions average slightly larger than the female. Average measurements for gray wolves in Minnesota are 60 inches (1.5 m) long from nose to tail, ranging from 52 to 72 inches (1.3 to 1.8 m). The tail is approximately 1¹/₂ feet (46 cm) long. Although weight varies depending on environmental conditions, age, and sex, an average wolf from the Great Lakes region weighs between 58 and 67 pounds (26.3 and 30.4 kg), with a maximum of about 100 pounds (45.4 kg). The gray wolf or eastern timber wolf (*C. l. lycaon*) subspecies occurs in Michigan.

Wolves can easily be confused with coyotes (*Canis latrans*), dog-wolf hybrids, and large dogs. General appearance and behavior differences are useful, but wolves are usually noticed by the sign they leave. Tracks are especially indicative of wolf presence. Wolf tracks can easily be distinguished from coyotes by the width of the track. Tracks that are greater than $2^3/4$ inches (7.0 cm) in width are not made by coyotes. The track size of a wolf generally ranges from $2^3/4$ to $3^1/4$ inches (7.0 cm to 8.3 cm) wide and $3^3/4$ to $5^1/2$ inches (9.5 to 14.0 cm) long, although larger prints do occur (Murie 1974). There may be some overlap in track size between a large dog and wolf. Two of Michigan's felines, the lynx (*Felis lynx*) and the cougar (*Felis concolor*), may have comparable tracks to the wolf, but only wolves (like other canids) leave wide, pronounced claw marks. To confirm a sighting, photograph the tracks using a standard object for scale or make a track cast using plaster of paris.

At a distance, a wolf can be distinguished from a coyote by its heavier build, longer and broader snout, and shorter ears. Size is generally less useful and more deceiving at a distance. Wolves often travel with their tail held down and have tufts of hair on their cheeks that give a broad appearance to their head. If a canid scat with a diameter of $1^1/2$ inches (3.8 cm) or more is discovered (particularly one with hairs from a large animal), it is probably from a wolf (Weaver and Fritts 1979).

Coyote howls are higher pitched than those of wolves and contain barks before, during, and after howling (yip-yap howls). Wolves usually produce a single, deep and long howl with silent pauses of one minute or more between each howl. Wolf pups can distort these differences considerably with their yapping. Wolf howls may carry 4 miles (6.4 km) or more depending on weather conditions. They will respond to simulated howls, particularly at night, from July through September (Harrington and Mech 1982). This is useful for locating individuals (Fuller and Sampson 1988) and can be used with other indices, such as radiotelemetry (Fuller and Snow 1988), to estimate densities (Crete and Messier 1987).

Total Range: The gray wolf has a circumpolar distribution and was historically widespread in North America from Canada and Alaska south to central Mexico. Now it occurs in viable populations only in Alaska, Minnesota, and much of Canada (numbers are increasing in the U.S. northern Rocky Mountains). Alaska's declining population is estimated to be 5,000; Canada's sizable population is estimated at 50,000. South of the Canada–United States border, wolves nearly have been exterminated, often by unjustified and systematic killings. Minnesota's population of over 1,200 individuals has dispersed into Wisconsin (Mech and Nowak 1981; Thiel and Welch 1981), which has approximately 50 individuals. Dispersing wolves into Michigan's Upper Peninsula are known to originate from Wisconsin and Ontario.

State History and Distribution: Historically, this species was found throughout Michigan. One year after its statehood in 1837, Michigan instituted a wolf bounty. Except for a 13-year lapse during a state trapper program, the bounty system remained intact until its repeal in 1960. By 1910, the wolf had been extirpated from the southern part of the Lower Peninsula and individuals in the northern Lower Peninsula were too scattered for packs to develop. Since the late 1950s, continued persecution of the wolf has prevented natural attempts to form a stable pack structure, a requirement for a viable population. By 1965, the wolf was officially declared a protected species, but it was too late to protect Michigan populations.

Hendrickson et al. (1975) estimated that a minimum of six individuals remained on the Upper Peninsula mainland in 1973.

It is now estimated that about 25 wolves (loners and small groups), survive throughout the Upper Peninsula mainland. Encouragingly, groups recently have been found in Chippewa, Gogebic, Iron, Mackinac, and Schoolcraft counties. Since the winter of 1986–87, several wolves from Wisconsin have been followed in Michigan's Iron County. And, for the first time since the late 1950s, pack formation was possible in 1987; unfortunately, the adult female was killed. In this same territory, seven wolf pups were produced in 1991 (the first mainland production since the mid-1950s) and one in 1992.

Michigan's second breeding wolf population occurs on Isle Royale National Park, Keweenaw County. In the winter of 1948–49, a pack crossed a 14-mile (22.5 km) ice bridge from Sibley Peninsula, Ontario, to Isle Royale. Since that crossing, the island has become internationally known for its well-studied wolf-moose (*Alces alces*) relationship. Annual intensive studies since 1959 have provided detailed information about wolf population dynamics and how this predator is an integral part of the ecosystem (Mech 1966; Wolfe and Allen 1973; Peterson 1977; Allen 1993; Peterson and Page 1988).

After many years of small population fluctuations, numbers on Isle Royale slowly increased until 1980, when they peaked at 50 individuals (around 1 wolf per 7 square miles [18 square km]) (Peterson and Page 1988). A declining food base ultimately caused a population decline to 14 individuals by 1982. The population rebounded from this low to a more stabilized total of 24 in 1983. Subsequently, the population dropped to 16 individuals in the winter of 1987–88, 12 in 1988–89, and 9 in 1989–90. Disease brought from the mainland was believed to be a contributing factor to the wolf decline. Consequently, researchers have intensified wolf studies through live-capture for radio-collaring and blood sampling to monitor the incidence of disease and determine the degree of inbreeding.

Habitat: The gray wolf does not have specific habitat requirements other than the need for extensive areas with minimal human disturbance and sufficient wild ungulate (e.g., deer) densities. It is not strictly a wilderness mammal; the wolf can survive in moderately developed regions, as long as there is limited contact with humans. Pimlott (1967) concluded that deer densities of ten individuals per square mile (2.6 square km) were sufficient to sustain wolf populations; although more recently, Keith (1983) found that only four deer per square mile (2.6 square km) were required. Regardless of specific densities, Upper Peninsula deer numbers are not a limiting factor.

Areas of 100 square miles (161 square km) with road densities less than one mile (1.6 km) of linear road per square mile (2.6 square km) are suitable wolf habitat (Bailey 1992; Thiel 1985). While hunting, wolves may frequent any habitat, although areas such as trails, roads, and ice-covered bodies of water are often utilized (Wolfe and Allen 1973). Dens are typically situated in underground burrows, often enlarging those excavated by other animals. Den sites include rock crevices and ledges, hollow logs, overturned stumps, and debris piles. Most dens are in well-drained locations usually with an open view of the surroundings and a readily available water supply.

Ecology and Life History: The gray wolf is a very social animal with strong family attachments. Its behavior, ecology, and management have been intensively studied in North America (Murie 1944; Mech 1970; Klinghammer 1978; Zimen 1981; Ballard et al. 1987), particularly in the Upper Great Lakes region (Van Ballenberghe et al. 1975; Peterson 1977; Allen 1993; Fritts and Mech 1981; Nelson and Mech 1981; Fuller 1989).

Packs are family groups, usually including the alpha (dominant) male and female, offspring of previous years, young-of-the-year, and occasionally unrelated individuals. In the Upper Great Lakes region, a pack typically consists of five to seven members. Harmony within the pack is maintained through a hierarchical system. Competition for dominance occurs, but major injuries through fighting usually are avoided by ritualized submissive postures and facial expressions. Pack size depends, in part, on the density and type of ungulate species present.

Normally, only the alpha male and female of a pack mate and reproduce (they usually pair for life), even when subordinate, sexually mature and receptive males and females are present. From late December to January, the alpha pair begin their relatively long courtship. In Michigan, the female wolf is usually receptive to mating in February and March (Baker 1983). After a gestation period of 63 days, a litter of four to ten pups (average of six) is born in April or May. The same den site is commonly used in successive years (Fuller 1989).

During the early pup-rearing period, the female cares for and remains with the young while the remainder of the pack hunts for the family cause. Pack structure typically loosens during the summer, with some individuals becoming more independent. Nearly one month after birth, the juveniles begin to explore outside the den; family groups frequently change den sites at this time (Fuller 1989). Approximately three weeks later, the pups are weaned. Soon after, the pack, accompanied by the young, seek out various above-ground areas called rendezvous sites that serve as nursery areas (e.g., beaver meadows). By September or October, the young-of-the-year can participate in hunts. Wolves are sexually mature in 2 to 3 years, and are potentially reproductive up to approximately 11 years of age (Mech 1988).

The pack is an important social unit that enables the killing of large prey through cooperation, increasing the probability of rearing the young successfully, and helps to defend hunting territories against other packs. Home ranges vary greatly from 50 square miles (130 square km) in Minnesota (Mech and Frenzel 1971) to 5,000 square miles (13,000 square km) in Alaska (Burkholder 1959). Territories are maintained by scent marking and are vigorously defended (Peters and Mech 1975). Howling provides a way of locating and assembling pack members and also serves as an advertisement of the pack's presence to other, neighboring packs. In response to food shortages, packs may trespass; individuals occasionally are killed during territorial disputes.

Wolves are highly dependent on ungulate populations. Although snowshoe hare (*Lepus americanus*)(Stebler 1944), beaver (*Castor canadensis*), and other mammals and birds are commonly taken, they only serve as supplements. Carrion and berries are eaten when available. In the upper Great Lakes region, the primary prey species is the white-tailed deer (*Odocoileus virginianus*) and, in some areas, moose. Of the large prey species, the weak, diseased, injured, young, and aged are more likely to be killed than healthy breeding individuals (Mech 1970; Mech and Frenzel 1971; Mech and Karns 1977; Nelson and Mech 1981). During long, severe winters with deep snow and starving deer, wolf predation may be at its highest (Nelson and Mech 1986). Wolf predation on livestock is typically low; less than 1% of domesticated animals is taken, even in areas with relatively high wolf densities (Fritts 1982).

Conservation/Management: One of the greatest threats to the remaining populations of wolves in North America is not habitat destruction, but an antiwolf attitude by the general public, which is usually based upon erroneous beliefs. Since the late 1950s, when the wolf was eliminated as a viable part of Michigan's fauna, pressure from a minority of the public has prevented pack formation and, therefore, reproduction (Hendrickson et al. 1975; Weise et al. 1975). Hook and Robinson (1982) found the most important factor contributing to an antipredator attitude was fear of the wolf, even though this fear is based on unreasonable assumptions. Providing information to hunters and local residents is the most effective way for gaining acceptance of the wolf.

The Upper Peninsula contains several areas that are suitable for supporting wolf packs. There are several cases indicating that wolves are increasing in numbers and expanding their Great Lakes range (Jensen et al. 1986; Thiel 1988; Thiel and Hammill 1988). Wolves can recover in Michigan's Upper Peninsula, without any introductions, if there is public support. Reestablishing a viable wolf population in Michigan will require a public education program, law enforcement to reduce illegal wolf harassment, and careful monitoring of individual wolves present in the Upper Peninsula.

LITERATURE CITED

Allen, D. L. 1993. *Wolves of Minong: Isle Royale's wild community.* Univ. Mich. Press, Ann Arbor.

Bailey, R. (Ed.). 1992. Recovery plan for the eastern timber wolf. *U.S. Fish Wildl. Serv.,* Twin Cities, MN.

Baker, R. H. 1983. *Michigan mammals.* Mich. State Univ. Press, E. Lansing.

Ballard, W. B., J. S. Whitman, and C. L. Gourdner. 1987. Ecology of an exploited wolf population in south-central Alaska. *Wildl. Monogr.* No. 98.

Burkholder, B. L. 1959. Movements and behavior of a wolf pack in Alaska. *J. Wildl. Mgmt.* 23:1–11.

Crete, M., and F. Messier. 1987. Evaluation of indices of gray wolf, *Canis lupus,* density in hardwood-conifer forests of southwestern Quebec. *Can. Field-Nat.* 101:147–52.

Fritts, S. H. 1982. Wolf predation on livestock in Minnesota. *U.S. Fish Wildl. Serv., Resour. Publ.* No. 145.

Fritts, S. H., and L. D. Mech. 1981. Dynamics, movements, and feeding ecology of a newly protected wolf population in northwestern Minnesota. *Wildl. Monogr.* No. 80.

Fuller, T. K. 1989. Population dynamics of wolves in north-central Minnesota. *Wildl. Monogr.* No. 105.

Fuller, T. K., and B. A. Sampson. 1988. Evaluation of a simulated howling survey for wolves. *J. Wildl. Mgmt.* 52:60–63.

Fuller, T. K., and W. J. Snow. 1988. Estimating winter wolf densities using radiotelemetry data. *Wildl. Soc. Bull.* 16:367–70.

Harrington, F. H. and L. D. Mech. 1982. An analysis of howling response parameters useful for wolf pack censusing. *J. Wildl. Mgmt.* 46:686–93.

Hendrickson, J., W. L. Robinson, and L. D. Mech. 1975. Status of the wolf in Michigan, 1973. *Am. Midl. Nat.* 94:226–31.

Hook, R. A., and W. L. Robinson. 1982. Attitudes of Michigan citizens toward predators. Pp. 382–94 *in* F. H. Harrington and A. C. Paquet (eds.), *Wolves of the world: Perspectives of behavior, ecology, and conservation.* Noyes Publ., Park Ridge, N.J.

Jensen, W. F., T. K. Fuller, and W. L. Robinson. 1986. Wolf (*Canis lupus*) distribution on the Ontario-Michigan border near Sault Ste. Marie. *Can. Field-Nat.* 100:363–66.

Keith, L. B. 1983. Population dynamics of wolves. Pp. 66–77 *in* L. Carbyn (ed.), Wolves of Canada and Alaska. *Can. Wildl. Serv. Rept. Ser.* No. 45.

Klinghammer, E. 1978. *The behavior and ecology of wolves.* Garland STPM Press, New York.

Mech, L. D. 1966. The wolves of Isle Royale. *U.S. Natl. Park Serv., Fauna Ser.* No. 7.

Mech, L. D. 1970. *The wolf: The ecology and behavior of an endangered species.* Natural History Press, Garden City, N.Y.

Mech, L. D. 1988. Longevity in wild wolves. *J. Mamm.* 69:197–98.

Mech, L. D., and L. D. Frenzel. 1971. Ecological studies of the timber wolf in northeastern Minnesota. *U.S. Dept. Agric. For. Serv., N. Cent. For. Exp. Sta. Res. Pap.* NC-52.

Mech, L. D., and P. D. Karns. 1977. Role of the wolf in a deer decline in the Superior National Forest. *U.S. Dept. Agric. For. Serv., N. Cent. For. Exp. Sta. Res. Pap.* NC-148.

Mech, L. D., and R. M. Nowak. 1981. Return of the gray wolf to Wisconsin. *Am. Midl. Nat.* 105:408–9.

Murie, A. 1944. The wolves of Mount McKinley. *U.S. Natl. Park Serv., Fauna Ser.* No. 5.

Murie, O. J. 1974. *A field guide to animal tracks.* Houghton Mifflin Co., Boston.

Nelson, M. E., and L. D. Mech. 1981. Deer social organization and wolf predation in northeastern Minnesota. *Wildl. Monogr.* No. 77.

Nelson, M. E., and L. D. Mech. 1986. Relationship between snow depth and gray wolf predation on white-tailed deer. *J. Wildl. Mgmt.* 50:471–74.

Peterson, R. O. 1977. Wolf ecology and prey relationships on Isle Royale. *U.S. Natl. Park Serv., Sci. Monogr. Serv.* No. 11.

Peterson, R. O., and R. E. Page. 1988. The rise and fall of Isle Royale wolves, 1975–1986. *J. Mamm.* 69:89–99.

Peters, R. P., and L. D. Mech. 1975. Scent-marking in wolves. *Am. Sci.* 63:628–37.

Pimlott, D. H. 1967. Wolf predation and ungulate populations. *Am. Zool.* 7:267–78.

Stebler, A. M. 1944. The status of the wolf in Michigan. *J. Mamm.* 25:37–43.

Thiel, R. P. 1985. The relationship between road densities and wolf habitat suitability in Wisconsin. *Am. Midl. Nat.* 113:404–7.

Thiel, R. P. 1988. Dispersal of a Wisconsin wolf into upper Michigan. *Jack-Pine Warbler* 66:143–47.

Thiel, R. P., and J. H. Hammill. 1988. Wolf specimen records in upper Michigan, 1960–1986. *Jack-Pine Warbler* 66:149–53.

Thiel, R. P., and R. J. Welch. 1981. Evidence of recent breeding activity in Wisconsin wolves. *Am. Midl. Nat.* 106:401–2.

Van Ballenberghe, V., A. W. Erickson, and D. Byman. 1975. Ecology of the timber wolf in northeastern Minnesota. *Wildl. Monogr.* No. 43.

Weaver, J. L., and S. H. Fritts. 1979. Comparison of coyote and wolf scat diameters. *J. Wildl. Mgmt.* 43:786–88.

Weise, T. F., W. L. Robinson, R. A. Hook, and L. D. Mech. 1975. An experimental translocation of the eastern timber wolf. *Audubon Conserv. Rept.* No. 5. U.S. Fish Wildl. Serv.,
Twin Cities, Minn.

Wolfe, M. L., and D. L. Allen. 1973. Continued studies of the status, socialization, and
relationships of Isle Royale wolves, 1967 to 1970. *J. Mamm.* 54:611–33.

Zimen, E. 1981. *The wolf.* Delacorte Press, New York.

Marten

· *Martes americana* (Turton)
· **State Threatened**

Map 4. Confirmed individual martens. Historically statewide (Baker 1983); ● = 1992 (Michigan DNR).

Status Rationale: Early in this century, North American marten populations declined due to widespread logging and trapping. Recently, natural colonization, reestablishment projects, habitat improvement, and trapping regulations have reversed this downward trend in the Great Lakes region. Michigan, in particular, has worked intensively to reestablish this species, with encouraging results.

Field Identification: This member of the weasel family has a yellowish brown to dark brown pelage, light-colored head contrasting with the body, white-tipped ears, and a distinctive pale buff patch on the breast (the size and color of this patch varies). These characters, along with the lack of a white chinstrap, should distinguish the marten from the smaller but similar mink (*Mustela vison*). The larger, but similar appearing, fisher lacks the buff breast patch and yellowish brown pelage of the marten; instead, it has a dark brown pelage with grizzled, silvery-tipped hairs on its head and shoulders. The average length of the adult marten ranges from 14 to 17 inches (35.6 to 43.2 cm) for the body, plus a 7 to 9 inch (17.8 to 22.9 cm) tail; males are 12% to 15% longer (Baker 1983). Most individuals weigh 1^1/2 to 2^3/4 pounds (0.68 to 1.25 kg); males are 40% to 80% heavier than females.

Tracks are frequently the best evidence of marten presence. Generally, the weasel family's five-toed tracks can be distinguished from most other similarly sized mammals by their tendency to form a twin-print pattern while running (hind feet register in front tracks). The tracks of a marten and mink are similar, and

53

identification is particularly difficult in marginal tracking conditions (e.g., deep snow). In snow, marten tracks are 1¹/2 to 1³/4 inches (3.8 to 4.5 cm) wide by 2¹/2 to 2³/4 inches (6.4 to 7.0 cm) long for females and 2 to 2¹/2 inches (5.1 to 6.4 cm) wide by 3 to 3¹/2 inches (7.6 to 8.9 cm) long for males. Scats associated with weasel-like tracks can be attributed to the marten's more omnivorous diet if berry pits and seeds are present (when in season). Compared to marten tracks, weasel (*Mustela* spp.) tracks are smaller, and fisher (*Martes pennanti*) tracks are larger.

Total Range: Prior to European settlement, the marten occurred in boreal forests from the southern Rocky Mountains, upper Great Lakes region, Appalachian Mountains of Pennsylvania and New York, and throughout the New England states, north to the Arctic treeline. By the early 1900s, it had disappeared from much of the United States and southern Canada. Today, the marten is returning to its southern range through natural colonization and release projects. It has now reclaimed much of the remaining suitable habitat in its historic U.S. range. Reestablishment programs in the Great Lakes region have been successful in Michigan, Ontario, and Wisconsin. Martens have increased and expanded in Minnesota because of protection (Mech and Rogers 1977; Berg 1982).

State History and Distribution: Historically, the marten was abundant in mature forests throughout Michigan. Intensive fur trapping practices, coupled with subsequent logging and fires, decimated the marten population and its habitat until it was extirpated in the Lower Peninsula by 1911 and in the Upper Peninsula by 1939. It disappeared from Isle Royale, Keweenaw County, in the early 1900s, the last being trapped in 1905 (Martin 1988).

In the winters of 1968–69 and 1969–70, a reestablishment effort released 99 martens into the Whitefish River area of Hiawatha National Forest in Alger and Delta counties. This release was originally thought to have failed, but numerous recent sightings near the area confirm that martens are present and reproducing. From 1978 to 1981, the Michigan Marten Reintroduction Program was responsible for an additional 148 martens released at three sites: the Huron Mountain Club, Marquette County; the McCormick Wilderness Area, Marquette and Baraga counties; and in the Iron River district of the Ottawa National Forest, Iron County (Churchill et al. 1981). Other releases from 1989–91 in Luce and Chippewa counties help continue its complete range recovery in the Upper Peninsula.

Since 1981, there have been reliable marten reports in 13 Upper Peninsula counties. Western Alger and Marquette counties probably support the highest numbers. Today, martens probably occur in all of the Upper Peninsula counties. In November and December 1985, 49 martens were released in the Pigeon River Country State Forest in the northern Lower Peninsula. To supplement this population, another 21 were released in the Huron-Manistee National Forest and 15 in Lake County's Pere Marquette State Forest in March 1986. Subsequent investigations of the 1986 release established that martens had remained near the release site and were reproducing (Irvine 1989); similar findings also are known for the 1985 release site.

Habitat: Throughout much of their range, martens are generally associated with mature, upland coniferous forest (Marshall 1951; de Vos 1952; Koehler and Hornocker 1977; Bateman 1986). However, marten populations in the Great Lakes region also utilize younger, mixed conifer-hardwood stands (Hagmeier 1956; Francis and Stephenson 1972; Mech and Rogers 1977) and northern white-cedar swamps (*Thuja occidentalis*) (de Vos and Guenther 1952). The use of second-growth mixed forests has been important to the success of the marten's reestablishment in Michigan, since this habitat type has replaced much of the original mature boreal forests. Recent observations in the western Upper Peninsula have found martens extensively using young, mixed northern hardwood stands. An eastern hemlock (*Tsuga canadensis*) component appears to be preferred, particularly in winter.

While large disturbances provide unsuitable marten habitat, small and scattered openings provide optimal foraging areas with abundant prey and fruiting shrubs. Martens typically avoid openings in the winter (Raine 1983), utilizing them more in the summer (DeBlaay 1980; Schupbach 1977). Windfalls, debris piles, hollow logs, and snags are important components of marten habitat. They provide habitat for small mammalian prey, crucial for winter survival (Steventon and Major 1982; Strickland et al. 1982), den sites (Wynne and Sherburne 1984), and winter cover and resting areas (Allen 1982; Buskirk et al. 1989). Other den types include tree cavities and abandoned buildings. Squirrel nests (Murie 1961) and "witches'-brooms" (Wynne and Sherburne 1984) are used for shelter during summer and winter.

Ecology and Life History: Adult martens are solitary, territorial animals. Home ranges vary considerably by region. The home range of the male marten is typically larger and may overlap several female home ranges. In Algonquin Provincial Park, Ontario (the origin of Michigan's 1978–81 martens), home ranges for males and females were 900 and 260 acres (364 and 105 ha), respectively (Francis and Stephenson 1972; Wynne and Sherburne 1984). The size of the home range depends on habitat quality, prey availability, and marten population density. Thompson and Colgan (1987) found that low prey availability was closely associated with larger home ranges.

The marten's long gestation is due to delayed implantation, a reproductive condition in which cell division of the fertilized egg is internally inhibited for 7 to 8 1/2 months. This reproductive modification permits courtship and mating in July or August and the birth of young the following spring (Baker 1983). In March or April, the female selects a den in a protective hollow for the two to three young. Only the female cares for the young until they become relatively self-sufficient at approximately 11 weeks. Juvenile males move greater distances than females and may disperse up to 25 miles (40 km) (Francis and Stephenson 1972; Taylor and Abrey 1982; Bateman 1986). By the second or third summer, female martens are sexually mature, but males are able to breed the first year (Jonkel and Weckwerth 1963). Rest-site preferences have been well studied, indicating use of aboveground sites in summer (Masters 1980; Martin and Barrett 1983; Buskirk et al. 1989) and ground-level sites in winter (Steventon and Major 1982; Spencer 1987; Buskirk et al. 1989).

The marten is generally active during twilight and nighttime hours throughout the year. The primary prey is voles, particularly red-backed voles (*Clethrionomys gapperi*) (Cowan and Mackay 1950; Quick 1955; Murie 1961; Weckwerth and Hawley 1962; Thompson and Colgan 1987). Martens are opportunistic hunters, feeding arboreally on flying squirrels (*Glaucomys* spp.) or on the ground for red squirrels (*Tamiasciurus hudsonicus*). Other small mammals, birds and their eggs, carrion, and seasonally available fruit and insects are also part of its diet (Murie 1961; Francis and Stephenson 1972; Strickland et al. 1982; Buskirk and MacDonald 1984). During population peaks, snowshoe hare (*Lepus americanus*) may comprise over half of the marten's diet (Raine 1983).

Conservation/Management: The marten's vulnerability to habitat destruction through logging and its susceptibility to trapping has led to its extirpation in many parts of its range, including Michigan. The destruction of its habitat—mature coniferous and mixed wilderness forests—ultimately led to its demise, and trapping pressure became an increasingly important decimating factor during the logging era. However, with a stable population in a suitable area of quality habitat, the marten can withstand regulated trapping pressure.

Recent reestablishment efforts have encountered some difficulties. By wandering from the release site, martens are more susceptible to being shot, trapped, or hit by automobiles. Dispersal, therefore mortality, may be reduced significantly by following these procedures: (1) conduct feasibility studies; (2) use a gentle release program (hold for at least one week); and (3) provide a readily available food source (e.g., carrion) (Berg 1982; Davis 1983). Michigan is now able to translocate individuals from established in-state populations to suitable habitat in other parts of the state.

The current, gradual succession from pioneer plant communities (e.g., aspen-birch) to climax shade-tolerant communities (e.g., hemlock–northern hardwoods and spruce-fir) is crucial for the marten's recovery. To improve marten habitat, forests must be managed for the retention of some old growth, including such shade-tolerant conifers as eastern hemlock. Use of selective cutting or small clearcuts in even-aged stands may enhance marten habitat and increase prey abundance (Koehler et al. 1975). Because martens are trapped easily (Yeager 1950; deVos 1951), the immediate release site should be closed to dry-set trapping (Davis 1983). With the recent successful reestablishment projects throughout much of northern Michigan, the marten's outlook for survival is now encouraging.

LITERATURE CITED

Allen, A. W. 1982. Habitat suitability index models: Marten. *U.S. Fish Wildl. Serv., Biol. Rept.* 82 (10.11).

Baker, R. B. 1983. *Michigan mammals.* Mich. State Univ. Press, E. Lansing.

Bateman, M. C. 1986. Winter habitat use, food habits and home range size of the marten, *Martes americana*, in western Newfoundland. *Can. Field-Nat.* 100:58–62.

Berg, W. E. 1982. Reintroduction of fisher, pine marten, and river otter. Pp. 158–73 *in* G. C. Sanderson (ed.), Midwest furbearer management. *Proc. Midwest Fish Wildl. Conf.*, Wichita, KS.

Buskirk, S. W., and S. D. MacDonald. 1984. Seasonal food habits of marten in south-central Alaska. *Can. J. Zool.* 62:944–50.

Buskirk, S. W., S. C. Forrest, M. G. Raphael, and H. J. Harlow. 1989. Winter resting site ecology of marten in the central Rocky Mountains. *J. Wildl. Mgmt.* 53:191–96.

Churchill, S. J., L. A. Herman, M. F. Herman, and J. P. Ludwig. 1981. Final report on the completion of the Michigan marten reintroduction program. *Ecol. Res. Serv., Inc.*, Iron River, MI, Unpubl. Rept.

Cowan, I., and R. H. Mackay. 1950. Food habits of the marten (*Martes americana*) in the Rocky Mountain region of Canada. *Can. Field-Nat.* 64:100–104.

Davis, M. H. 1983. Post-release movements of introduced marten. *J. Wildl. Mgmt.* 47:59–66.

DeBlaay, T. J. 1980. *A survey of marten (Martes americana) habitat and prey availability in Michigan's upper peninsula.* M.S. thesis, Mich. Tech. Univ., Houghton.

de Vos, A. 1951. The ecology and management of fisher and marten in Ontario. *Ontario Dept. Lands For., Tech. Bull.*

de Vos, A. 1952. Ecology and management of fisher and marten in Ontario. *Ontario Dept. Lands For., Tech. Bull.*

de Vos, A., and S. E. Guenther. 1952. Preliminary live-trapping studies of marten. *J. Wildl. Mgmt.* 16:207–14.

Francis, G. R., and A. B. Stephenson. 1972. Marten ranges and food habits in Algonquin Provincial Park, Ontario. Ontario Ministry Nat. Resour., *Res. Rept. (Wildlife)*, No. 91.

Hagmeier, E. M. 1956. Distribution of marten and fisher in North America. *Can. Field-Nat.* 70:149–68.

Irvine, G. W. 1989. Evaluation of marten translocations on the Manistee National Forest and the Pere Marquette State Forest. *Mich. Dept. Nat. Resour.*, Unpubl. Rept.

Jonkel, C. J., and R. P. Weckwerth. 1963. Sexual maturity and implantation of blastocysts in the wild pine marten. *J. Wildl. Mgmt.* 27:93–98.

Koehler, G. M., and M. G. Hornocker. 1977. Fire effects on marten habitat in the Selway-Bitterroot wilderness. *J. Wildl. Mgmt.* 41: 500–505.

Koehler, G. M., W. R. Moore, and A. R. Taylor. 1975. Preserving the pine marten: Management guidelines for western forests. *West. Wildlands Mag.* 2(3): 31–36.

Marshall, W. H. 1951. Pine marten as a forest product. *J. Forestry* 49:899–905.

Martin, C. 1988. The history and reestablishment potential of marten, lynx, and woodland caribou on Isle Royale. *Natl. Park Serv., Resour. Mgmt. Rept.* 88-2.

Martin, S. K., and R. H. Barrett. 1983. The importance of snags to pine marten habitat in the northern Sierra Nevada. Pp. 114–16 *in* Proc. of symposium on snag habitat management. *U.S. For. Serv., Gen. Tech. Rept.* RM-99.

Masters, R. D. 1980. Daytime resting sites of two Adirondack pine martens. *J. Mamm.* 61:157.

Mech, L. D., and L. L. Rogers. 1977. Status, distribution and movements of martens in northeastern Minnesota. U.S. Dept. Agric., *For. Serv. N. Cent. For. Exp. Sta. Res. Pap.* NC-143.

Murie, A. 1961. Some food habits of the marten. *J. Mamm.* 42:516–21.

Quick, H. F. 1955. Food habits of marten (*Martes americana*) in northern British Columbia. *Can. Field-Nat.* 69:144–47.

Raine, R. M. 1983. Winter habitat use and responses to snow cover of fisher (*Martes pennanti*) and marten (*Martes americana*) in southeastern Manitoba. *Can. J. Zool.* 61:25–34.

Schupbach, T. A. 1977. *History, status, and management of the pine marten in the Upper Peninsula of Michigan.* M.S. thesis, Mich. Tech. Univ., Houghton.

Spencer, W. D. 1987. Seasonal rest-site preferences of martens in the northern Sierra Nevada. *J. Wildl. Mgmt.* 51:616–21.

Steventon, J. D., and J. T. Major. 1982. Marten use of habitat in a commercially clearcut forest. *J. Wildl. Mgmt.* 46:175–82.

Strickland, M. A., C. W. Douglas, M. Novak, and N. P. Hunziger. 1982. Marten. Pp. 599–612 in J. A. Chapman and G. A. Feldhammer (eds.), *Wild mammals of North America.* Johns Hopkins Univ. Press, Baltimore.

Taylor, M. E., and N. Abrey. 1982. Marten, *Martes americana,* movements and habitat use in Algonquin Provincial Park, Ontario. *Can. Field-Nat.* 96:439–47.

Thompson, I. D., and P. W. Colgan. 1987. Numerical responses of martens to a food shortage in northcentral Ontario. *J. Wildl. Mgmt.* 51:824–35.

Weckwerth, R. P., and V. D. Hawley. 1962. Marten food habits and population fluctuations in Montana. *J. Wildl. Mgmt.* 26:55–74.

Wynne, K. M., and J. A. Sherburne. 1984. Summer home range use by adult marten in northwestern Maine. *Can. J. Zool.* 62:941–43.

Yeager, L. E. 1950. Implications of some harvest and habitat factors of pine marten management. *Trans. 15th N. Am. Wildl. Conf.,* 319–34.

Cougar

- *Felis concolor* Linnaeus
- **Federally and State Endangered**

Map 5. Valid reported observations of cougars. Historically statewide (Baker 1983); ▲ = 1962 to 1992, individuals; ● = 1962 to 1992, family groups (Michigan DNR).

Status Rationale: The cougar is a critically endangered mammal in eastern North America. Since the 1950s, however, an increasing number of confirmed sightings have been recorded. Today, several areas throughout its former range, including northern Michigan, may support small populations of cougars.

Field Identification: The cougar, or mountain lion, is the only native Michigan cat with a long tail. Its large size, short tawny pelage, lanky build, small head, and long curled tail (one-third of its total body length) are diagnostic. The long tail is especially indicative; cougars have a tendency to hold the tail low, flicking it frequently with the tip turned upward. Males and females average a total length of 7 1/2 feet and 6 1/2 feet (2.3 and 2.0 m), respectively. Males weigh an average of 190 pounds (86 kg) and females 115 pounds (52 kg).

The best method for positive identification is to investigate sign. Adult cougars typically have round tracks, 3 inches (7.6 cm) in width and length (Murie 1974). Front tracks are commonly 1/2 inch wider (1.3 cm). Large coyotes (*Canis latrans*), small gray wolves (*Canis lupus*), and lynx (*Felis lynx*) and many domestic dogs also have similarly sized tracks.

Unlike canids, felid tracks typically show no claw marks; if they do, the impressions are very thin. Cougar tracks tend to have widely spaced toes (1/4 to 1/2 inch [0.6 to 1.3 cm] apart) shaped like teardrops, with a small outer toe, whereas members of the dog family have narrowly spaced (less than 1/4 inch, 0.6 cm, apart),

59

equal-sized, rounded toes. Other distinctive features of the cougar track include the presence of a forwardly raised, flat, three-lobed heel pad, which has an abrupt, squared-off appearance in the front. With canids, the cross section of the heel pad shows a distinctively higher center lobe and the track slopes down gradually to the front of the heel pad (Downing and Fifield 1985). In winter, tracks on snow-covered logs or rocks indicate a felid, since canids rarely utilize these slippery vantage points. Also, look for the cougar's long tail marks in the snow.

Like Michigan's other native (but short-tailed) cats, the bobcat (*Felis rufus*) and lynx, cougars commonly cover food caches to help conceal and preserve uneaten prey by neatly scratching leaf litter and other debris within a 2 to 3 foot (0.6 to 0.9 m) radius. Other cougar signs include the scrape and mound. Scrapes, made of forest debris, are usually 8 to 10 inches (20.3 to 25.4 cm) wide and 3 to 6 inches (7.6 to 15.2 cm) high and are used as urine scent posts to mark territories. Mounds are much larger than scrapes, usually several inches deep and 2 to 4 feet (0.6 to 1.2 m) across, and are apparently used by the young to await the return of a hunting female cougar. Scats can be used to help identify species by the fecal bile acid (Major et al. 1980; Johnson et al. 1984).

Total Range and Taxonomic Status: Once ranging from the Canadian Yukon through South America, the cougar has experienced a general decline in abundance and distribution over much of its range. This has been particularly evident in the eastern United States (Downing 1981a). Since Canada's boreal forests historically provided low densities of ungulates, cougars only ranged north to the extreme southern reaches of Canada (although still encompassing the entire Great Lakes region). Since the 1950s, reports of the cougar have been increasing in many areas of its eastern range (e.g., Dixon 1982; Lawrence 1983); and it now may be reestablished at several historical locations, such as the Canadian Maritimes (Wright 1971; van Zyll de Jong and Van Ingen 1978), southern Appalachians (Culbertson 1977; Downing 1981b), Manitoba (Nero and Wrigley 1977), and the upper Great Lakes region (Berg 1984; Gerson 1988). Numbers in the United States and Canada are estimated at 16,000 individuals (Nowak 1976).

Of the 14 mountain lion subspecies recognized north of Mexico (Hall 1981), the 3 eastern subspecies are rare or declining. These are the federally endangered Florida panther (*F. c. coryi*) with 30 to 50 remaining in southern Florida and additional small populations possibly inhabiting parts of the southeastern United States (Downing 1981b); the federally endangered eastern cougar (*F. c. couguar*), which appears to be making a labored comeback in parts of its eastern U.S. range (Wright 1972; Frome 1979; Downing 1981a; Currier 1983); and the taxonomically controversial Wisconsin puma (*F. c. schorgeri*), which formerly occurred throughout much of the Midwest (Hall 1981). All cougars occurring throughout Michigan are now protected at both the federal and state level.

State History and Distribution: Historically, the cougar occurred throughout the state. Following European settlement, the cougar declined due to lower prey abundance, unrelenting hunting pressure, and general human persecution. By the turn of the century, the cougar was considered extirpated in Michigan, although there are a few documented Upper Peninsula records several years afterward (Manville 1948). In recent years, reports of cougars have become increasingly more frequent and believable in Michigan and the upper Great Lakes region. This upward trend may be partially attributed to public awareness and understanding, increasing favorable habitat conditions for prey, lessening human persecution, and small, outlying populations reestablishing their former range in Michigan.

In 1984, confirmation of a Michigan cougar (through a blood sample) was obtained for the first time in decades in Menominee County (Zuidema 1985). There also are encouraging signs that the Michigan cougar is not transient but occurs in a self-sustaining population—based on several reliable sightings of adult cougars with kittens (Zuidema 1985). Since 1982, sightings of cougars have been reported from most Upper Peninsula counties with breeding evidence reported from Delta, Marquette, Menominee, and Schoolcraft counties. Despite these reports, there has not been a cougar specimen recovered during this century in Michigan. Some cougar sightings may be the result of the release of captive animals.

Habitat: The adaptable cougar is not particularly dependent on a specific habitat; rather, it requires minimally human-impacted areas and sufficiently high ungulate populations. Cougars, particularly adults, avoid habitats frequented by people and rarely use sites logged within 6 years (Van Dyke et al. 1986b). Although historical habitat use is relatively unknown, the eastern cougar probably shared the same areas used by its primary prey, the white-tailed deer (*Odocoileus virginianus*). These habitats include areas with a mosaic of vegetation types and ages, as well as northern white cedar (*Thuja occidentalis*) swamps and other lowland areas in the winter. Studies of the southern Manitoba population recorded 40% of the cougar reports from wilderness areas with few roads, 30% from disturbed areas interspersed with large tracts of forest, and 30% from farmland areas with little forest cover (Nero and Wrigley 1977). Seldom-used roadways are frequented by cougars for travel routes and hunting forays.

Ecology and Life History: Adult cougars are solitary, except during the breeding season. Males have relatively large home ranges, which may overlap several of the typically smaller female ranges; these are maintained through mutual avoidance (Hornocker 1969). Average summer and winter home ranges for males can vary considerably (e.g., 114 and 47 square miles [296 and 122 km], respectively) depending on the density of prey, geographic location, and individual physical characteristics (Seidensticker et al. 1973).

In eastern North America, cougars probably court and mate during the winter months. Females typically locate their dens under fallen logs and brush. After a three-month gestation period, an average litter of two to four kittens is born in the summer. In optimal conditions, females may breed every year (Robinette et al. 1961). The young are weaned two to three months after birth and are dependent for up to two years. Afterward, immatures disperse to locate unoccupied habitat and establish a home range.

Deer are important prey for the cougar (Robinette et al. 1959; Hornocker 1970; Toweill and Meslow 1977). Nero and Wrigley (1977) found that the white-tailed deer was the primary prey base in the Great Lakes region. Cougars kill proportionally more old and young deer, thereby producing a healthier deer population and a dampening of prey oscillations (Hornocker 1970). Deer densities of ten or more individuals per square mile (2.6 square km) usually provide conditions suitable for winter survival of cougars, while less than five deer per square mile (2.6 square km) could jeopardize a cougar population's local survival or establishment. Although deer and other ungulates are primary food items, cougars prey on other mammals, such as the porcupine (*Erethizon dorsatum*) and snowshoe hare (*Lepus americanus*) (Robinette et al. 1959; Wright 1959; Spalding and Lesowski 1971).

Conservation/Management: According to the United States Fish and Wildlife Service Recovery Plan for the eastern cougar (Downing 1981a), recovery of the eastern subspecies will require at least three self-sustaining populations, each having at least 50 breeding adults, in its original U.S. range. While sightings are important indicators of mountain lion presence, observations of this secretive cat are independent of time spent outdoors (Van Dyke and Brocke 1987) and cannot measure population size. For example, western mountain lion hunters average only one sighting per nine years or every 2,000 days afield. Systematic surveys such as scent stations and intensive road surveys are needed to evaluate Michigan's cougar numbers and distribution. For every 310 square miles (806 square km), no more than 224 linear miles (360 km) need to be searched for cougar confirmation: 56 linear miles (90 km) suffice in ideal tracking conditions (Van Dyke et al. 1986a).

The existence of the cougar in Michigan has only been recently confirmed. Whether individuals are from small, remnant populations that survived human pressures through the last two centuries, transients from the western Great Lakes region, or privately released (or escaped) western subspecies, the cougar needs to be recognized, protected, and studied in Michigan's Upper Peninsula.

LITERATURE CITED

Baker, R. H. 1983. *Michigan Mammals*. Mich. State Univ. Press, E. Lansing.

Berg, W. E. 1984. Mountain lions in Minnesota? *Minn. Volunteer* 47:1–7.

Culbertson, N. 1977. Status and history of the mountain lion in the Great Smokey Mountain National Park. *Natl. Park Serv., Mgmt. Rept.* 15.

Currier, M. J. 1983. Mammalian species: *Felis concolor*. *Am. Soc. Mamm., Mamm. Ser.* No. 200.

Dixon, K. R. 1982. Mountain lion. Pp. 711–27 *in* Chapman, J. A., and G. A. Feldhammer (eds.), *Wild mammals of North America: Biology, management, and economics*. Johns Hopkins Univ. Press, Baltimore.

Downing, R. L. 1981a. Eastern cougar recovery plan. *U.S. Fish Wildl. Serv.*, Clemson, SC.

Downing, R. L. 1981b. The current status of the cougar in the southern Appalachian. Pp. 142–51 *in* Proc. Nongame Endangered Wildl. Symp., Athens, GA.

Downing, R. L., and V. L. Fifield. 1985. Differences between tracks of dogs and cougars. Worcester Sci. Center, Worcester, MA, Unpubl. Rept.

Frome, M. 1979. Panthers wanted: Alive, back East where they belong. *Smithsonian* 10:82–88.

Gerson, H. B. 1988. Cougar, *Felis concolor*, sightings in Ontario. *Can. Field-Nat.* 102:419–24.

Hall, E. R. 1981. *The mammals of North America*, vol. 2. John Wiley and Sons, New York.

Hornocker, M. G. 1969. Winter territoriality in mountain lions. *J. Wildl. Mgmt.* 33:457–64.

Hornocker, M. G. 1970. An analysis of mountain lion predation upon mule deer and elk in the Idaho Primitive Area. *Wildl. Monogr.* No. 21.

Johnson, M. K., R. C. Belden, and D. R. Aldred. 1984. Differentiating mountain lion and bobcat scats. *J. Wildl. Mgmt.* 48:239–44.

Lawrence, R. D. 1983. *The ghost walker.* Holt, Rinehart and Winston, New York.

Major, M., M. K. Johnson, W. S. Davis, and T. F. Kellogg. 1980. Identifying scats by recovery of bile acids. *J. Wildl. Mgmt.* 44:290–93.

Manville, R. H. 1948. The vertebrate fauna of the Huron Mountains, Michigan. *Am. Midl. Nat.* 39:615–40.

Murie, O. J. 1974. *A field guide to animal tracks.* Houghton Mifflin Co., Boston.

Nero, R. W., and R. E. Wrigley. 1977. Status and habits of the cougar in Manitoba. *Can. Field-Nat.* 91:28–40.

Nowak, R. M. 1976. The cougar in the United States and Canada. *U.S. Fish Wildl. Serv.,* Washington, D.C.

Robinette, W. L., J. S. Gashwiler, and O. W. Morris. 1959. Food habits of the cougar in Utah and Nevada. *J. Wildl. Manage.* 23:261–73.

Robinette, W. L., J. S. Gashwiler, and O. W. Morris. 1961. Notes on cougar productivity and life history. *J. Mamm.* 42:204–17.

Seidensticker, J. C., M. G. Hornocker, W. V. Wiles, and J. P. Messick. 1973. Mountain lion social organization in the Idaho Primitive Area. *Wildl. Monogr.* No. 35.

Spalding, D. J., and J. Lesowski. 1971. Winter food of the cougar in south-central British Columbia. *J. Wildl. Mgmt.* 3:378–81.

Toweill, D. E., and E. C. Meslow. 1977. Food habits of cougars in Oregon. *J. Wildl. Mgmt.* 41:576–78.

Van Dyke, F. G., and R. H. Brocke. 1987. Sighting and track reports as indices of mountain lion presence. *Wildl. Soc. Bull.* 15:251–56.

Van Dyke, F. G., R. H. Brocke, and H. G. Shaw. 1986a. Use of road track counts as indices of mountain lion presence. *J. Wildl. Mgmt.* 50:102–9.

Van Dyke, F. G., R. H. Brocke, H. G. Shaw, B. B. Ackerman, T. P. Hemker, F. G. Lindzey. 1986b. Reactions of mountain lions to logging and human activity. *J. Wildl. Mgmt.* 50:95–102.

van Zyll de Jong, C. G., and E. Van Ingen. 1978. Status report on Eastern Cougar (*Felis concolor cougar*) in Canada. *COSEWIC Rept.*

Wright, B. S. 1959. *The ghost of North America: The story of the eastern panther.* Vantage Press, New York.

Wright, B. S. 1971. The cougar in New Brunswick. Pp 108–19 *in* S. E. Jorgensen and L. D. Mech (eds.), *Proc. of a symposium on the native cats of North America, their status and management,* U.S. Fish Wildl. Serv., Twin Cities, MN.

Wright, B. S. 1972. *The eastern panther: A question of survival.* Clarke, Irwin and Co. Ltd., Toronto.

Zuidema, M. R. 1985. Cougars in the U.P. *Timber Producer* 5:56–66.

Lynx

· *Felis lynx* Linnaeus
· **State Endangered**

Map 6. Confirmed individual lynx. Historically statewide (Baker 1983); ● = 1961 to 1989 (Harger 1965; Martin 1989).

Status Rationale: The lynx has yet to recover in Michigan or the Great Lakes region. Its specific range limits and population status are difficult to define due to its natural cyclic fluctuations, secretive behavior, and low numbers. However, current conditions, including complete protection and forest regeneration, may aid in its recovery in Michigan's Upper Peninsula.

Field Identification: The lynx is morphologically similar to the more common bobcat (*Felis rufus*), but may be distinguished by its longer, prominent black ear tufts (2 inches [5.1 cm] in length); grizzled, silvery gray pelage; larger, furred feet; and stubbier tail, which is completely black at the tip. The bobcat's longer tail has several dark bands prior to the tail tip, which is only black on the top half. Males are slightly larger than females. Total body length ranges from 30 to 42 inches (76.2 to 106.7 cm) (Baker 1983) and the typical weight ranges from 22 to 30 pounds (10.0 to 13.6 kg). The subspecies (*F. l. canadensis)* occurs in Michigan.

Since lynx are rarely seen, the best way to identify their presence is by tracks. The foreprints of the lynx are usually twice the size as those of the bobcat. Average track measurements are $3^3/4$ inches (9.5 cm) wide and $4^1/2$ inches (11.4 cm) long for the front and 3 inches (7.6 cm) wide and $3^1/8$ inches (7.9 cm) long for the hind foot pads (Murie 1974). Surveys in Colorado attribute felid tracks greater than $3^3/8$ inches (8.6 cm) wide with a straddle less than $7^1/4$ inches (18.4 cm) and a stride less than $17^1/4$ inches (43.8 cm) to those of a lynx (Murray 1987).

The cougar (*Felis concolor*) has tracks similar to the lynx. The heel pads of the lynx are usually round and unlobed, while the cougar has irregular, three-lobed heel pads. In winter, the well-furred "snowshoe" feet of the lynx are especially distinctive. In deep snow, the tail marks of the cougar are characteristic. In comparison with canids, felid tracks generally have no claw marks. Lynx can be vocal, particularly during the mating season, when its caterwauling can be heard at night.

Total Range: The lynx occurs in boreal regions worldwide. Historically, its North American distribution included much of the northern contiguous United States south to Oregon, Colorado, Iowa, and West Virginia north to the Canadian-Alaskan treeline. Since the fur trade and heavy logging of forests, its distribution in the United States and other areas has been severely reduced. Today, lynx still occur throughout much of Alaska and Canada and, since the early 1960s, have reclaimed some of their former range in Canada (van Zyll de Jong 1973; McCord and Cardoza 1982). In the contiguous United States, it still may be found in scattered areas of the northern Rockies (south through Utah), extreme northern portions of the New England states (e.g., the White Mountains of New Hampshire), and the upper Great Lakes region. Minnesota may be the only Great Lakes state with a self-sustaining population.

State History and Distribution: As with many of Michigan's furbearers, the lynx was formerly widespread, although occurring only locally in the southern regions. Because of logging, agricultural clearing, shooting, and trapping, this deep forest denizen was extirpated from the Lower Peninsula. Its Upper Peninsula history is less distinct, although by the late 1920s it had probably vanished as a viable reproducing species. However, due to their wandering habit and cyclic population upswings, there has been a periodic flow of immigrants into the eastern Upper Peninsula from Ontario. The paucity of recent reports probably does not reflect actual lynx status because the secretive cat is rarely encountered. The lynx was considered to have naturally spread throughout the Upper Peninsula originating from Ontario populations, but lynx records continue to be very scarce.

Once common on Isle Royale, the lynx disappeared in the early 1930s (Mech 1966; Martin 1989). Since then, lynx have been observed only during peak irruptive years, including recent sightings in the late 1980s. Unless reestablished, the lynx can only reach this national park by crossing the ice bridge that occasionally forms between the island and Ontario's mainland (the last ice bridge was in 1976). High wolf densities also may preclude lynx presence (Pulliainen 1965).

Habitat: This cat is generally an inhabitant of remote, dense, boreal conifer forests. Lynx can neither tolerate extensive clear-cuts, nor can they survive solely in large stands of younger forest. Overall lynx population declines in the early 1900s are associated with the deforestation caused by logging and fires (de Vos and Matel 1952). Limited second-growth habitats, however, may enhance lynx habitat by supporting higher prey populations, particularly in winter.

Ecology and Life History: Except for females with young, lynx are typically solitary predators (Barash 1971; Parker et al. 1983). Courtship and mating occurs in February through May. The female lynx selects large hollow logs, overturned stumps, or thick brush for den sites. After a two-month gestation period, an average of two to three kittens are born, fewer if prey is scarce (Brand and Keith 1979). During years of high food availability, lynx become more prolific, producing three to six young. At two months, juveniles accompany the female on hunting forays and remain dependent through the first winter. This time is used to refine hunting techniques. Juvenile survival is strongly associated with hare abundance (Brand et al. 1976; Brand and Keith 1979). Juvenile lynx disperse at approximately one year and usually first reproduce at two years of age.

Lynx territories are well defined in undeveloped areas. In Minnesota, male lynx occupy large home ranges (55 to 92 square miles [143 to 239 square km]) that may be adjacent to or partially encompass several female home ranges (19 to 46 square miles [49 to 120 square km]) (Mech 1980). Territories are vigorously defended and marked by scent posts that are used as communication stations. Established males will usually drive out young intruders. Therefore, in saturated populations, dispersing lynx may travel long distances to find unoccupied, suitable habitat. Lynx are known to disperse 300 miles (483 km) in the Great Lakes region (Mech 1977), and even greater movements, of nearly 500 miles (805 km), are known in northern Canada. Lynx densities can dramatically fluctuate, from one individual per 16 square miles (42 square km) during years of low populations and low prey availability, to one individual for every 4 square miles (10 square km) in peak years (Brand et al. 1976).

The lynx is ideally adapted for hunting in deep snow with its proportionally long legs, snowshoelike foot pads, and exceptional leaping abilities. This cat's major prey is the snowshoe hare (*Lepus americanus*) (Parker et al. 1983). The lynx is a specialist on this prey species, to the extent that populations fluctuate in direct response to hare abundance (Elton and Nicholson 1942; Wing 1953; Nellis et al. 1972; Brand and Keith 1979). Carrion from white-tailed deer (*Odocoileus virginianus*) occasionally is eaten, but rarely is one killed. Ruffed grouse (*Bonasa umbellus*) (van Zyll de Jong 1966), other birds, small mammals, and even fish (Saunders 1963) serve as occasional food supplements. If snow conditions significantly affect lynx hunting success, this adept tree climber may temporarily shift its focus to arboreal prey (Nellis and Keith 1968). A population in Canada was found to rely heavily on red squirrels (*Tamiasciurus hudsonicus*) (More 1976).

Conservation/Management: After near extirpation by the middle of this century, the lynx apparently staged a temporary comeback in Michigan's Upper Peninsula through natural immigration during the 1960s. Although current factors are favorable for reestablishing a lynx population, reproduction still has not been recently confirmed in Michigan. Continual immigration into Michigan from Ontario may improve this situation, and populations from the western Great Lakes region also may enter Michigan. The lynx is now fully protected by law, and the pressures resulting in its extirpation are less intense (e.g., hunting, trapping, logging). The current timber harvesting methods of smaller clear-cuts should be beneficial as long as large, mature conifer stands are left intact.

The following threats still cast a shadow over the lynx's tentative recovery in Michigan: (1) illegal poaching for furs and overexploitation of Canadian populations, because pelt prices remain high; (2) natural population lows combined with continued human-induced mortality; (3) mismanagement of mature coniferous forests; and (4) incidental trapping, to which lynx are particularly vulnerable (Quinn and Thompson 1985). If harvest of this species in the upper Great Lakes region were to be curtailed three to five years immediately after snowshoe hare numbers decline, it would avoid delaying or suppressing a natural increase of the prey population and aid in the recovery of the lynx (Berrie 1974; Brand and Keith 1979).

The wandering nature of dispersing lynx, cyclic populations, proximity of lynx populations in Ontario, and maturing forests all enhance the lynx's chances of establishing populations in northern Michigan.

LITERATURE CITED

Baker, R. H. 1983. *Michigan mammals*. Mich. State Univ. Press, E. Lansing.

Barash, D. P. 1971. Cooperative hunting in the lynx. *J. Mamm.* 52:480.

Berrie, P. M. 1974. Ecology and status of the lynx in interior Alaska. Pp.4–41 *in* R. L. Eaton (ed.), *The World's Cats*. World Wildlife Safari, Winston, OR.

Brand, C. J., and L. B. Keith. 1979. Lynx demography during a snowshoe hare decline in Alberta. *J. Wildl. Mgmt.* 43:827–49.

Brand, C. J., L. B. Keith, and C. A. Fischer. 1976. Lynx responses to changing snowshoe hare densities in central Alberta. *J. Wildl. Mgmt.* 40:416–28.

de Vos, A., and S. E. Matel. 1952. The status of the lynx in Canada, 1929–1952. *J. Forestry.* 50:742–45.

Elton, C., and M. Nicholson. 1942. The ten-year cycle in numbers of lynx in Canada. *J. Animal Ecol.* 11:215–44.

Harger, E. M. 1965. The status of the Canada lynx in Michigan. *Jack-Pine Warbler* 43:150–53.

McCord, C. M., and J. E. Cardoza. 1982. Bobcat and lynx. Pp. 728–66 *in* J. A. Chapman and G. A. Feldhammer (eds.), *Wild mammals of North America*. Johns Hopkins Univ. Press, Baltimore.

Martin, C. 1989. The history and reestablishment potential of marten, lynx, and woodland caribou on Isle Royale. *Natl. Park Serv., Resour. Mgmt. Rept.* 88-2.

Mech, L. D. 1966. *The wolves of Isle Royale*. Nat. Park Serv. Fauna Ser. No. 7. Washington, D.C.

Mech, L. D. 1977. Record movement of a Canadian lynx. *J. Mamm.* 58:676–77.

Mech, L. D. 1980. Age, sex, reproduction, and spatial organization of lynxes colonizing northeastern Minnesota. *J. Mamm.* 61:261–67.

More, G. 1976. Some winter food habits of lynx (*Felis lynx*) in the southern Mackenzie District, Northwest Territories. *Can. Field-Nat.* 90:499–500.

Murie, O. 1974. *A field guide to animal tracks*. Houghton Mifflin Co., Boston.

Murray, J. A. 1987. *Wildlife in peril: The endangered mammals of Colorado*. Roberts Rinehart, Boulder, CO.

Nellis, C. H., and L. B. Keith. 1968. Hunting activities and success of lynxes in Alberta. *J. Wildl. Mgmt.* 32:718–22.

Nellis, C. H., S. P. Wetmore, and L. B. Keith. 1972. Lynx-prey interactions in central Alberta. *J. Wildl. Mgmt.* 36:320–29.

Parker, G. R., J. W. Maxwell, L. D. Morton, and G. E. Smith. 1983. The ecology of the lynx (*Lynx canadensis*) on Cape Breton Island. *Can. J. Zool.* 61:770–86.

Pulliainen, E. 1965. Studies of the wolf in Finland. *Ann. Zool. Fenn.* 2:215–59.

Quinn, N. W., and J. E. Thompson. 1985. Age and sex of trapped lynx (*Felis canadensis*), related to period of capture and trapping technique. *Can. Field-Nat.* 99:267–69.

Saunders, J. K. 1963. Food habits of the lynx in Newfoundland. *J. Wildl. Mgmt.* 27:384–90.

van Zyll de Jong, C. G. 1966. Food habits of the lynx in Alberta and the Mackenzie District, N.W.T. *Can. Field-Nat.* 80:18–23.

van Zyll de Jong, C. G. 1973. The status and management of the Canada Lynx in Canada. Pp 16–21 *in* S. E. Jorgensen and L. D. Mech (eds.), *Proc. of a symposium on the native cats of North America: Their status and management.* U.S. Fish Wildl. Serv., Twin Cities, MN.

Wing, L. W. 1953. Cycles of lynx abundance. *J. Cycles Res.* 2:28–51.

Prairie Vole

· *Microtus ochrogaster* (Wagner)
· **State Threatened**

Map 7. Confirmed individual prairie voles. ▲ = before 1962 (Baker 1983). No records known since 1962.

Status Rationale: Considered a species of the prairies, the prairie vole is recorded in Michigan and probably inhabited the Prairie Peninsula that once extended into the state's southwestern corner. It is still abundant throughout most of its North American range but now may be gone from Michigan.

Field Identification: As with most voles, this species has a stout body, small, partly concealed ears, relatively short tail, and small, black eyes. Adults average 4 1/2 inches (11.4 cm) in length with a tail of 1 1/2 inches (3.8 cm)—a third of the body-head length (Baker 1983). The pelage varies from gray to dark brown, sometimes resulting in a grizzled appearance. The belly usually has a buff-yellow cast. The nominate subspecies occurs in Michigan.

Prairie voles make use of surface runways (1 to 3 1/2 inches [2.5 to 8.9 cm], in diameter) formed through the base of vegetation. Runways are the most conspicuous sign of vole presence, although they may not be visible in sparse vegetation; mole tunnels also may be used. Although highly variable, mice (e.g., *Peromyscus*) tracks generally show a leaping four-print pattern while voles leave a double-print pattern (Murie 1974). Two other closely related *Microtus* species may be easily confused with the prairie vole: the meadow vole (*M. pennsylvanicus*) and woodland vole (*M. pinetorum*). The latter species may be separated by its shorter tail (2/3 to 1 inch [1.9 to 2.5 cm]), more rufous-colored fur, and tendency for more forested habitats. The upper body color of the frequently associated meadow vole

69

lacks the grizzled-brown fur and is generally darker; undersides tend to be white. The tail of the meadow vole is longer (more than $1^1/2$ inches [3.8 cm]). The locally distributed southern bog lemming (*Synaptomys cooperi*), which also is commonly found with the prairie vole and may even share runway systems with it (Rose and Spevak 1978), is easily identified by its short tail (less than 1 inch [2.5 cm]) and white belly fur.

Total Range: The prairie vole has a middle North American distribution, primarily confined to the Great Plains from south-central Canada south to Oklahoma. Its range continues eastward to southwestern Ohio. Some expansion of this vole's range occurred outside the original prairies following the conversion of forests to agriculture (Choate and Williams 1978).

State History and Distribution: As indicated by Baker (1983), fossil records for the prairie vole exist since postglacial times, including a specimen found in Leelanau County. Its presence just prior to European settlement is debatable, although upland prairie and oak savanna habitat did extend well into southwest Michigan (Chapman 1984). Certainly, Michigan was at the extreme northeastern periphery of its distribution. The last individual recorded in Michigan was in Kalamazoo County (northeast of Schoolcraft), even though systematic efforts were made in the late 1970s and early 1980s to document prairie voles in historical Berrien County sites.

Populations are contiguous south of Michigan and remain at relatively high densities (e.g., northern Indiana). Prairie voles are known to occupy habitats along artificial corridors, such as railroad grades and roadways (Getz et al. 1978). Therefore, if these populations are viable, it is possible that prairie voles are, at least, occasionally present in Michigan.

Habitat: As its name implies, this vole occupies open habitats, such as abandoned fields, managed rights-of-way, and old field borders, as well as hayfields and other grass-associated agriculture lands. A known Berrien County population (now extirpated) utilized an abandoned orchard. The prairie vole prefers areas with thick ground cover of herbaceous vegetation (Martin 1956). In fact, the amount of vegetative cover may control prairie vole population numbers and cycles at certain threshold levels (Birney et al. 1976). Any lack of cover can be detrimental to vole populations and may even cause declines under high predation pressure (Getz 1970). The prairie vole may be displaced from areas of dense vegetation if there is a high meadow vole density, (Klatt and Getz 1987).

Prairie voles often share the same areas with meadow voles (Krebs 1977) and southern bog lemmings (Rose and Spevak 1978), although typically little interaction occurs (Keller and Krebs 1970; Getz et al. 1978). In areas containing these species, the meadow vole and lemming are typically found in more moist situations. From most recent Michigan records, prairie voles tend to favor upland sites, similar to those sites studied outside the state (Findley 1954; DeCoursey 1957; Keller and Krebs 1970; Birney et al. 1976).

Ecology and Life History: Prairie voles are social throughout the year, except during the breeding season and at population peaks when interaction with other individuals only occurs when necessary (Krebs 1970; Rose and Gaines 1976; Getz and Hofmann 1986). Like many microtines, this vole regularly may exhibit substantial local population fluctuations, reaching peak densities every two to four years (Gaines and Rose 1976; Beasley 1978). Densities in some Midwest populations are less than 10 individuals per acre (0.4 ha); although, within populations at their peak cycles, densities may reach more than 145 individuals per acre (0.4 ha) (Baker 1983). During local population peaks, individuals require accessible areas for dispersal (Krebs et al. 1969). Interestingly, the prairie vole generally has monogamous breeding behavior at both low and high densities (Getz and Hofmann 1986; Getz et al. 1987).

Unlike the meadow vole, which typically makes its nest above ground or slightly elevated in clumps of vegetation, the prairie vole nests underground. This breeding behavior is probably related to the different vegetative height preferences of the two species. In the Midwest, the prairie vole breeds primarily in the summer (Krebs et al. 1969); although, in the Great Plains (Rose and Gaines 1978) and during population peaks (Cole and Batzli 1978), it may produce young in any season. Prairie voles are prolific and may have up to five litters of two to six young per litter following a gestation period of three weeks (Baker 1983). Growth rates of individuals from the spring litters are high (Sauer and Slade 1986); subsequently, young-of-the-year may breed at five to six weeks of age. Only about one-third of the young survive the first 30 days (Getz and Hofmann 1986), and few individuals survive more than a year.

The prairie vole is a grazer and subsists on seeds and the leaves and stems of herbaceous vegetation. Zimmerman (1965) found Midwest prairie vole diets contained primarily vegetation, particularly Canada bluegrass (*Poa compressa*). However, forbs (nongrass, nonwoody plants) are usually the preferred forage (Cole and Batzli 1978; Klatt and Getz 1987). In spring and summer, individuals are active day and night (Danielson and Swihart 1987).

The species home range varies from 1/4 to 1/2 acre (0.1 to 0.2 ha) (Martin 1956; Harvey and Barbour 1965; Danielson and Swihart 1987); live-trapping studies may underestimate home range size or vary from direct measurements (Ambrose 1969; Desey et al. 1989). Unlike the meadow vole, home ranges of the prairie vole do not differ according to sex (Gaulin and Fitzgerald 1988).

Conservation/Management: Although abundant over much of its North American range, the prairie vole apparently has disappeared in Michigan. The original upland prairies and oak savannas that were present in most of southwestern Michigan, which probably provided suitable habitat, are now virtually gone (Chapman 1984). However, studies have shown that this vole easily accommodates to hot, dry summers and cold, snowy winters (Rose and Gaines 1978), climate similar to that of southwestern Michigan. In addition to its marginal Michigan distribution, limiting factors likely include intensive land-use patterns, such as conversion to row crops and the high use of chemicals. An increase in exotic predator composition (e.g., domestic cats) and densities has been associated with high rodent mortality (George 1974; Warner 1985). Grazing by large domesticated herbivores in, at least, tallgrass communities (Grant et al. 1982) and possible displacement by other ubiquitous rodents, such as meadow voles (Getz et al. 1978, Klatt and Getz 1987), also may have led to its decline.

As shown by several studies (e.g., Birney et al. 1976; Klatt and Getz 1987), the degree of vegetative cover for the prairie vole is critical for a population's success. This is particularly evident for declining *Microtus* populations in areas with sparse vegetative cover during winter (Baker and Brooks 1982). Since prairie voles are social, home ranges are small and general habitat and forage availability are not limiting. Managing connected odd areas (e.g., sites difficult for maneuverability of farm equipment) with low, thick plant growth will benefit this species. Surveys are needed to locate extant populations, enabling management of the sites and protection of this small yet crucial component of our native grassland fauna.

LITERATURE CITED

Ambrose, H. W. 1969. A comparison of *Microtus ochrogaster* home ranges as determined by isotype and live trapping methods. *Am. Midl. Nat.* 81:535–55.

Baker, J. A., and R. J. Brooks. 1982. Impact of raptor predation on a declining vole population. *J. Mamm.* 63:297–300.

Baker, R. H. 1983. *Michigan mammals*. Mich. State Univ. Press, E. Lansing.

Beasley, L. E. 1978. *Demography of southern bog lemmings (Synaptomys cooperi) and prairie vole (Microtus ochrogaster) in southern Illinois.* Ph.D. diss., Univ. IL.

Birney, E. C., W. E. Grant, and D. D. Baird. 1976. Importance of vegetative cover to voles of *Microtus* populations. *Ecology* 57:1043–51.

Chapman, K. A. 1984. *An ecological investigation of native grassland in southern lower Michigan.* M.S. thesis, Western Mich. Univ., Kalamazoo.

Choate, J. R., and S. L. Williams. 1978. Biogeographic interpretation of variation within and among populations of the prairie vole, *Microtus ochrogaster*. *Texas Tech. Univ., Mus. Occ. Pap.* No. 49.

Cole, F. R., and G. O. Batzli. 1978. Influence of supplemental feeding on a vole population. *J. Mamm.* 59:809–19.

Danielson, B. J., and R. K. Swihart. 1987. Home range dynamics and activity patterns of *Microtus ochrogaster* and *Synaptomys cooperi* in snytopy. *J. Mamm.* 68:160–65.

DeCoursey, G. E. 1957. Identification, ecology and reproduction of *Microtus* in Ohio. *J. Mamm.* 38:44–52.

Desey, E. A., G. O. Batzli, and L. Jike. 1989. Comparison of vole movements assessed by live trapping and radiotracking. *J. Mamm.* 70:652–56.

Findley, J. S. 1954. Competition as a possible limiting factor in the distribution of *Microtus. Ecology* 35:418–20.

Gaines, M. S., and R. K. Rose. 1976. Population dynamics of *Microtus ochrogaster* in eastern Kansas. *Ecology* 57:1145–61.

Gaulin, S. J., and R. W. Fitzgerald. 1988. Home-range size as a predictor of mating systems in *Microtus. J. Mamm.* 69:311–19.

George, W. G. 1974. Domestic cats as predators and factors in winter shortages of raptor prey. *Wilson Bull.* 86:384–96.

Getz, L. L. 1970. Influence of vegetation on the local distribution of the meadow vole in southern Wisconsin. *Univ. Conn. Occ. Pap.* 1:213–41.

Getz, L. L., and J. E. Hofmann. 1986. Social oragnization in free-living prairie voles, *Microtus ochrogaster. Behav. Ecol. Sociobiol.* 18:275–82.

Getz, L. L., F. R. Cole, and G. L. Gates. 1978. Interstate roadsides as dispersal routes for *Microtus pennsylvanicus. J. Mamm.* 59:208–12.

Getz, L. L., J. E. Hofmann, C. J. Carter. 1987. Mating system and population fluctuations of the prairie vole, *Microtus ochrogaster. Am. Zool.* 27:909–20.

Grant, W. E., E. C. Birney, N. R. French, and D. M. Swift. 1982. Structure and productivity of grassland small mammal communities related to grazing-induced changes in vegetative cover. *J. Mamm.* 63:248–60.

Harvey, H. J., and R. W. Barbour. 1965. Home range of *Microtus ochrogaster* as determined by a modified minimum area method. *J. Mamm.* 46:398–402.

Keller, B. L., and C. J. Krebs. 1970. *Microtus* populations biology; Part 3. Reproductive changes in fluctuating populations of *M. ochrogaster* and *M. pennsylvanicus* in southern Indiana, 1965–67. *Ecol. Monogr.* 40:263–94.

Klatt, B. J., and L. L. Getz. 1987. Vegetation characteristics of *Microtus ochrogaster* and *M. pennsylvanicus* habitats in east-central Illinois. *J. Mamm.* 68:569–77.

Krebs, C. J. 1970. *Microtus* population biology: Behavioral changes associated with the population cycle in *M. ochrogaster* and *M. pennsylvanicus. Ecology* 51:34–52.

Krebs, C. J. 1977. Competition between *Microtus pennsylvanicus* and *Microtus ochrogaster. Am. Midl. Nat.* 97:42–49.

Krebs, C. J., B. L. Keller, and R. H. Tamarin. 1969. *Microtus* population biology: Demographic changes in fluctuating populations of *M. ochrogaster* and *M. pennsylvanicus* in southern Indiana. *Ecology* 50:587–607.

Martin, E. P. 1956. A population study of the prairie vole (*Microtus ochrogaster*) in northeastern Kansas. *Univ. Kansas Publ., Mus. Nat. Hist.* 8:361–416.

Murie, O. J. 1974. *A field guide to animal tracks.* Houghton Mifflin Co., Boston.

Rose, R. K., and M. S. Gaines. 1976. Levels of aggression in fluctuating populations of the prairie vole, *Microtus ochrogaster*, in eastern Kansas. *J. Mamm.* 57:43–57.

Rose, R. K., and M. S. Gaines. 1978. The reproductive cycle of *Microtus ochrogaster* in eastern Kansas. *Ecol. Monogr.* 48:21–42.

Rose, R. K., and A. M. Spevak. 1978. Aggressive behavior in two sympatric microtine rodents. *J. Mamm.* 59:213–16.

Sauer, J. R., and N. A. Slade. 1986. Field-determined growth rates of prairie voles (*Microtus ochrogaster*): Observed patterns and environmental influences. *J. Mamm.* 67:61–68.

Warner, R. E. 1985. Demography and movements of free-ranging domestic cats in rural Illinois. *J. Wildl. Mgmt.* 49:340–46.

Zimmerman, E. G. 1965. A comparison of habitat and food of two species of *Microtus. J. Mamm.* 46:605–12.

Wolverine

· *Gulo gulo* (Linnaeus)
· **State Extirpated**

The secretive wolverine, which formerly occupied much of the northern hemisphere, is the largest member of the weasel family (Mustelidae). In North America, its northward retreat occurred early during European settlement (Baker 1983). Historically a resident throughout Canada and Alaska south to northern California, Colorado, the Great Lakes, and northern New England, it has retreated from much of its southern range.

In the contiguous United States, the wolverine is now confined to California, Oregon, Washington, Idaho, Montana, Wyoming, and Colorado (Chapman and Feldhammer 1982; Nowak 1973). Minnesota (e.g., Birney 1974) and North Dakota may occasionally harbor wandering individuals. Most of these states protect this species and recent increasing populations suggest that it is making a moderate comeback in several western states (Nowak 1973). Canadian populations still have not fully recovered, particularly in the eastern and southern areas (van Zyll de Jong 1975).

Michigan's original wolverine population probably occupied the northern Lower and Upper peninsulas. Specimen records are lacking, but Baker (1983) considered the wolverine to occur in at least 16 counties, ranging throughout the Upper Peninsula south to Kent and Sanilac counties. The wolverine probably disappeared from the northern Lower Peninsula in the 1880s, possibly surviving past the turn of the century in the Upper Peninsula (Manville 1948; Baker 1983).

Today, wolverine populations reside in areas well north of Michigan in northwestern Ontario, and, although long-distance movements of over 200 miles (nearly 400 km) are known (Gardner et al. 1986), the natural reestablishment of this species is not likely. Restoration projects may be possible in heavily forested areas with few roads, such as those found in the Huron Mountains, Marquette County, and the Betsy Lake Natural Area, Luce and Chippewa counties.

Once considered to be an aggressive and fierce animal, the wolverine is now known to be a secretive and solitary species. Wolverines average 30 pounds (13.6 kg) (slightly less for females). They are a stout, muscular animal with an overall shaggy appearance. The pelage is coarse with a dark mid-back surrounded by buffy brown, with dark underparts.

Mating occurs in the summer. After a several month period of dormancy, the fertilized eggs are implanted (Rausch and Pearson 1972). Usually two to three young are born under a fallen tree between February and April (Baker 1983), and they are weaned after seven to eight weeks. After weaning, the young will remain with their mother for up to two years. The wolverine is omnivorous and has a diverse diet, but a large part is dependent upon carrion, which is likely its primary access to most large prey.

The wolverine typically exists in large territories at extremely low densities (Krott 1959). In Montana, Hornocker and Hash (1981) found one individual per 25 square miles (65 square km). Individuals may defend territories of 77 to 163 square miles (200 to 424 square km) (Hornocker and Hash 1981; Quick 1953). Males maintain a territory that commonly overlaps several smaller female territories (Krott 1959). The wolverine is sensitive to human disturbance (e.g., large-scale logging and trapping), which was partially responsible for its disappearance in many areas. These threats are now reduced or controlled. With the reforestation of the Upper Peninsula, perhaps this species could once again inhabit Michigan.

LITERATURE CITED

Baker, R. H. 1983. *Michigan mammals*. Mich. State Univ. Press, E. Lansing.
Birney, E. C. 1974. Twentieth century records of wolverine in Minnesota. *Loon* 46:78–81.
Chapman, J. A., and G. A. Feldhammer (eds.). 1982. *Wild mammals of North America: Biology, Management, and Economics*. Johns Hopkins Univ. Press, Baltimore.
Gardner, C. L., W. B. Ballard, and H. Jessup. 1986. Long distance movement by an adult wolverine. *J. Mamm.* 67:603.
Hornocker, M. G., and H. S. Hash. 1981. Ecology of the wolverine in northwestern Montana. *Can. J. Zool.* 59:1286–1301.
Krott, P. 1959. Demon of the north. Alfred A. Knopf, New York.
Manville, R. H. 1948. The vertebrate fauna of the Huron Mountains, Michigan. *Am. Midl. Nat.* 39:615–40.
Nowak, R. M. 1973. Return of the wolverine. *Natl. Parks Conserv. Mag.* 47:20–23.
Quick, H. F. 1953. Wolverine, fisher, and marten studies in a wilderness region. *Trans. 18th N. Am. Wildl. Conf.*, 513–32.
Rausch, R. A., and A. M. Pearson. 1972. Notes on the wolverine in Alaska and the Yukon Territory. *J. Wildl. Mgmt.* 36:249–68.
van Zyll de Jong, C. G. 1975. The distribution and abundance of the wolverine (*Gulo gulo*) in Canada. *Can. Field-Nat.* 89:431–37.

Woodland Caribou

- *Rangifer tarandus caribou*
  (Linnaeus)
- **State Extirpated**

Like most of North America's ungulates and larger mammals, the caribou's range and numbers have declined markedly since European settlement, particularly during the 1800s and early 1900s. This species historically had an extensive holarctic distribution. However, southern populations in much of this range now have disappeared. In North America, the wide-ranging caribou once was distributed from the Alaskan-Canadian tundra south to the alpine tundra and boreal forest of the northern Rockies, upper Great Lakes region, and northern New England and adjacent southeastern Canadian provinces (Chapman and Feldhammer 1982).

Today, the contiguous United States contains only a small and federally endangered caribou population in the Selkirk Mountains of northern Idaho (Servheen and Lyon 1989). Original caribou populations lingered in Maine until at least 1908 (Palmer 1938) and in the U.S. Great Lakes region into the early 1940s (Nelson 1947).

Great Lakes Canadian populations also have retreated northward (de Vos and Peterson 1951), subsequently providing fewer chances for recruitment to Minnesota, Wisconsin, and Michigan. Nevertheless, wandering individuals are occasionally observed in extreme northern Minnesota (Mech et al. 1982).

In Michigan, the postglacial caribou population initially comprised the tundra subspecies, the barren-ground caribou (*R. t. granti*); due to changing habitat, it was replaced by the woodland caribou (*R. t. caribou*) that was present during European settlement (Baker 1983). The woodland caribou probably was never common in Michigan (Schorger 1942); however, Baker (1983) found records of woodland caribou in 12 counties and considered this species to have ranged throughout the Upper Peninsula and nothern Lower Peninsula, even extending as far south as Lapeer County.

The woodland caribou disappeared from the Lower Peninsula in the mid-1800s and from the Upper Peninsula mainland by 1912. Caribou disappeared from Isle Royale about 1928. Although the woodland form is less migratory than the tundra form, seasonal ranges for males and females average 460 and 208 square miles (1,196 to 541 square km), respectively (Fuller and Keith 1981). The woodland caribou is a fairly large ungulate, with males weighing an average of 375 pounds (170 kg) and females an average of 230 pounds (104 kg). Both sexes grow the curved, palmate antlers, although those of females are smaller. The pelage is generally brownish gray with varying shades found throughout the body. In autumn, the male caribou gathers a small harem of females and, after a 32-week gestation period, a single calf is born to each female between late May and early June. Due to synchronized breeding, 80% to 90% of the calves are born in a $1^1/_2$–week period (Dauphine and McClure 1974; Bergerud 1975). The calf begins to eat solid food in 2 weeks and develops sexual maturity in its second or third year. Woodland caribou are highly dependent on lichens (Holleman et al. 1979), however the diet is generally more varied in summer (Cringhan 1957; Bergerud 1972).

The decline of this species in the Great Lakes region and other areas is dependent on several limiting factors, including habitat destruction, overhunting, and competition with unnaturally high white-tailed deer (*Odocoileus virginianus*) populations. This latter factor was probably responsible for this species' decline in many areas (Bergerud 1974), including Michigan. The woodland caribou requires boreal habitat containing an abundance of such openings as bogs (Baker 1983; Fuller and Keith 1981). Several areas in the northern half of the Upper Peninsula currently provide suitable boreal habitat, particularly within the Lake Superior drainage basin.

The closest self-sustaining populations are along the north shore of Lake Superior; recent reestablishment projects on Michipicoten and Montreal islands have been successful (Bergerud and Mercer 1989). Although an evaluation of the feasibility of reestablishing caribou on Isle Royale was not encouraging due to formerly high wolf (*Canis lupus*) densities (Martin 1988), according to Bergerud and Mercer (1989) the current low wolf population provides "an excellent opportunity to reintroduce caribou to this National Park." Caribou populations acclimated to living on islands are relatively close to Isle Royale, occurring in the Lake Nipigon area (Cumming and Beange 1987) and the Slate islands (Mech et al. 1982).

Results from releases in other states and provinces, totaling 33 between 1924 and 1985 (13 were successful), indicate the following requirements: (1) a low frequency of meningeal or brainworm (*Parelaphostrongylos tenuis*) infections carried by white-tailed deer, (2) low predator densities; wolf numbers cannot exceed ten per 1,000 square km (Bergerud 1985), and (3) spring and summer ranges that provide adequate escape cover and lichen food supplies (Bergerud and Mercer 1989). At this time, the relatively low reproductive rate in caribou combined with losses due to brainworm and predation probably preclude the reestablishment of caribou in Michigan.

LITERATURE CITED

Baker, R. H. 1983. *Michigan mammals*. Mich. State Univ. Press, E. Lansing.

Bergerud, A. T. 1972. Food habits of Newfoundland caribou. *J. Wildl. Mgmt.* 36:912–23.

Bergerud, A. T. 1974. Decline of caribou in North America following settlement. *J. Wildl. Mgmt.* 38:757–70.

Bergerud, A. T. 1975. The reproductive season in Newfoundland caribou. *Can. J. Zool.* 53:1213–21.

Bergerud, A. T. 1985. Antipredator strategies of caribou dispersion along shorelines. *Can. J. Zool.* 63:1324–29.

Bergerud, A. T., and W. E. Mercer. 1989. Caribou introductions in eastern North America. *Wildl. Soc. Bull.* 17:111–20.

Chapman, J. A., and G. A. Feldhammer (eds.). 1982. *Wild mammals of North America: Biology, Management, and Economics.* Johns Hopkins Univ. Press, Baltimore.

Cringhan, A. T. 1957. History, food habits and range requirements of the woodland caribou of continental North America. *Trans. 22nd. N. Am. Wildl. Conf.,* 485–500.

Cumming, H. G., and D. B. Beange. 1987. Dispersion and movements of woodland caribou near Lake Nipigon, Ontario. *J. Wildl. Mgmt.* 51:69–79.

Dauphine, T. C., Jr., and R. L. McClure. 1974. Synchronous mating in Canadian barren-ground caribou. *J. Wildl. Mgmt.* 38:54–66.

de Vos, A., and R. L. Peterson. 1951. A review of the status of woodland caribou (*Rangifer caribou*) in Ontario. *J. Mamm.* 32:329–37.

Fuller, T. K., and L. B. Keith. 1981. Woodland caribou population dynamics in northeastern Alberta. *J. Wildl. Mgmt.* 45:197–213.

Holleman, D. F., J. R. Luick, and R. G. White. 1979. Lichen intake estimates for reindeer and caribou during winter. *J. Wildl. Mgmt.* 43:192–201.

Martin, C. 1988. The history and reestablishment potential of marten, lynx, and woodland caribou of Isle Royale. Isle Royale Natl. Park, *Resour. Mgmt. Rept.* 88-2.

Mech, L. D., M. E. Nelson, and H. F. Drabik. 1982. Reoccurrence of caribou in Minnesota. *Am. Midl. Nat.* 108:206–8.

Nelson, U. C. 1947. Woodland caribou in Minnesota. *J. Wildl. Mgmt.* 11:283–84.

Palmer, P. S. 1938. Late records of caribou in Maine. *J. Mamm.* 19:37–43.

Schorger, A. W. 1942. Extinct and endangered mammals and birds of the Upper Great Lakes Region. *Trans. Wisc. Acad. Sci., Arts Ltrs.* 34:23–44.

Servheen, G., and L. J. Lyon. 1989. Habitat use by woodland caribou in the Selkirk Mountains. *J. Wildl. Mgmt.* 53:230–37.

Bison

· *Bison bison* (Linnaeus)
· State Extirpated

The bison once had an extensive range throughout much of North America and influenced the settlement of North America more than any other native species (Roe 1970). Prior to European settlement, the bison had an estimated population of 40 to 60 million and occurred from the Appalachian Mountains west to the Rocky Mountains, and from northwestern Mexico to Great Slave Lake, Northwest Territories, Canada (Hall 1981; McDonald 1981; Chapman and Feldhammer 1982). However, in less than 100 years this abundant animal was brought to the edge of extinction, with approximately 1,000 individuals surviving after the turn of the century (Sandoz 1954; Dary 1974; McHugh 1972). Today, there are more than 65,000 bison in North America (Jennings 1978), but only a small proportion survive in free-ranging herds.

The bison is divided into two subspecies: (1) the plains bison (*B. b. bison*), which is the most common one found in zoological parks, wildlife reserves, and in National Parks (e.g., Yellowstone National Park), and (2) the federally endangered wood bison (*B. b. athabascae*). The wood bison was believed extinct in 1940 but was rediscovered in 1957, surviving in an isolated population near the Nyarling River in the Northwest Territories (Banfield and Novakowski 1960).

The plains bison subspecies once occurred in the southern part of the Lower Peninsula (and possibly western Upper Peninsula)(Baker 1983). Records of their actual distribution are scarce because it had probably disappeared from Michigan by the end of the eighteenth century. Documentation of its presence exists for Berrien, Cass, Jackson, Kalamazoo, Monroe, and Wayne counties (Baker 1983). These presettlement herds probably preferred southern Michigan's small prairies and extensive oak savanna regions. Bison remained in the Great Lakes region in Minnesota until the early 1880s (Schorger 1942).

Bison are unmistakable in appearance, characterized by a high shoulder hump, short hairy neck, and massive head with curved horns (found on both sexes). Mature male bison typically weigh approximately 1,400 to 1,600 pounds (635 to 726 kg) with the smaller females weighing 900 to 1,100 pounds (408 to 499 kg) (Baker 1983). The breeding season extends from July to October. One calf (rarely two calves) is produced between April and June after a 9- to 9^1/2–month gestation period. The young are weaned in several months and are sexually mature in one year, although actual mating usually does not occur for two to three years. The bison is a grazer, mainly dependent on various grasses, sedges, and other herbaceous plant material.

The nearly complete transformation of the bison's native Michigan habitat into agricultural lands severely limits its reestablishment, particularly free-ranging herds. The tremendous acreages required to prevent bison-human conflicts and the bison's tendency to wander long distances when released into new range both result in poor relocation success and public protest.

LITERATURE CITED

Baker, R. H. 1983. *Michigan mammals.* Mich. State Univ. Press, E. Lansing.

Banfield, A. W., and N. S. Novakowski. 1960. The survival of the wood bison (*Bison bison athabascae* Rhoads) in the Northwest Territories. *Natl. Mus. Can. Hist. Pap.* No. 8.

Chapman, J. A., and G. A. Feldhammer (eds.). 1982. *Wild mammals of North America: Biology, management, and economics.* Johns Hopkins Univ. Press, Baltimore.

Dary, D. A. 1974. The buffalo book: The saga of an American symbol. *Avon Books/Swallow Press,* Chicago.

Hall, E. R. 1981. *The mammals of North America.* Vol. 2. John Wiley Sons, New York.

Jennings, D. C. 1978. *Buffalo history and husbandry.* Pine Hill Press, Freeman, ND.

McDonald, J. N. 1981. *North American bison: Their classification and evolution.* Univ. Calif. Press, Berkeley.

Roe, F. G. 1951. *The North American buffalo: A critical study of the species in the wild state.* Univ. Toronto Press, Toronto.

Sandoz, M. 1954. *The buffalo hunters.* Hastings House, New York.

Schorger, A. W. 1942. Extinct and endangered mammals and birds of the upper Great Lakes region. *Wisc. Acad. Sci., Arts, Ltrs.,* 23–44.

Eastern Elk

- *Cervus elaphus canadensis* Erxleben
- Extinct

This extinct, eastern North American subspecies historically roamed from southern Ontario and New York south through the Allegheny Mountains to Georgia and Louisiana west to the edge of the Rocky Mountains (Hall 1981). Due to widespread settlement and intensive hunting in this region, the eastern elk was greatly reduced by the mid 1800s. Although there is some question about whether the few survivors into the twentieth century were of native stock (e.g., in Minnesota), this subspecies apparently became extinct in the late 1800s with late records of 1877 in Pennsylvania (Allen 1942) and 1890 in Minnesota (Schorger 1942).

In Michigan, it probably occurred throughout the state but was most abundant in the Lower Peninsula (Hall 1981; Baker 1983). Here, it experienced a steady decline through the 1800s. The last population sites were in Huron, Sanilac, and Tuscola counties (Murie 1951). It became extinct in the state around 1871 (Wood and Dice 1923).

Efforts by the state to reestablish the elk (or wapiti) were successful with the reintroduction of seven Rocky Mountain wapiti (*C. e. nelsoni*) in Cheboygan County in 1918 (Moran 1973). This subsequently led to the formation of Michigan's current self-sustaining population (Ruhl 1940). Today's increasing population of over 1,000 individuals resides largely in seven northern lower Michigan counties. The principal range of 600 square miles (1,560 square km) includes the area encompassed by Indian River, Onaway, Atlanta, and Gaylord (Baker 1983).

LITERATURE CITED

Allen, G. M. 1942. Extinct and vanishing mammals of the western hemisphere. *Am. Comm. Int. Wildl. Protection, Spec. Bull.* No. 11.

Baker, R. H. 1983. *Michigan mammals.* Mich. State Univ. Press, E. Lansing.

Hall, E. R. 1981. *The mammals of North America.* Vol. 2. John Wiley and Sons, New York.

Moran, R. J. 1973. The Rocky Mountain elk in Michigan. *Mich. Dept. Nat. Resour. Wildl. Div., Resour. Dev. Rept.* No. 267.

Murie, O. J. 1951. *The elk of North America.* Stackpole Books, Harrisburg, PA.

Ruhl, H. D. 1940. Game introductions in Michigan. *Trans. 5th N. Am. Wildl. Conf.,* 424–27.

Schorger, A. W. 1942. Extinct and endangered mammals and birds of the upper Great Lakes region. *Wisc. Acad. Sci., Arts, Ltrs.,* 23–44.

Wood, N. A., and L. R. Dice. 1923. Records of the distribution of Michigan mammals. *Mich. Acad. Sci., Arts, and Ltrs.* 3:424–69.

Birds

Endangered and Threatened

Common loon (*Gavia immer*)
Least bittern (*Ixobrychus exilis*)
Osprey (*Pandion haliaetus*)
Bald eagle (*Haliaeetus leucocephalus*)
Red-shouldered hawk (*Buteo lineatus*)
Merlin (*Falco columbarius*)
Peregrine falcon (*Falco peregrinus*)
Yellow rail (*Coturnicops noveboracensis*)
King rail (*Rallus elegans*)
Piping plover (*Charadrius melodus*)
Caspian tern (*Sterna caspia*)
Common tern (*Sterna hirundo*)
Barn owl (*Tyto alba*)
Long-eared owl (*Asio otus*)
Short-eared owl (*Asio flammeus*)
Loggerhead shrike (*Lanius ludovicianus*)
Yellow-throated warbler (*Dendroica dominica*)
Prairie warbler (*Dendroica discolor*)
Kirtland's warbler (*Dendroica kirtlandii*)

Extirpated

Trumpeter swan (*Cygnus buccinator*)
Greater prairie chicken (*Tympanuchus cupido*)
Whooping crane (*Grus americana*)*
Eskimo curlew (*Numenius borealis*)*
Lark sparrow (*Chondestes grammacus)*

Extinct

Passenger pigeon (*Ectopistes migratorius*)

---

*These species are not officially state listed but may occur in Michigan and are federally listed.

An Introduction to Birds

David N. Ewert
The Nature Conservancy

Three hundred years ago, Michigan was a rich mosaic of forests and openings. Spruce clustered along Lake Superior, the northern shores of Lakes Michigan and Huron, and south into the bogs of Calhoun and Washtenaw counties. Moist, well-drained areas supported cathedrallike stands of eastern hemlock (*Tsuga canadensis*), American beech (*Fagus grandifolia*) and sugar maple (*Acer saccharum*), intermingled with younger stands of the same species, yellow birch (*Betula alleghaniensis*) and white birch (*Betula papyrifera*), and other early succession species where windfalls had occurred. Pines and oaks flourished on sandy soils and outcrops where episodic fires ensured their survival. In the south, oak and hickory dominated much of the landscape. Where fire was frequent, especially in southwestern Michigan, savannas and patches of prairie replaced the extensive forest. Michigan's forests were largely unfragmented and highly variable in structure and species composition. Natural openings—whether bog, prairie, fen, or areas in different stages of early succession—were concentrated in extremely wet or dry locations.

Presettlement Avifauna

The avifauna reflected these conditions. Had there been a threatened or endangered species list of birds in 1690, it might have included field sparrow, eastern meadowlark, brown-headed cowbird, and common grackle (Brewer 1988) but not the then-abundant passenger pigeon or the widespread common loon, king rail, and red-shouldered hawk. Today much of the Michigan landscape has become highly fragmented, urbanized, suburbanized, and altered by intensive agriculture. Vast stretches of unmanaged forest are few and far between, and old-growth forests have virtually disappeared; the present avifauna reflects these dramatic changes in habitat.

Michigan's bird life has not been stable during the last few thousand years. Whatever preglacial avifauna existed was completely eliminated or displaced by glaciers that retreated 10,000 to 15,000 years ago. Our birds, then, are newcomers that reached the state at different times and whose numbers and distribution have varied with ecosystem modification induced by climatic change during the last 13,000 years.

History of Michigan's Avifauna

Written accounts by settlers, early land surveyors, geologists, and others provided the first extensive information on the distribution of many species of birds in the state. The first statewide summaries appeared in Gibbs 1879 and Cook 1893. As interest in birds increased, our knowledge of Michigan birds has expanded rapidly. Twentieth-century statewide treatments of Michigan birds include those by Barrows (1912), Wood (1951), Zimmerman and Van Tyne (1959), Payne (1983), and Brewer et al. (1991). Several regional treatments have appeared, most notably those by Cuthbert (1963) for Isabella County; Pettingill (1974) for the Straits of Mackinac region; McWhirter and Beaver (1977) for the Lansing area; Kelley (1978) for the Detroit area; Kenaga (1983) for the Saginaw Bay region; and Mlodinow (1984) for Berrien County and southern Lake Michigan. Many other regional summaries have appeared in the *Jack-Pine Warbler* and other journals. Many other publications have focused on Michigan's birds, especially the Kirtland's warbler, of which the books by Mayfield (1960) and Walkinshaw (1983) are internationally known. Supplementing these studies have been Christmas bird counts, breeding bird censuses, and breeding bird surveys sponsored by the United States Fish and Wildlife Service. Michigan's bird life has received much attention, and the distribution and relative abundance of members of our avifauna, including threatened and endangered species, is relatively well known.

Our Rare Species: Who, Where, and Why

Settlement of Michigan by native Americans, and then Europeans, drastically altered our bird life, both directly and indirectly. Of the 238 species known to have bred in the state, 1 species is extinct (passenger pigeon), and 2 native species are considered extirpated, the greater prairie chicken, trumpeter swan, and lark sparrow. Seven species are endangered, 12 species are threatened, and 20 species are potential candidates for listing (special concern species).

Populations of Michigan's 39 bird species that are of special concern, threatened, or endangered seem to have decreased, in large part, because of direct and indirect effects of habitat fragmentation, degradation, and contamination. Other factors, such as loss of winter habitat, are becoming increasing concerns, especially for species that winter in the neotropics. The introduction of exotics, such as the mute swan and European starling, has resulted in increased competition for nest sites with some native species threatened in the state.

Historically, large, conspicuous, social species declined rapidly because of overhunting and habitat destruction. Among the early targets was the passenger pigeon, which vanished from Michigan in 1898 (Barrows 1912). The wild turkey, the sandhill crane, and many migratory shorebirds, including the Eskimo curlew, probably declined when market hunting became commonplace (Barrows 1912; Wolinski 1988). Of these, the passenger pigeon suffered a similar fate throughout its range, and it is the only Michigan bird species to have become extinct. On the brighter side, the sandhill crane has recovered well and continues to increase

(Walkinshaw and Hoffman 1974; Hoffman 1989), and wild turkeys have been successfully reintroduced by the Michigan Department of Natural Resources and are prospering in many parts of the state.

Hunting probably has little effect on threatened and endangered species of birds in Michigan today. Of far greater consequence is the change in habitat, loss of habitat, and introduction of environmental contaminants. Even the disappearance of the passenger pigeon was at least partly attributed to habitat loss.

The deforestation of Michigan in the 1800s was amazingly rapid. Virtually the entire state was logged and cleared in 100 years. Large, conspicuous species, such as the red-shouldered hawk and barred owl, became more locally distributed and their centers of distribution retreated northward. Pileated woodpeckers and common ravens were also nearly eliminated from the state (Barrows 1912). It is likely that many songbirds, especially those now known to dwell in the forest interior (e.g., Cerulean warbler and the blackburnian warbler), experienced undocumented declines. For forest birds in Michigan, the late 1800s and early 1900s must have been a disaster.

Of the four species of forest-dwelling birds endangered or threatened in Michigan, at least two, the Kirtland's warbler and red-shouldered hawk, are seriously affected by the loss of forest. A third species, the yellow-throated warbler (which is at the northern limit of its breeding range in Michigan), recolonized the state as floodplain forests matured. Concern has developed over the Cerulean warbler, an area-sensitive species that breeds in floodplain forests, because of the loss of forest in Michigan and other breeding areas, and in the Andes Mountains, where it winters.

One of the changes associated with the loss and fragmentation of forest was the spread of species into the state that were formerly rare or locally distributed, such as the parasitic brown-headed cowbird, an invader from the prairies (Mayfield 1977). Increased rates of nest parasitism by cowbirds and predation of the nests of songbirds are likely to be increasing problems in fragmented landscapes (Temple and Cary 1988). If cowbird control measures had not been implemented, the Kirtland's warbler might now be extinct (Mayfield 1975 and 1977). Had the forests not been cleared, the cowbird would not have dispersed northward through southern Michigan, and the two species probably would not have met. Other cowbird host species (not yet listed as endangered or threatened) may be candidates for the threatened and endangered list if the cowbird continues to increase.

The red-shouldered hawk also has been severely affected by forest loss and fragmentation in southern Michigan. Populations have declined precipitously and shifted northward, where large forest tracts remain. However, reproduction is low. Reasons for declines are unclear but may include negative interactions with other species of raptors, the adverse effects of contaminants, and lower prey abundance and availability (Evers et al. in Eastman 1989).

Chemical contaminants likely played a major role in the decline of many of our listed species, especially those near the top of the food chain nesting near or within aquatic environments. The peregrine falcon, red-shouldered hawk, bald eagle, osprey, Cooper's hawk, northern harrier, common tern, and double-crested cormorant may have reached low population levels in the 1960s because of such environmental contaminants as organochlorines and PCBs (Adams et al. 1988). Ingestion of lead pellets may reduce populations of the king rail, while mercury-contaminated food may be responsible for local mortality in common loons. Although chemical and heavy metal contamination continues to be a threat to Michigan's avifauna, it is reassuring that at least some contaminants have decreased sufficiently to permit implementation of the peregrine falcon reintroduction program.

The indirect effect of habitat loss is painfully evident for the Kirtland's warbler, but the absolute loss of habitat has seriously affected many more of our threatened and endangered species. For example, common loon populations have plummeted and nearly disappeared from the southern Lower Peninsula, a region that probably supported the highest number of nesting pairs (McPeek and Evers 1989). They have also declined dramatically in the eastern Upper Peninsula since the mid-1980s (Evers 1990).

Intense foot and vehicle traffic on beaches of the Great Lakes has reduced the population of piping plovers from an estimated 200 pairs at settlement (Russell 1983) to fewer than 17 in 1992. Loss of prairie and oak savanna to agriculture and other development has meant the demise or near demise of the greater prairie chicken, lark sparrow, and barn owl in Michigan; fewer than 5 pairs of loggerhead shrikes remain in the state as grasslands and pastures have succumbed to "clean farming." Similarly, loss of wetlands, especially in southeastern Michigan, may be responsible for the low numbers of king rails, black terns, and short-eared owls recorded in recent decades. For some species with specialized requirements, specific cultural practices precipitate further declines. For example, it is estimated that almost 500 loons drown in fish nets each year in Michigan; these trap nets are being evaluated to see if they can be designed to capture fish while allowing loons to escape.

Fire is another factor that can determine species distribution and abundance. Black-backed woodpeckers and Kirtland's warblers colonize coniferous habitats created by fire; without fire, or an ecological equivalent, these species would be without appropriate habitat. Fire suppression, with its obvious benefit to humans, has consequences to species that benefit from fire.

Several species that are of special concern, threatened, or endangered in Michigan may always have been rare in the state. It is difficult to assess if populations have changed and if so, why. In this group are the merlin and yellow rail; Michigan is at the southern limit of their breeding range and populations are at least stable. Similarly the Wilson's phalarope, Forster's tern, and yellow-headed blackbird are at or near their eastern range limit in Michigan, and typically occur in large marshes, especially those bordering the Great Lakes. Dickcissels, which occasionally invade Michigan grasslands and alfalfa fields from the prairie states, may become widespread and common in some years (e.g., 1988). Nonetheless, it appears that many of these species are declining elsewhere, so efforts should be made to maintain their habitats in Michigan.

Current Conservation Efforts

The list of special concern, threatened, and endangered bird species comprises only 16% of the 238 species that have documented breeding records in the state. Because of their precarious survival in the state, special programs are in place to protect these species. Much progress is being made. Increased numbers of double-crested cormorants (Ludwig 1984), raptors such as the bald eagle, osprey, merlin, and Cooper's hawk (Evers 1989), and yellow-throated warblers are being found in Michigan and elsewhere. Common loons have not declined below 300 pairs, even though development pressure on the lakes in which they breed is intensifying. However, protection efforts by the Michigan Department of Natural Resources (including the Natural Heritage Program), United States Forest Service, National Park Service, and Michigan Loon Preservation Association are also intensifying. Piping plovers and Kirtland's warblers—the subjects of protection and management programs by government agencies and conservation organizations—are at dangerously low levels, although declines have been at least temporarily arrested. Once-rare species now common in Michigan, such as the wood duck, Canada goose, and common raven (Evers et al. in Eastman 1989) provide encouragement that recovery of other species will be successful.

The Future

Reducing the list of endangered and threatened bird species will take careful planning and coordination between agencies and major landowners, not only in Michigan but also in such overseas locations as the Bahamas (Kirtland's warbler winter habitat) and Peru (Cerulean warbler winter habitat). Special attention must be devoted to habitat protection as well as identification of the essential requirements of each species for breeding, migration, and for overwintering if successful conservation strategies are to be implemented. Because much work is being done with the 39 species of immediate concern, we hope that we can ensure their survival.

But what of the other 360 species of birds that inhabit Michigan? It will be even more difficult to prevent many more species from being threatened or endangered as the human population grows and the demand on resources increases. Clearly, protecting our birds, as well as other animals and plants, requires that we anticipate future land development and management trends. We must design preserves and implement management techniques that provide the resources necessary to maintain viable populations of our remaining species throughout the year. Protection planning for Michigan's birds will require intrastate, interstate, and international cooperation. Implementation of plans will become increasingly complex, urgent, and challenging.

LITERATURE CITED

Adams, R. J., Jr., G. A. McPeek, and D. C. Evers. 1988. Bird population changes in Michigan, 1966–1985. *Jack-Pine Warbler* 66:71–86.

Barrows, W. B. 1912. *Michigan bird life.* Mich. Agric. Coll. Spec. Bull., E. Lansing.

Brewer, R. 1988. An early Michigan bird list. *Jack-Pine Warbler* 66: 47–53.

Brewer, R., G. A. McPeek, and R. J. Adams, Jr. 1991. *The atlas of breeding birds of Michigan.* Mich. State Univ. Press, E. Lansing.

Cook, A. J. 1893. *Birds of Michigan.* 2d ed. Mich. Agric. Exp. Sta. Bull. 94.

Cuthbert, N.L. 1963. *The birds of Isabella County, Michigan.* Edwards Brothers, Ann Arbor, MI.

Evers, D. C. 1989. Michigan raptor populations may be on the upswing. *Eyas* 12:10–15.

Evers, D. C. 1990. Northern Great Lake common loon monitoring program: 1990 final report. *Mich. Dept. Nat. Resour.,* Unpubl. Rept.

Evers, D. C., G. A. McPeek, and R. A. Adams, Jr. 1989. Michigan's changing bird populations. Pp. 19–29 in John Eastman (ed.), *Enjoying Birds in Michigan.* CES Publications, Grand Rapids, MI.

Gibbs, M. 1879. Annotated list of the birds of Michigan. *U.S. Geol. Geog. Survey Bull.* 5:481–97.

Hoffman, R. H. 1989. Status of the sandhill crane population in Michigan's lower peninsula, 1986–87. *Jack-Pine Warbler* 67:18–28.

Kelley, A. H. 1978. *Birds of southeastern Michigan and southwestern Ontario.* Cranbrook Inst. Science, Bloomfield Hills, MI.

Kenaga, E. E. 1983. *Birds, birders, and birding in the Saginaw Bay area.* Chippewa Nature Center, Midland, MI.

Ludwig, J. D. 1984. Decline, resurgence and population dynamics of Michigan and Great Lakes double-crested cormorants. *Jack-Pine Warbler* 62:91–102.

McPeek, G., and D. Evers. 1989. Guidelines for the protection and management of the common loon in southwestern Michigan. *Mich. Loon Preserv. Assoc.,* Unpubl. Rept.

McWhirter, D. W., and D. L. Beaver. 1977. Birds of the Capital Count Area of Michigan with seasonal and historical analysis. *Mich. State Univ., Mus. Biol. Ser.* No. 5:353–442.

Mayfield, H. 1960. Kirtland's warbler. *Cranbrook Inst. Science Bull.* No. 40.

Mayfield, H. 1975. The numbers of Kirtland's warblers. *Jack-Pine Warbler* 53:38–47.

Mayfield, H. 1977. Brown-headed cowbird: Agent of extinction? *Am. Birds* 31:107–13.

Mlodinow, S. 1984. *Chicago area birds.* Chicago Review Press, Chicago.

Payne, R. B. 1983. A distributional checklist of the birds of Michigan. *Univ. Mich. Mus. Zool. Misc. Publ.* No. 164.

Pettingill, O. S., Jr. 1974. *Ornithology at the University of Michigan Biological Station and the birds of the region.* Kalamazoo Nature Center Spec. Publ. No. 1.

Russell, R .P., Jr. 1983. The piping plover in the Great Lakes region. *Am. Birds* 37:951–55.

Temple, S. A., and J. R. Cary. 1988. Modeling dynamics of habitat-interior bird populations in fragmented landscapes. *Conserv. Biology* 2:340–47.

Walkinshaw, L. H. 1983. Kirtland's Warbler: The natural history of an endangered species. *Cranbrook Inst. Science Bull.* No. 58.

Walkinshaw, L. H., and R. H. Hoffman. 1974. Southern Michigan sandhill crane survey. 1971–1973. *Jack-Pine Warbler* 52:103–14.

Wolinski, R. A. 1988. Some bird population changes in Michigan: 1900 to 1965. *Jack-Pine Warbler* 66:55–69.

Wood, N. A. 1951. The birds of Michigan. *Univ. Mich. Mus. Zool. Misc. Publ.* No. 75.

Zimmerman, D. A., and J. Van Tyne. 1959. A distributional checklist of the birds of Michigan. *Univ. Mich. Mus. Occ. Pap.* No. 608.

Common Loon

· *Gavia immer* (Brunnich)
· **State Threatened**

Map 8. Confirmed common loon nesting pairs. Historically statewide (Barrows 1912); ● = 1983 to 1992 (Michigan DNR).

Status Rationale: The common loon is threatened by many human-related activities within its nesting and wintering grounds. Although many of the loon's northernmost populations are relatively stable, others have contracted in the lower Great Lakes region and throughout the southern periphery of its historical nesting range. Nesting loons have nearly disappeared from the southern half of the Lower Peninsula, while the rest of Michigan supports a population estimated at fewer than 300 nesting pairs.

Field Identification: The common loon is a large, long- and heavy-bodied, diving bird with a daggerlike black bill and a sloped forehead. It averages 32 inches (81.3 cm) in length with a wingspan of 5 feet (1.5 m). In breeding plumage, the common loon can be identified by its checkered, black-and-white back, white breast, and dark head. Winter adults lose their black-and-white pattern and become drab gray above and white below, plumage similar to that of the juvenile's. The loon has four distinctive calls, including the most familiar laughing tremolo, the territorial yodel (used only by the male), the lonesome wail (used for long-distance communication), and the simple "talking"calls (hoots). The yodel call can be used as a voice imprint to identify individual males (Miller 1988). These four calls are frequently mixed. While flying, only the tremolo is used.

In summer, the common loon may be confused with the double-crested cormorant (*Phalacrocorax auritus*) or with Michigan's two largest mergansers, the common (*Mergus merganser*) and red-breasted (*Mergus serrator*). Cormorants are black and sit on the water with their bill tilted upwards, while the black-and-white plumaged loons maintain a flat-headed profile. Both mergansers have a thinner bill and a solid, dark back. Two migratory loon species, both uncommon in Michigan, are easily distinguished from the common loon when in their breeding plumage. The smaller, red-throated loon (*Gavia stellata*) lacks a broken white collar and has a light gray head with a red throat. The Pacific loon (*Gavia arctica*) (formerly the arctic loon) has a gray head with a black throat patch. Winter plumage differences between these three loons are more subtle.

Total Range: Unlike the other four loon species that nest in the tundra regions throughout the Northern Hemisphere, the common loon is restricted to North America (including Greenland) and Iceland. It breeds from the northern reaches of the continent south to the western Canadian border and northern Rocky Mountains of the contiguous United States, east through the northern Great Lakes region, New York, and much of New England.

The U.S. Great Lakes region currently contains an estimated 13,000 breeding individuals. This does not include a large, unknown number of nonbreeders and subadults that summer on the Great Lakes. The Wisconsin loon population is apparently increasing and expanding (Strong 1988b). Great Lakes populations (including Michigan) generally migrate south for the winter to the Atlantic and Gulf coasts from Newfoundland south to Texas. Recent banding programs in Michigan (Evers 1992) may soon provide information on specific wintering areas for Michigan nesting loon populations.

State History and Distribution: Michigan's breeding common loon population once covered the entire state, an area encompassing more than 2,000 lakes larger than 50 acres. In the early 1900s, the loon began to experience the pressure of lakeshore development along the southern extremities of the state. During the 1920s and 1930s, the common loon declined dramatically in southern Michigan; by the early 1940s, many southern counties had lost their last nesting pairs. Southeastern Michigan's breeding population disappeared by the late 1970s.

During the 1980s, the common loon was largely restricted to northern lower Michigan (north of Mecosta, Clare, and Iosco counties) and the Upper Peninsula. A remnant population of 2 pairs in Barry County and 2 to 3 pairs in southeastern Newaygo and northwestern Montcalm counties represent the last of Michigan's southern population (McPeek and Evers 1988). In 1989, estimates from several surveys placed Michigan's population between 275 and 300 nesting pairs (Robinson et al. 1988; Evers 1992). These surveys, sponsored by the DNR, began in 1983 and proved to be a valuable monitoring tool (Robinson et al. 1988). Survey results have shown the nesting loon population of the western Upper Peninsula and Isle Royale to be substantial and possibly increasing. Seney National Wildlife Refuge, Schoolcraft County, in the central Upper Peninsula also has a relatively dense nesting loon population. However, the proportion of suitable lakes occupied in the eastern Upper Peninsula is lower than expected, even though productivity rates are relatively high and constant, averaging more than 1.5 young per successful nest

(Robinson 1988; Evers 1992). Substantial numbers of nonbreeding loons summer on the Great Lakes, which provides a deceiving picture of Michigan's actual breeding population.

The common loon may be seen throughout the state in spring and autumn migration. Several peninsulas that extend into the Great Lakes concentrate migrating loons, particularly Whitefish Point on eastern Lake Superior (Ewert 1982). Between 4,000 and 9,000 common loons generally are counted each spring by the Whitefish Point Bird Observatory. Lakes Michigan and Huron are important staging areas, where loons congregate in spring and autumn to prepare for long migration treks. Wintering individuals occasionally are sighted in the Great Lakes (Payne 1983).

Habitat: Common loons prefer to nest in large, inland lakes (between 40 to 70 acres [16.2 to 28.3 ha]) containing several islands and minimal shoreline development. The minimal lake size is typically 30 acres (12.1 ha), although loons have nested on lakes of 11 acres (4.5 ha) in Michigan (Miller and Dring 1988). Loon territories may include lakes other than the primary nesting lake. Research in southwestern Michigan shows that territorial pairs frequently use two to six different lakes (Evers 1989; McPeek and Evers 1989).

Nests are usually placed on sheltered islands less than 4 feet (1.2 m) from shore (Vermeer 1973) or on lake shorelines with steep underwater slopes (McIntyre 1983). Since it is difficult for a loon to walk on land, this facilitates a quick underwater exit. Natural island (Olson and Marshall 1952; Vermeer 1973; McIntyre 1975; Jung 1987) and bog-mat nest sites (Reiser 1988) are preferred.

Ecology and Life History: In March, loons begin to move north from the ocean overwintering grounds to Michigan's northern forested lakes. Southern Michigan's populations arrive immediately following ice-off, typically in late March (Evers and McPeek 1987). Weather conditions (e.g., heavy rain and fog) and ice breakup on the Great Lakes have a bearing on the time and magnitude of loon migration (Ewert 1982). The heaviest migration concentrations at Whitefish Point usually occur from early to mid-May between 5:00 and 7:00 A.M., sometimes with peaks of more than 700 birds per hour.

Male loons may arrive before the females, alone or in groups of two or more, to establish their territories on suitable lakes. Breeding adult loons are highly site faithful but are known to frequently switch mates between and within years (Evers 1992). Courtship is simple, mainly involving ritualized bill dipping and splash diving by the pair (Sjolander and Agren 1972). After a territory is established, the pair constructs a nest of debris and vegetation within a few feet from the water's edge. Nesting areas are used annually (Strong et al. 1987a), even after persistent nest failures (Titus and Van Druff 1981; Strong et al. 1987a). Loons commonly renest if the initial nest is destroyed early in the breeding season.

Both parents incubate the two eggs (sometimes one and rarely three) for approximately one month. After hatching, the black, down-covered chicks almost immediately take to the water, sometimes swimming on their own, but frequently riding on a parent's back. Adult loons rear the young in specific nursery (chick-rearing) areas, typically situated close to shore in small, secluded coves with shallow water (McIntyre 1983; Strong 1985). After 2 weeks, the young become excellent swimmers and divers. In 8 to 12 more weeks, they are capable of sustained flight (Barr 1973; McIntyre 1975).

Loons typically congregate in large bodies of water in late summer, prior to the major migratory movements in September and October. Upon reaching the ocean wintering grounds, juveniles often remain along the coast until they reach three years of age. Some loons in nonbreeding plumage (immatures) oversummer on the Great Lakes, and McIntyre (1988) suggests a northward movement of these loons along the Atlantic Coast to the Canadian Maritimes. Overwintering adults remain territorial during the day but group together at night (McIntyre 1978).

Loons are opportunistic feeders and depend primarily on a variety of fish up to 8 inches (20.3 cm) long (Barr 1973). Crayfish, frogs, aquatic insects, and aquatic plants are also eaten. The young initially are fed plant material and invertebrates with fish comprising an increasingly greater part of the diet as they mature. Breeding adult loons will use neighboring lakes as auxiliary feeding areas (Miller and Dring 1988); although, unlike the red-throated loon, the adults usually do not bring food back to the young.

Conservation/Management: The common loon is still present in much of its northern breeding range. In Michigan and many other parts of its southern breeding range, however, the loon is vulnerable to a wide variety of serious threats that are compounded by the loon's late sexual maturity (probably 6–7 years) and behavior typical of long-lived birds. One of the most direct and conspicuous of these threats is increased shoreline development and recreational use of lakes (Vermeer 1973; Ream 1976; Plunkett 1979; Jung 1987; Strong 1988a). While on its breeding territory, the loon is very sensitive to human disturbance, which is accompanied by noise and water pollution, boating, and other recreational activities, as well as habitat destruction.

On the Great Lakes, loons face major threats from commercial fishing and botulism outbreaks. Nets, particularly trap nets used for commercial fishing, unintentionally entrap loons underwater during routine fishing operations. Michigan loon mortality related to trap nets are estimated to be higher than 500 individuals per year. Through the bioaccumulation of mercury in aquatic food chains, common loons may become weakened and susceptible to avian botulism (Type E), which causes large dieoffs.

Another potential threat to loon populations is the effect of acidification. Lakes with lower pH levels may increase methyl mercury levels, reduce their biological carrying capacity (e.g., by depleting fish numbers), and therefore significantly decrease loon reproductive success (Haines 1981; Alvo and Russell 1983; Parker and Brocke 1985; Alvo 1986). Some studies have shown little relationship between acid rain levels and loon breeding success (McNichol et al. 1987; Parker 1988). However, Schindler (1988) and Mitchell (1989) have shown that acid rain can be a major contributor to the deterioration of aquatic ecosystems and bird populations dependent on them. Because Michigan's lakes typically have a limestone bedrock foundation and a high pH (which buffers the acid rain effects), nesting loons have not been as affected as those areas with low pH levels, such as in the Adirondack Mountains, New York.

Other hazards to loons include increases in raccoons (*Procyon lotor*), which destroy nests (Wood in Sutcliffe 1979), fluctuating water levels (Lange and Lange 1987), and predation of chicks by large fish, turtles, and gulls. On inland lakes along Lake Michigan, mute swan (*Cygnus olor*) populations are increasingly in contact with nesting loons and may affect fledging success (McPeek and Evers 1989). On their wintering grounds, the common loon must also contend with oil spills, toxic chemicals, commercial fishing nets, and heavy metal contamination of their diet. Young birds in particular are vulnerable, because they may remain on the ocean until three years of age.

In 1986, relief for Michigan's struggling loon population was initiated by the Loon Registry Program (Ewert 1988). Because boating activity is a major threat to loons, in 1987 the Michigan Loon Preservation Association began providing needed support for the common loon in the state (replacing the registry program with project LoonWatch in 1989). Designated loon rangers are responsible for monitoring each known active breeding pair in Michigan. On developed lakes, signs and buoys can help protect nesting loons from disturbance by unsuspecting boaters and canoeists. Since nest sites, as well as nursery areas, are frequently reused, their protection is crucial for loon breeding success (Strong et al. 1987a). On lakes lacking suitable island and/or shoreline nesting habitat, artificial floating platforms can be anchored in sheltered areas of the lake. If properly placed, a nest should be protected from predation and waves produced by boats and storms.

Conflicts between commercial fisheries and loons must be lessened. One reason loons have been captured accidentally more often in recent years is because of the increased use of trap nets in Michigan. This net keeps the struggling loon above the water surface, which then attracts and catches other loons. During a single week at Whitefish Point, one trap net captured more than 50 loons with at least half of these dying as a result of the encounter. As summer progresses and fish typically move away from shore, incidences of loon capture decrease. Two alternatives are to set trap nets 6 feet (1.8 m) below the water's surface or change to pound nets (used in other states). State officials have also developed a trap design to allow loons (but not fish) to escape through larger mesh at the top of the trap's heart.

Although the common loon is a symbol of the wilderness quality of lakes, it can tolerate limited recreational development (Heimberger et al. 1983). If the lake's ecological community is respected, common loons and humans can coexist.

Comments: Loons are superbly built for life on the water. The loon's powerful, webbed feet (which are placed well back on the body) provide underwater mobility and aid in capturing fish. To reduce buoyancy, loons have the ability to compress air from the feathers and lungs, and their bones are solid (unlike the spongy bones of most birds). Once in the air, the loon is a strong, swift flyer. However, to become airborne it needs at least a 75-foot (23-m) runway of water and, depending on wind direction, up to a quarter mile. The common loon cannot take off from land. Reflections from highways simulate bodies of water and occasionally fool loons into landing. If found, these stranded loons should be immediately taken to a large body of water.

Literature Cited

Alvo, R. 1986. Lost loons of the northern lakes. *Nat. Hist.* 90:59–64.

Alvo, R., and D. J. Russell. 1983. *Ontario lakes loon survey.* Long Point Bird Observatory. Unpubl. Progr. Rept. No. 2.

Barr, J. F. 1973. *Feeding biology of the common loon (Gavia immer) in oligotrophic lakes of the Canadian Shield.* Ph.D. diss. Univ. Guelph, Guelph, Ont.

Barrows, W. B. 1912. *Michigan bird life.* Mich. Agric. Coll. Spec. Bull., E. Lansing.

Evers, D. C. 1989. Multiple-lake usage by common loons in declining and non-declining populations. *N. Am. Loon Fund,* Unpubl. Rept.

Evers, D. C. 1992. Northern Great Lake common loon monitoring program. *Mich. Dept. Nat. Resour.,* Unpubl. Rept.

Evers, D. C., and G. McPeek. 1987. Conservation of common loons in southwestern Michigan. *N. Am. Loon Fund,* Unpubl. Rept.

Ewert, D. N. 1982. Spring migration of common loons at Whitefish Point, Michigan. *Jack-Pine Warbler* 60:135–43.

Ewert, D. N. 1988. Status of the Michigan loon registry program in 1987. Pp. 103–4 *in* P. I. Strong (ed.), *Papers from the 1987 conference on loon research and management.* N. Am. Loon Fund, Meredith, NH.

Haines, T. A. 1981. Acidic precipitation and its consequences for aquatic ecosystems: A review. *Trans. Am. Fish Soc.* 110:669–707.

Heimberger, M., D. Euler, and J. Barr. 1983. The impact of cottage development on common loon reproductive success in central Ontario. *Wilson Bull.* 95:431–39.

Jung, R. E. 1987. An assessment of human impact on the behavior and breeding success of the common loon (*Gavia immer*) in the northern Lower and eastern Upper Peninsulas of Michigan. *Univ. Mich. Biol. Stat.,* Unpubl. Rept.

Lange, J. P., and J. A. Lange. 1987. Breeding loon survey and habitat assessment/improvement project in Michigan's eastern Upper Peninsula and northern Lower Peninsula. *North Lakes Loon Project,* Unpubl. Rept.

McIntyre, J. W. 1975. *Biology and behavior of the common loon (Gavia immer) with reference to its adaptability in a man-altered environment.* Ph.D. diss., Univ. of Minn., Minneapolis.

McIntyre, J. W. 1978. Wintering behavior of common loons. *Auk* 95:396–403.

McIntyre, J. W. 1983. Nurseries: a consideration of habitat requirements during the early chick-rearing period in common loons. *J. Field Ornith.* 54:247–53.

McIntyre, J. W. 1988. *The common loon: Spirit of northern lakes.* Univ. Minn. Press, Minneapolis.

McNichol, D. K., B. E. Bendell, and R. K. Ross. 1987. Studies of the effects of acidification on aquatic wildlife in Canada: Waterfowl and trophic relationships in northern Ontario. *Can. Wildl. Serv. Occ. Pap.* No. 62.

McPeek, G., and D. C. Evers. 1988. Protection and management of southwestern Michigan's declining common loon population. *Mich. Loon Preserv. Assoc.,* Unpubl. Rept.

McPeek, G., and D. C. Evers. 1989. Guidelines for the protection and management of the common loon in southwest Michigan. *Mich. Loon Preserv. Assoc.,* Unpubl. Rept.

Miller, E. 1988. Collection of yodel calls for individual identification of male common loons. Pp. 44–52 *in* P. I. Strong (ed.), *Papers from the 1987 conference on loon research and management.* N. Am. Loon Fund, Meredith, NH.

Miller, E., and T. Dring. 1988. Territorial defense of multiple lakes by common loons: A preliminary report. Pp. 1–14 *in* P. I. Strong (ed.), *Papers from the 1987 conference on loon research and management*. N. Am. Loon Fund, Meredith, NH.

Mitchell, B. A. 1989. Acid rain and birds: How much proof is needed? *Am. Birds* 43:234–41.

Olson, S. T., and W. M. Marshall. 1952. The common loon in Minnesota. *Occ. Pap. Minn. Mus. Nat. Hist.* 5.

Parker, K. E. 1988. Common loon reproduction and chick feeding on acidified lakes in the Adirondack Park, New York. *Can. J. Zool.* 66:804–10.

Parker, K. E., and R. H. Brocke. 1985. Common loon chick feeding on acidified lakes in the Adirondack Park. *N. Am. Loon Fund*, Unpubl. Rept.

Payne, R. P. 1983. A distributional checklist of the birds of Michigan. *Univ. Mich. Mus. Zool. Misc. Publ.* No. 164.

Plunkett, R. L. 1979. Major elements of a five-year comprehensive plan of research and management for the Great Lakes and northeastern United States populations of the common loon, *Gavia immer*. Pp. 154–62 in S. A. Sutcliffe (ed.), *The common loon*. Natl. Audubon Soc., New York.

Ream, C. H. 1976. Loon productivity, human disturbance and pesticide residues in Alberta. *Wilson Bull.* 88:427–32.

Reiser, M. H. 1988. Productivity and nest site selection of common loons in a regulated lake system. Pp. 15–18 *in* P. I. Strong (ed.), *Papers from the 1987 conference on loon research and management*. N. Am. Loon Fund, Meredith, NH.

Robinson, W. L. 1988. A survey of breeding common loons in the eastern Upper Peninsula of Michigan, 1988. *Mich. Dept. Nat. Resour.*, Unpubl. Rept.

Robinson, W. L., J. H. Hammill, H. R. Hill, and T. A. deBruyn. 1988. The status of the common loon in Michigan. Pp. 132–44 *in* P. I. Strong (ed.), *Papers from the 1987 conference on loon research and management*. N. Am. Loon Fund, Meredith, NH.

Schindler, D. W. 1988. Effects of acid rain on freshwater ecosystems. *Science* 239:149–57.

Sjolander, S., and G. Agren. 1972. Reproductive behavior of the common loon. *Wilson Bull.* 84:296–308.

Strong, P. I. V. 1985. *Habitat selection by common loons*. Ph.D. thesis, Univ. Maine, Orono.

Strong, P. I. V. (ed.). 1988a. *Papers from the 1987 conference on loon research and management*. N. Am. Loon Fund, Meredith, NH.

Strong, P. I. V. 1988b. Changes in Wisconsin's common loon population. *Passenger Pigeon* 50:287–90.

Strong, P. I. V., J. A. Bissonette, and J. S. Fair. 1987a. Reuse of nesting and nursery areas by common loons. *J. Wildl. Mgmt.* 51:123–27.

Strong, P. I. V., S. A. LaValley, and R. C. Burke. 1987b. A colored plastic leg band for common loons. *J. Field Ornith.* 58:218–21.

Sutcliffe, S. A. (ed.). 1979. *The common loon. Proc. second N. Am. Conf. on common loon research and mgmt.*

Titus, J. R., and L. W. Van Druff. 1981. Response of the common loon (*Gavia immer*) to recreational pressures in the Boundary Waters Canoe Area, northeastern Minnesota. *Wildl. Monogr.* No. 79.

Vermeer, K. 1973. Some aspects of the nesting requirements of common loons. *Wilson Bull.* 85:429–35.

Least Bittern

· *Ixobrychus exilis* (Gmelin)
· **State Threatened**

Map 9. Breeding evidence of least bitterns. Historically statewide, rare in the Upper Peninsula (Barrows 1912); ● = 1983 to 1992 (Brewer et al. 1991).

Status Rationale: The emergent wetland habitat required by breeding and migratory least bitterns has been severely reduced and altered in Michigan. Although least bittern breeding population trends are difficult to monitor and quantify, this heron's decline has been obvious in several regions.

Field Identification: The least bittern averages 11 to 14 inches (28 to 36 cm) in length with a 16- to 18-inch (41 to 46 cm) wingspan. Adults are characterized by their overall rufous and chestnut-colored appearance contrasting with a darker back. The dark cap, nape, and back of the male is distinctly darker than that of the female. Optimal lighting will reveal the male's cap and back as a dark blue-greenish. Immatures are similar to females, but have heavier streaking on the throat and breast. Both mature and immature birds have two white lines bordering the wing shoulders (i.e., scapulars). Although it is difficult to flush, the pale buff wing patches of the least bittern in flight are diagnostic. Its quick wingbeats, outstretched neck, and dangling legs are also characteristic of the species. A rare dark phase, known as Cory's bittern, has chestnut-colored rather than pale buff wing patches. Other Michigan members of the heron family are considerably larger. The relatively common and widespread green heron (*Butorides striatus*), which may be twice as large, has dark wings.

The least bittern usually climbs or walks through its herbaceous habitats. Because of its secretive behavior, its presence is typically only known from a fast series of three to five low, dovelike *coo* notes, frequently repeated. These male territorial calls are most often heard during the breeding season at crepuscular hours. Females respond with ticking calls. Various cackles and *tut-tut-tut* calls are given when it is agitated or disturbed.

Total Range: This wide-ranging heron occurs in parts of North, Central, and South America, south to northern Argentina and southern Brazil. In North America, the least bittern is generally restricted to the eastern United States, ranging from the Great Plains states eastward to the Atlantic Coast and north to the Great Lakes region and the New England states. The species is on the Blue List, the National Audubon Society's watch for declining species, and, since 1979, the least bittern has been reported to be "down or greatly down" and extirpated from some regions (Tate 1986). Populations in the United States are migratory, overwintering along the U.S. Gulf Coast south through Mexico and the Carribean islands into northern South America.

State History and Distribution: The least bittern was known as "an abundant bird in all suitable places in the state" at the turn of the century (Barrows 1912). By the early 1980s, nesting was confirmed in 27 counties (Payne 1983). Since that time, the species has declined dramatically in all its former strongholds. It is now nearly absent from the Great Lakes coastal marshes of southeastern Michigan, the southernmost tier of counties, and Seney National Wildlife Refuge. Kalamazoo County populations exhibited more than a 50% decline since the mid 1970s (Adams in Brewer et al. 1991). The lack of recent observations in the St. Clair Flats is perplexing, since this relatively intact ecosystem formerly supported a large breeding population (Wood 1951) and Canadian surveys in the mid-1980s found high densities in several Lake St. Clair locations (Woodliffe in Cadman et al. 1987).

Today, the least bittern is an uncommon species in the southern Lower Peninsula, and rarely found in the northern Lower Peninsula except along marshes associated with Lakes Michigan and Huron. Disjunct interior populations remain within the Houghton Lake wetland complex in Roscommon County. Upper Peninsula nesting pairs regularly occur in Portage Marsh, Delta County, along the Portage Channel, Houghton County, and Munuscong Bay in the St. Marys River system.

Habitat: The least bittern's breeding habitat requires emergent wetlands interspersed with open water, forming a mosaic of habitat transition zones. In Michigan, these habitats are most commonly found along the Great Lakes shoreline in bays, coves, and associated wetlands. Isolated emergent wetlands in the state's interior are also occupied (particularly in the southern Lower Peninsula) as well as those associated with lakes, rivers, floodings, and sewage ponds. Brown and Dinsmore (1986) suggest this may be an area-dependent breeding species, requiring at least 12 acres (4.9 ha) of suitable habitat.

Nest sites typically are associated with dense stands of cattail (*Typha* spp.) and bulrush (*Scirpus* spp.) (Weller 1961). They are placed above shallow water (less than 3 feet or 1.1 m deep), suspended 6 to 30 inches (15 to 76 cm) above water in herbaceous vegetation (Hands et al. 1989). Occasionally, nests are placed several feet off the ground in other herbaceous vegetation, such as *Phragmites* (Dillon 1959) and sedge (*Carex* spp.) (Graber et al. 1978), and in shrubs (Walkinshaw 1978), such as buttonbush (*Cephalanthus occidentalis*) (Baker 1940; Trautman 1940).

Foraging habitat during migration is similar to that occupied during the breeding season. In spring, least bitterns prefer average water depths of 9 inches (23 cm), vegetation densities of 27 stems per square foot, and stem heights of 25 inches (64 cm); in autumn, they occur in similar water depths, thinner vegetation densities of 15 stems per square foot, and taller stem heights of 47 inches (119 cm) (Reid in Hands et al. 1989).

Ecology and Life History: Returning individuals arrive in late April and early May to the southern Lower Peninsula and soon after in northern Michigan. Males call frequently with their low, cooing sounds during the prenesting period. Territorial defense is characterized by two behaviors: (1) a common bitternlike freeze posture with neck and bill stretched upward, and (2) a more aggressive posture with wings spread and curved forward, possibly to give the illusion that it is a larger bird (Hancock and Hugh 1984).

Nest building is done primarily by the males (Davidson 1944; Weller 1961). Nesting pairs sometimes associate with others, forming loose colonies. Colonial nesting may be a response to exceptional, local feeding areas (Kushlan 1973). The nest is constructed in dead, matted vegetation, providing a platform and a protective canopy. In Michigan, egg laying occurs from mid-May through July. Later records are from second nesting attempts. The usual clutch size is four to five eggs; nests with six eggs are renests (Weller 1961). Double brooding has not been confirmed (Hands et al. 1989), but is suspected (Weller 1961).

Incubation by both sexes (more so by the female) lasts 17 to 20 days with eggs hatching asynchronously over a 3-day period. Males will continue incubation if the female is lost (Aniskowicz 1981). The brooding period is shortened by the young's tendency to leave the nest early, sometimes within one week; however, the nest remains the central site for the family unit for several more weeks. Males are the main food providers for the young (Weller 1961). Young will assume the freeze position if threatened, frequently swaying, imitating wind-blown vegetation. Southward migration probably begins in late August, peaking in early to mid-September, and ending by early October.

The least bittern is nearly an exclusive forager in emergent wetlands. It usually feeds at the edges of the open water-vegetation transition areas, clasping and clinging to vegetation with both feet (Allen 1915; Eastwood 1932; Sutton 1936). Prey are primarily aquatic, including small fish, such large insects as damselflies, tadpoles, and other amphibians, and crayfish. Small mammals and birds are occasionally eaten. Feeding strategies include (1) stand-and-wait, (2) walking slowly, and (3) wing flicking (Kushlan 1976). Feeding platforms are frequently constructed and may be important for providing adults and young a resting place.

Conservation/Management: Habitat loss and degradation are the primary causes for declining least bittern populations. Since European settlement, more than half of the wetlands in the contiguous United States and nearly three-quarters in Michigan have been lost (Tip of the Mitt Watershed Council 1987; Hands et al. 1989). A decline of the least bittern has followed. Many existing emergent wetlands are threatened by changing water levels, often related to human-induced disturbances. Within Great Lakes emergent wetlands, breeding habitat has been reduced by high water levels, particularly in the mid 1980s. These coastal wetlands also are exposed to high environmental contaminant loads, chemicals that severely affected reproductive success in such other Great Lakes birds as Caspian terns (*Sterna caspia*) and double-crested cormorants (*Phalacrocorax auritus*) (Kurita et al. 1987) and bald eagles (*Haliaeetus leucocephalus*). Organochlorines and other pesticides have been detected in several herons (see, e.g., Heinz et al. 1984 and 1985; Heithammer et al. 1985), including dieldrin in least bitterns (Causey and Graves 1969). No critical levels are known, however, for this heron.

Interior wetlands are also threatened by water level fluctuations, contaminants, and general degradation. Changing water levels caused by nearby draining, channeling, and general changes in the area's hydrological system can severely reduce least bittern breeding success by flooding or drying potential nest sites. The least bittern may be most successful in semicolonial conditions, thereby requiring large, contiguous wetlands; solitary pairs probably need at least 12 acres (4.9 ha) of suitable habitat for breeding (Brown and Dinsmore 1986). Human-related disturbances adjacent to emergent wetlands also reduce habitat quality by increasing erosion and sedimentation, which is often followed by the invasion of woody vegetation.

Predators such as raccoons (*Procyon lotor*)(Hansen 1984), and mink (*Mustela vison*)(DeVore 1968), and other wetland inhabitants) can severely affect productivity. Higher raccoon densities and decreasing contiguous wetland areas increase the probability of predation.

Protection and management of large, emergent wetlands are vital for this heron's survival. Legal protection for its habitat has been implemented through the Michigan Wetlands Protection Act of 1979. The optimal, long-term management plan would actively rotate a complex of several distinct areas at least 12 acres (4.9 ha) or more in size. A hemimarsh stage (half open water and half vegetated) is the preferred marsh condition. Nests are typically situated 6 inches to 20 feet (15 cm to 6.1 m) from open-water edges (Nero 1950; Weller 1961), which can be natural openings (e.g., created by muskrats [*Ondatra zibethica*]) or artificially created ones.

Finding least bitterns is difficult, but playing recordings can enhance surveys (Johnson et al. 1981; Manci and Rusch 1988; Swift et al. 1988). Recommended field techniques are (1) using at least five minutes of playback of the *cooing* call at each survey point for a minimum of three visits, (2) calling between mid-May and mid-June, and (3) calling during morning hours with calm winds (Swift et al. 1988). Under these favorable conditions, a relatively high response rate is expected for individuals within 100 feet (30 m). By applying this technique annually at selected sites, biweekly during the peak calling period, indices of distribution and abundance can be obtained (Manci and Rusch 1988).

By protecting the remaining 25% of Michigan's original wetlands and identifying the marshes occupied by the least bittern, the process of actively managing populations can begin.

LITERATURE CITED

Allen, A. A. 1915. The behavior of the least bittern. *Bird Lore* 17:425–30.

Aniskowicz, B. T. 1981. Behavior of a male least bittern incubating after loss of mate. *Wilson Bull.* 93:395–97.

Baker, B. W. 1940. Some observations of the least bittern. *Jack-Pine Warbler* 18:113–14.

Barrows, W. B. 1912. *Michigan bird life.* Mich. Agric. Coll. Spec. Bull., E. Lansing.

Brewer, R., G. A. McPeek, and R. J. Adams, Jr. 1991. *The atlas of breeding birds of Michigan.* Mich. State Univ. Press, E. Lansing.

Brown, M., and J. J. Dinsmore. 1986. Implications of marsh size and isolation for marsh bird management. *J. Wildl. Mgmt.* 50:392–97.

Cadman, M. D., P. F. J. Eagles, and F. M. Helleiner. 1987. *Atlas of the breeding birds of Ontario.* Univ. Waterloo Press, Waterloo, Ont.

Causey, M. K., and J. B. Graves. 1969. Insecticide residues in least bittern eggs. *Wilson Bull.* 81:340–41.

Davidson, F. S. 1944. Nesting of the least bittern. *Flicker* 16:19–21.

DeVore, J. E. 1968. A nesting study of the king rail and least bittern. *Migrant* 39:53–58.

Dillon, S. T. 1959. Breeding of the least bittern in Manitoba. *Auk* 76:524–25.

Eastwood, S. K. 1932. Notes on the feeding of the least bittern. *Wilson Bull.* 44:240.

Graber, J. W., R. R. Graber, and E. L. Kirk. 1978. *Illinois birds: Ciconiiformes. Ill. Nat. Hist. Surv., Urbana, Biol. Notes* No. 109.

Hancock, J., and E. Hugh. 1984. *The herons handbook.* Harper Row, New York.

Hands, H. M., R. D. Drobney, and M. R. Ryan. 1989. Status of the least bittern in the northcentral United States. *Missouri Coop. Fish Wildl. Res. Unit,* Univ. Missouri, Columbia.

Hansen, S. C. 1984. Breeding of the least bittern in Pottawatomie County, Kansas. *Kansas Ornith. Soc. Bull.* 35:37–39.

Heinz, G. H., D. M. Swineford, and D. E. Katsma. 1984. High PCB residues in birds from the Sheboygan River. *Environ. Monit. Assess.* 4:155–61.

Heinz, G. H., T. C. Erdman, S. D. Haseltine, and C. Stafford. 1985. Contaminant levels in colonial waterbirds from Green Bay and Lake Michigan, 1975–1980. *Environ. Monit. Assess.* 5:223–36.

Heithammer, K. R., R. D. Atkinson, T. S. Baskett, and F. D. Samson. 1985. Metals in riparian wildlife of the lead mining district of southeastern Missouri. *Archeol. Environ. Contam. Toxicol.* 14:213–23.

Johnson, R. R., B. T. Brown, L. T. Haight, and J. M. Simpson. 1981. Playback recordings as a special avian censusing technique. Pp. 68–75 *in* C. J. Ralph and J. M. Scott, (eds.), *Estimating numbers of territorial birds. Stud. Avian Biol.* 6.

Kurita, H., J. P. Ludwig, and M. E. Ludwig. 1987. Results of the 1987 Michigan colonial waterbird monitoring project on Caspian terns and double-crested cormorants: egg incubation and field studies of colony productivity, embryologic mortality and deformities. *Ecol. Res. Serv. Inc.,* Bay City, MI, Unpubl. Rept.

Kushlan, J. A. 1973. Least bittern nesting colonially. *Auk* 90:685–86.

Kushlan, J. A. 1976. Feeding behavior of North American herons. *Auk* 93:86–94.

Manci, K. M., and D. H. Rusch. 1988. Indices to distribution and abundance of some inconspicuous waterbirds on Horicon Marsh. *J. Field Ornith.* 59:67–75.

Nero, R. W. 1950. Notes on a least bittern nest and young. *Passenger Pigeon* 12:3–8.

Payne, R. B. 1983. A distributional checklist of the birds of Michigan. *Univ. Mich. Mus. Zool. Misc. Publ.* No. 164.

Sutton, G. M. 1936. Food capturing tactics of the least bittern. *Auk* 53:74.

Swift, B. L., S. R. Orman, and J. W. Ozard. 1988. Response of least bittern to tape-recorded calls. *Wilson Bull.* 100:496–99.

Tate, J., Jr. 1986. The blue list for 1986. *Am. Birds* 40:227–36.

Tip of the Mitt Watershed Council. 1987. Michigan wetlands: Yours to protect. *Tip of the Mitt Watershed Council*, Conway, MI.

Trautman, M. B. 1940. The birds of Buckeye Lake. *Univ. Mich. Mus. Zool. Misc. Publ.* No. 44.

Walkinshaw, L. H. 1978. *Birds of the Battle Creek, Calhoun County, Michigan area.* Univ. Microfilms International, Ann Arbor, MI.

Weller, M. W. 1961. Breeding biology of the least bittern. *Wilson Bull.* 73:11–35.

Wood, N. A. 1951. The birds of Michigan. *Univ. Mich. Mus. Zool. Misc. Publ.* No. 75.

Osprey

- *Pandion haliaetus* (Linnaeus)
- **State Threatened**

Map 10. Confirmed osprey nesting pairs. Historically statewide (Barrows 1912); ● = 1983 to 1992 (Michigan DNR).

Status Rationale: The distribution and abundance of Michigan's osprey are increasing at a steady rate, rebounding from historical lows in the 1960s. Since the mid-1980s, nesting has been confirmed in 31 counties, and, in 1992, 226 pairs occupied nests. The osprey is responding favorably to protection and reductions in DDT contamination.

Field Identification: The osprey is distinctly marked with dark brown above, white below, and a white head with a broad dark cheek band. It often flies with an angle or crook in the long, narrow wings, showing the diagnostic black wrist (carpal) patches on the underside of its 5 to 5¹/₂ foot (1.5 to 1.7 m) wingspan. Adults are 22 to 25 inches (56 to 64 cm) long. Sexes are similar in plumage, although the larger female usually shows a more pronounced breast band or necklace. Males lack or have limited upper breast markings. Young-of-the-year are recognized by the scaly appearance on their back, formed by white-edged feathers. While perched, the osprey's long legs are obvious and the wing tips extend beyond the tail. A repeated succession of high, clear whistles are given when alarmed or near the nest. This widespread species is divided into four subspecies (Prevost in Bird 1983), with *P. h. carolinensis* occurring in North America.

The wing crook, obvious while soaring or gliding, distinguishes the osprey from most large birds. Distant flight profiles of gulls are similar, but lack carpal patches, and gulls have shorter wings. Soaring turkey vultures (*Cathartes aura*) have a dihedral silhouette, and eagles have a flattened wing profile. The osprey's hunting technique of hovering and diving for fish is diagnostic.

Total Range: The osprey is nearly cosmopolitan. In North America, it breeds primarily from northwestern Alaska and central Canada south to the northern United States, and along both the Atlantic and Pacific coasts to the Gulf States and Baja California. Henny (1983) estimated the total nesting osprey population in the contiguous United States at about 8,000 nesting pairs, primarily divided into five regions. These include, in order of abundance, the Atlantic Coast, Florida and Gulf Coast, Pacific Northwest, Western Interior, and Great Lakes region. Another 10,000 to 12,000 pairs are estimated for Canada and Alaska (Poole 1989).

Much of the North American population migrates to Central and northern South America for the winter. Eastern and Midwest populations do not funnel through Mexico, but cross the Caribbean Sea; 36% overwinter in South America north of the equator, 24% south of the equator, and 40% in the Caribbean islands, Central America, and the United States (Poole and Agler 1987). After migrating up to 5,000 miles (8,000 km), individuals overwinter along the ocean and gulf coasts and on major river systems.

State History and Distribution: In Michigan, the osprey was formerly widespread, "but apparently nowhere abundant" (Barrows 1912). Today, it rarely occurs in southern Michigan, but does continue to nest throughout northern Michigan. Nesting success in Michigan's ospreys plummeted in the 1950s and 1960s following widespread use of DDT and other organochlorine pesticides. Osprey reproductive success started to improve in the late 1960s and approached normal levels by 1971 (Postupalsky 1977). Over 100 occupied nests were known by 1978, and, in 1985, the 145 pairs produced a record 226 young (Postupalsky 1986). In 1992, 226 pairs were counted producing 187 young.

Confirmed nesting in the Lower Peninsula since 1983 is known for 19 counties (Brewer et al. 1991). This region is represented by two exemplary breeding concentrations, one at Fletcher Pond in Alpena and Montmorency counties (a nearly 9,000 acre [3,600 ha] water storage reservoir established in the early 1930s) and the other in the Houghton Lake area in Roscommon County. These pairs are mostly dependent on artificial nesting platforms. In the late 1980s, over 20 pairs nested on a dozen small floodings in the Houghton Lake area (Postupalsky 1989).

Southernmost nesting pairs occur along the Muskegon River in Muskegon County and in Mecosta County (e.g., Haymarsh State Game Area). Recent observations show nesting pairs in Huron County (Brewer et al. 1991) and along the St. Joseph River near the Michigan-Indiana border (Evers 1989). Individuals are frequently observed in other southern Lower Peninsula sites during the summer, but these birds are probably nonbreeders. Artificial nesting platforms in suitable habitat may attract nesting pairs (e.g., Allegan State Game Area).

The osprey is widespread in the Upper Peninsula, but its strongholds are in the eastern portion, including the Manistique lakes system (Postupalsky in Ogden 1977; Evers 1989) and the St. Marys River system. On Isle Royale, Keweenaw

County, the first successful nesting since the early 1970s was reported in 1984. Two nesting pairs were known there in 1989 (Fettig 1989).

Habitat: The osprey occurs primarily in forested regions adjacent to lakes, large rivers, and floodings. A Habitat Suitability Index (HSI) Model was developed and specifically outlines osprey habitat requirements (Vana-Miller 1987). In general, though, optimal foraging requirements include clear, shallow water that is relatively free of floating and emergent vegetation.

Preferred nesting areas are snags, dead-top pines, and tamaracks near bodies of water. Postupalsky (1977) considered two principal habitat types used by Michigan ospreys for nest sites; these include lowland conifer swamps and dead timber in natural or artificial floodings. In the Great Lakes region, nests generally are placed in the tops of live conifers, averaging 30 to 60 feet (9 to 18 m) high (Peck and James 1983), although pairs have a wide tolerance of nest trees and sites (see, e.g., Edwards and Collopy 1988). Ospreys adapt well to nesting on platforms and other artificial structures, such as navigational light towers, microwave towers (Postupalsky and Stackpole 1974), power transmission towers, duck blinds and channel markers (Reese 1970), and wooden utility poles (Sindelar in Ogden 1977). On islands without mammalian predators, nests frequently are built on the ground (see, e.g., Spitzer and Poole 1980).

Ecology and Life History: Ospreys return to Michigan in early April and begin breeding in late April. After a vocal, aerial courtship display, mating occurs within a few hours, usually at the same nest site used the preceding year. Between the time of pair formation and egg laying, the male typically feeds the female, possibly to ensure mate fidelity (Poole 1985). Members of nesting pairs return to the same nesting area (or site) independently for successive years (Spitzer et al. 1983); in effect, returning adults mate for life. When ospreys are numerous in an area, they may breed in loose colonies or aggregations (e.g., Fletcher Pond).

Both sexes participate in nest construction. To obtain the necessary nest material (sticks), they search for dead tree branches, grabbing them from the ground or from trees. The resulting eyrie (nest) is large, measuring several feet wide and deep, although typically smaller than an eagle's. The clutch size ranges from three to four eggs, usually three. Incubation is generally by the female, although the male may share in incubation up to 50% of the time (Poole 1989). Eggs hatch in about 37 days and young fledge at seven to eight weeks.

Postbreeding and postfledging dispersal begins in August (Poole and Agler 1987). Foraging techniques are partially learned behaviors. After several weeks, young are relatively self-sufficient (Szaro 1978). Most individuals leave the Upper Peninsula by early October and are south of the state's border by early November.

Sexual maturity is reached at three years of age. Natal dispersal is low for juveniles. Of those that return, nearly three-quarters nest within 19 miles (30 km) and the other quarter between 19 and 77 miles (30 to 125 km) of their hatching site (Henny 1977). One-year-old ospreys from northern populations remain on the wintering grounds during the second summer. By the third summer, one-quarter to one-half of the two year olds return (Henny and Van Velzen 1972), and the remainder are on the breeding grounds by the fourth summer. Approximately half of the young ospreys die in their first year, with a 10% to 18% mortality rate thereafter (Henny and Wight 1969; Poole 1989). The adult mortality rate for Michigan birds is about 15% (Postupalsky 1989).

The osprey typically perches on a favorite snag overlooking water, searching for fish. Osprey use two types of feeding strategies before diving into the water (usually completely submerging): a gliding "interhover" and a hover. Under a variety of weather regimes, dives from hovering are significantly more successful than "interhovering" (Grubb 1977a). Diving and capture rates are reduced during cloudy and windy conditions (Grubb 1977b). Fish are caught by specialized talons (i.e., both outer toes are reversible) and are carried to a perch.

The osprey is dependent on live fish; dead fish are rarely taken. On rare occasions, the osprey takes nonfish prey, including mammals (see, e.g., McCoy 1966; Tait et al. 1972), birds (see, e.g., Kuser 1929; Sindelar and Schulter 1968), and reptiles (Postupalsky and Kleiman 1965), which comprise a small but regular part of a nesting pair's diet (Wiley and Lohrer 1973). Swenson (1979) found benthic-feeding fish (bottom dwellers) more vulnerable to osprey attacks than limnetic-feeding fish (open water dwellers). In Michigan, commonly caught fish species include small northern pike (*Esox lucius*), bullheads (*Ictalurus* spp.) sunfish (*Lepomis* spp.), crappie (*Pomoxis* spp.), and yellow perch (*Perca flavescens*). Dunstan (1974) found sunfish and crappie to be favored prey; pike were usually taken during spawning season or after winter and summer fish kills.

Conservation/Management: The osprey is sensitive to organochlorine pesticides. These persistent pollutants (e.g., DDE) are responsible for major declines of this species since 1950, as a result of thinning eggshells and severely lowered reproductive success (Ames 1966; Henny 1972; Keith 1966; Hickey and Anderson 1968; Wiemeyer et al. 1975; Henny et al. 1977; Spitzer et al. 1978). Osprey populations in the Northeast and Great Lakes region were most severely affected, but are now recovered since the 1972 ban of DDT. Pollution, such as organochlorines and acid rain (Mitchell 1989), remains a potential threat to osprey populations.

Human activity, such as logging, wetland destruction and alteration, and shoreline development, currently hampers further expansion of the osprey in Michigan. Unless a nesting pair is acclimated to human presence, continued intrusion at a nest site displaces adults and subsequently causes overexposure of eggs and young. Nesting osprey populations need to fledge an average of at least 0.8 young per active nesting pair each year to maintain stable populations (Henny and Wight 1969; Henny and Ogden 1970; Spitzer 1980). Shoreline and wetland development degrades suitable nesting and foraging habitat through cutting potential nest trees and reducing water clarity and access to prey.

Sufficient isolation of the nest site and territory is required for optimal nesting success. In areas with relatively intense development and related human disturbance, a safety zone of one-quarter mile (0.4 km) radius around the nest will provide the needed protection. In forested wetland nesting areas, safety zones can be successfully incorporated into timber practices with minimal costs (Smith and Nevers 1984).

Continued monitoring of populations statewide is needed, as well as active management (e.g., nesting platforms). Censusing consists of two aerial surveys in May and July, accompanied by ground checks. Surveys document current populations, but, more important, will detect developing problems. Although certain environmental pollutants are banned in the United States (e.g., DDT), they are still widely used in the osprey's Central and South American wintering range. Great Lakes organochlorine (e.g., DDE, dieldrin), PCBs, dioxin, and heavy metal contaminant loads also remain relatively high, necessitating large-scale pollutant controls on the breeding grounds.

Due to the loss of more than three-fourths of Michigan's wetlands (Tip of the Mitt Watershed Council 1987), floodings and impoundments created throughout the state are now important for maintaining osprey populations. Artificial nesting sites further assist population recovery (Henny et al. 1977). By erecting artificial nesting platforms, marginal and formerly suitable nesting habitat can be enhanced (Rhodes 1972; Postupalsky 1978; Johnson 1979; Scott and Stuart 1985). The tripod-type nesting platforms that are commonly used need to be placed 12 to 15 feet (4.6 m) high over shallow water, away from taller trees and areas of regular human use. Sticks should be arranged on the wooden top to simulate a nest, or nearby natural nests from decaying trees can be transferred.

Mortality of nestlings is reduced on properly placed platforms (Postupalsky and Stackpole 1974), since they are less susceptible to mammalian predation and wind. Foraging perches should be maintained nearby (e.g., large, live trees and snags near bodies of water). For artificial platforms to be most successful, they should be placed close to optimal foraging sites, since males typically supply food for the entire brood.

LITERATURE CITED

Ames, P. L. 1966. DDT residues in the eggs of the osprey in the northeastern United States and their relation to nesting success. *J. Appl. Ecol.* 3:87–97.

Barrows, W. B. 1912. *Michigan bird life.* Mich. Agric. Coll. Spec. Bull., E. Lansing.

Bird, D. M., (chief ed.). 1983. *Biology and management of bald eagles and ospreys.* Harpell Press, Ste. Anne de Bellevue, Quebec.

Brewer, R., G. McPeek, and R. J. Adams, Jr. 1991. *The atlas of breeding birds of Michigan.* Mich. State Univ. Press, E. Lansing.

Dunstan, T. C. 1974. Feeding activities of ospreys in Minnesota. *Wilson Bull.* 86:74–76.

Edwards, T. C., Jr., and M. W. Collopy. 1988. Nest tree preference of ospreys on northcentral Florida. *J. Wildl. Mgmt.* 52:103–7.

Evers, D. C. 1989. State report: Michigan raptor populations may be on the upswing. *Eyas* 12:10–15.

Fettig, S. 1989. Raptor inventory: 1989 update and analysis. Isle Royale Nat. Park, *Resour. Mgmt. Rept.* 89-2, Unpubl. Rept.

Grubb, T. G., Jr. 1977a. Why ospreys hover. *Wilson Bull.* 89:149–50.

Grubb, T. G., Jr. 1977b. Weather-dependent foraging in ospreys. *Auk* 94:146–49.

Henny, C. J. 1972. An analysis on population dynamics of selected avian species with special references to changes during the modern pesticide era. *U.S. Fish Wildl. Serv.,Wildl. Resour. Rept.* No. 1.

Henny, C. J. 1977. Research, management and status of the osprey in North America. Pp.

199–222 *in* R. D. Chancellor (ed.), *Proc. World Birds of Prey Conf. Internat. Counc. Bird Preserv.*, Vienna.

Henny, C. J. 1983. Distribution and abundance of nesting ospreys in the United States. Pp. 175–86 *in* D. M. Bird (chief ed.), *Biology and management of bald eagles and ospreys.* Harpell Press, Ste. Anne de Bellevue, Quebec.

Henny, C. J., M. A. Byrd, J. A. Jacobs, P. D. McLain, M. R. Todd, B. F. Halla. 1977. Mid-Atlantic coast osprey population: Present numbers, productivity, pollutant contamination, and status. *J. Wildl. Mgmt.* 41:254–65.

Henny, C. J., and J. C. Ogden. 1970. Estimated status of osprey populations in the United States. *J. Wildl. Mgmt.* 34:214–17.

Henny, C. J., and W. T. Van Velzen. 1972. Migration patterns and wintering localities of American ospreys. *J. Wildl. Mgmt.* 36: 1133–41.

Henny, C. J., and H. M. Wight. 1969. An endangered Osprey population: Estimates of mortality and production. *Auk* 86:188–98.

Hickey, J. J., and D. W. Anderson. 1968. Chlorinated hydrocarbons and eggshell changes in raptorial and fish-eating birds. *Science* 162: 271–73.

Johnson, F. L. 1979. Osprey nesting platforms in northern central Wisconsin. *Passenger Pigeon* 41:145–48.

Keith, J. O. 1966. Insecticide contaminations in wetland habitats and their effects on fish eating birds. *J. Appl. Ecol.* 3:71–85.

Kuser, C. D. 1929. An osprey feeds on ducks. *Bird Lore* 31:260–61.

Levenson, H., and J. R. Koplin. 1984. Effects of human activity on productivity of nesting ospreys. *J. Wildl. Mgmt.* 48:1374–77.

McCoy, J. 1966. Unusual prey for osprey. *Chat* 30:108–9.

Mitchell, B. D. 1989. Acid rain and birds: How much proof is needed? *Am. Birds* 43:234–41.

Ogden, J. C. (ed.). 1977. Transactions of the N. Am. Osprey Res. Conf., 1972. *U.S. Natl. Park Serv. Trans. Proc.* No. 2.

Peck, G. K., and R. D. James. 1983. Breeding birds of Ontario: Nidology and distribution. *Royal Ont. Mus. Life Sci. Misc. Publ.*, Toronto.

Poole, A. F. 1985. Courtship feeding and osprey reproduction. *Auk* 102: 479–92.

Poole, A. F. 1989. *Ospreys: A natural and unnatural history.* Cambridge Univ. Press, Cambridge.

Poole, A. F., and B. Agler. 1987. Recoveries of ospreys banded in the United States, 1914–84. *J. Wildl. Mgmt.* 51:148–55.

Postupalsky, S. 1977. Status of the osprey in Michigan. Pp. 153–65 *in* J. C. Ogden (ed.), N. Am. Osprey Res. Conf., 1972. *U.S. Natl. Park Serv. Trans. Proc.* No. 2.

Postupalsky, S. 1978. Artificial nesting platforms for ospreys and bald eagles. Pp. 35–45 *in* S. A. Temple (ed.), *Endangered birds: Management techniques for preserving threatened species.* Univ. Wisc. Press, Madison.

Postupalsky, S. 1986. Osprey population research in Michigan, 1985. *Mich. Dept. Nat. Resour.*, Unpubl. Rept.

Postupalsky, S. 1989. Osprey. Pp. 297–313 *in* I. Newton (ed.), *Lifetime reproduction in birds.* Academic Press, New York.

Postupalsky, S., and J. P. Kleiman. 1965. Osprey preys on turtle. *Wilson Bull.* 77:401–2.

Postupalsky, S., and S. M. Stackpole. 1974. Artificial nesting platforms for ospreys in Michigan. *Raptor Res. Rept.* 2:105–18.

Reese, J. G. 1970. Reproduction in a Chesapeake Bay osprey population. *Auk* 87:747–59.

Rhodes, L. I. 1972. Success of osprey nest structures at Martin National Wildlife Refuge. *J. Wildl. Mgmt.* 36:1296–99.

Scott, F., and C. S. Stuart. 1985. Success of osprey nest platforms near Loon Lake, Saskatchewan. *Blue Jay* 43:238–42.

Sindelar, C., and E. Schulter. 1968. Osprey carrying bird. *Wilson Bull.* 80:103.

Smith, C. F., and H. P. Nevers. 1984. Incorporating osprey nest production into timber management practices: New Hampshire's approach. Pp. 292–98 in *Proc. of the Workshop on Management of Nongame Species and Ecological Communities*, Univ. of Kentucky, Lexington.

Spitzer, P. R. 1980. *Dynamics of a discrete coastal breeding population of ospreys (Pandion haliaetus) in the northeastern United States during a period of decline and recovery, 1969–1978.* Ph.D. diss., Cornell Univ., Ithaca, NY.

Spitzer, P. R., and A. Poole. 1980. Coastal ospreys between New York City and Boston: A decade of reproductive recovery 1969–1979. *Am. Birds* 34:234–41.

Spitzer, P. R., A. F. Poole, and M. Scheibel. 1983. Initial population recovery of breeding ospreys in the region between New York City and Boston. Pp. 231–42 *in* D. M. Bird (ed.), *Biology and management of bald eagle and osprey.* Harpell Press, Ste. Anne de Bellevue, Quebec.

Spitzer, P. R., R. W. Risebrough, W. Walker, R. Hernandez, A. Poole, D. Puleston, and I. C. T. Nisbet. 1978. Productivity of ospreys in Connecticut–Long Island increases as DDE residue declines. *Science* 202:333–34.

Swenson, J. E. 1979. The relationship between prey species ecology and dive success in ospreys. *Auk* 96:408–12.

Szaro, R. C. 1978. Reproductive success and foraging behavior of the osprey at Seahorse Key, Florida. *Wilson Bull.* 90:112–18.

Tait, W. W., H. M. Johnson, and W. D. Courser. 1972. Osprey carrying a mammal. *Wilson Bull.* 84:341.

Tip of the Mitt Watershed Council. 1987. Michigan wetlands: Yours to protect. *Tip of the Mitt Watershed Council*, Conway, MI.

Vana-Miller, S. L. 1987. Habitat suitability index models: Osprey. *U.S. Fish Wildl. Serv., Biol. Rept.* 82 (10.154).

Wiemeyer, S. N., P. R. Spitzer, W. C. Krantz, T. C. Lamont, and E. Cromartie. 1975. Effects of environmental pollutants on Connecticut and Maryland ospreys. *J. Wildl. Mgmt.* 39:124–39.

Wiley, J. W., and F. E. Lohrer. 1973. Additional records of non-fish prey taken by ospreys. *Wilson Bull.* 85:468–70.

Bald Eagle

· *Haliaeetus leucocephalus*
  (Linnaeus)

· **Federally and State Threatened**

Map 11. Confirmed bald eagle nesting pairs. Historically statewide (Barrows 1912); ● = 1983 to 1992 (Michigan DNR).

Status Rationale: The bald eagle is an example of a species nearly extirpated in the United States by ecological degradation but recovered due to intense management and conservation efforts. Since the ban of DDT and the development of increased public understanding, eagle populations have slowly increased and returned to some of their former range. In Michigan, eagles have increased from 84 pairs in 1980 to 215 occupied nests in 1992.

Field Identification: The conspicuous white head and tail contrasting with the dark brown body, and the 6 to $7^{1}/_{2}$ foot (1.8 to 2.3 m) wingspan make identification of the adult bald eagle unmistakable. The large, strong bill and feet are yellow. Females are generally larger than males. When alarmed or near their nests, adults frequently give a series of cackling calls (e.g., *kak-kak-kak*), sometimes interspersed with guttural, grunting sounds.

   Juvenile and subadult plumages are highly variable and lack the diagnostic white head and tail of the adult. Aging techniques using plumage are described by McCollough (1989), using head, bill, and tail, and molting sequences. Head and tail plumages for the first $2^{1}/_{2}$ years are dark brown, increasing with white on the forehead and tail each year. At an age of $3^{1}/_{2}$ years, the white head and tail are prominent but exhibit some brown markings. The adult plumage is attained at $4^{1}/_{2}$ to $5^{1}/_{2}$ years. The dark brown undersides and back of subadults are mottled with white; this is most pronounced in 3-year olds (Southern 1967). Adult underwings

111

are brown with some white markings visible at the base of the primaries. Subadult underwing plumage varies, but usually has prominent white wing linings contrasting with the dark brown flight feathers.

The golden eagle (*Aquila chrysaetos*) is similar to the subadult bald eagle, and increasing numbers of these migrants are being observed in Michigan (Evers 1989). Several field marks provide positive identification, particularly the pattern of white in the underwing (Clark 1983). Subadult bald eagles have white on the wing linings and flight feathers, while golden eagles have white confined to the flight feathers. The smaller head and bill, completely feathered legs, and distinctly white tail base (in subadults) further separate the golden eagle from the bald eagle.

A reliable method of distinguishing distant bald eagles from other large soaring birds, such as the osprey (*Pandion haliaetus*) and turkey vulture (*Cathartus aura*) is by its flattened-wing flight silhouette. Ospreys have a distinctive wing crook, and turkey vultures soar with a pronounced dihedral. Eagles have a characteristic slow and shallow wingbeat.

The large nest is a useful indicator of a breeding pair's presence. Differences between an eagle's nest and the similar osprey's nest are sometimes subtle. Bald eagles typically use live trees in upland areas; their larger, cone–shaped nest is placed within the tree crown. The osprey often uses dead or topped spruces (*Picea* spp.) or tamaracks (*Larix larcina*) in lowland conifer zones with the smaller, rounded nest placed at the top of a tree (Mathisen 1968).

Total Range: The breeding range of the bald eagle formerly extended from the tree line of Alaska and Canada south to the southern United States and Baja California in Mexico. Currently, major breeding population centers in the contiguous United States are around the western Great Lakes basin, Florida, Chesapeake Bay, Maine, and the Pacific Northwest. Small populations remain in Arizona (with a few pairs recently found in adjacent Sonora, Mexico), the northern U.S. Rocky Mountains, and scattered groups or pairs increasingly occur elsewhere in the original breeding range. Canadian and Alaskan populations are large; an estimated 30,000 to 35,000 nesting pairs currently inhabit Alaska.

Rather than undergoing a true long-distance migration to specific wintering grounds, eagle populations generally just shift southward in late autumn. The three major concentrations of wintering eagles in the contiguous United States are the Mississippi River system (the largest) (Steenhof 1978), Pacific Northwest, and Rocky Mountain states. This eagle generally remains north of Mexico throughout the winter.

In the early 1960s, fewer than 500 pairs were known of the southern bald eagle (King 1981), the lowest population level of the bald eagle since European settlement. From the precipitous decline in the 1960s and 1970s (Sprunt et al. 1973; Grier 1982), it has increased to over 2,660 nesting pairs in 1989 (Anonymous 1990). This probably translates to between 10,000 to 12,000 individuals. Current populations are stable or increasing in many regions (e.g., New England [Nickerson in Pendleton 1989]).

The bald eagle is federally endangered throughout the contiguous United States except in Washington, Oregon, Minnesota, Wisconsin, and Michigan, where it is federally threatened. In four of the five recovery regions (i.e., Pacific Northwest, Southwest, Northern states, and Chesapeake Bay) the bald eagle population has attained the recovery goals set for each region (Anonymous 1990). Populations in the Southeast recovery region are increasing but have yet to recover. In the Northern states recovery region, the western Great Lakes population (Minnesota, Wisconsin, and Michigan) has increased from 306 known occupied nests in 1973 (Engel 1986) to nearly 900. Current midwinter counts indicate population increases. In the early 1980s, an estimated 70,000 bald eagles occurred in North America; approximately 22,000 were found outside their Alaskan and British Columbia strongholds (Gerrard in Bird 1983).

The bald eagle is divided into two subspecies, the northern form (*H. l. alascanus*) and the smaller, rarer southern form (*H. l. leucocephalus*). The geographic limits of each subspecies' breeding range are arbitrary, although it is generally recognized that the northern form occurs primarily north of a line from northern California to Maryland (or north of 40° north latitude) (American Ornithologists' Union 1957). The northern form breeds in Michigan, the southern form is an irregular visitor.

State History and Distribution: Although nowhere abundant, the bald eagle formerly nested throughout the state. Due to habitat destruction and human persecution, it dramatically declined in the southern Lower Peninsula in the early to mid 1900s. During the 1970s, bald eagle populations reached an all-time low and its state range was restricted to a few northern localities. Due to concerns over this decline, detailed annual censuses of the eagle's distribution, abundance, and reproductive success have been made since 1961. In 1981, the number of occupied nests started increasing, but leveled off at an annual average of 108 occupied nests until 1984. Further increases now place the number of occupied nests, in 1992, at 215, producing approximately 218 young. Michigan's interim bald eagle recovery goal is 200 nesting pairs by 2000.

Bald eagles now nest in 22 counties of the northern Lower Peninsula and in all counties of the Upper Peninsula (Brewer et al. 1991). The western half of the Upper Peninsula (particularly Gogebic and Iron counties) is the most important breeding center for Michigan bald eagles. Populations in the eastern Upper Peninsula and northern Lower Peninsula are productive and expanding (nearly 90% since the 1970s). Great Lakes shoreline production has increased since the 1970s, but is not as high as in the interior. This may be due to continued high levels of organic contaminants in the Great Lakes.

The expansion of Michigan's bald eagle populations is evident throughout the state. Bald eagle nesting attempts in the southern Lower Peninsula have taken place on the Shiawassee State Game Area and National Wildlife Refuge in Saginaw County, Pt. Mouilee State Game Area in Monroe County, and Allegan State Game Area in Allegan County. The first occupied bald eagle nest since 1969 at Isle Royale, Keweenaw County, was found in 1984.

Although immature eagles migrate south, most Michigan breeding adults remain in the upper Great Lakes region (Stalmaster 1987), scattered in isolated locations or gathered in small roosts near open water or other food sources. Adults with established territories apparently are permanent Michigan residents, depending on winter food availability. Michigan's immature and subadult eagles migrate into the southern United States (Postupalsky 1976).

A Michigan study carried out during January, 1979–81, found a minimum of 44 individual bald eagles throughout the state, most were adults (Lerg and Pierce 1981). In contrast, a January, 1988, winter eagle survey documented 192 sightings in the state. The county with the highest total was Delta. Expectedly, standardized counts of spring migrant eagles at Whitefish Point, Chippewa County, show an upward trend, a threefold increase between 1983 and 1988 (Evers 1989).

Habitat: Bald eagles prefer to nest in the tallest tree available (Mathisen in Bird 1983) adjacent to the shorelines of lakes, large rivers, floodings, and other bodies of water where prey is available throughout the breeding season (Hensel and Troyer 1964; Gerrard et al. 1975; Grier 1977). Large bodies of water are preferred over smaller ones (Whitfield et al. 1974; Stocek and Pearce 1981), but those greater than 3.8 square miles (9.6 square km) of surface area probably do not enhance habitat suitability for a nesting pair (Peterson 1986).

Nest trees usually have a clear flight path toward areas of water. This open forest structure is one of the most important nest selection criteria (Anthony et al. 1982). Michigan eagles nest primarily in white and red pines (*Pinus strobus* and *P. resinosa*, respectively), yellow birch (*Betula lutea*), maple (*Acer* spp.), oak (*Quercus* spp.), and aspen (*Populus* spp.). Live trees are generally favored over dead ones.

Food availability is the most crucial component of breeding and wintering habitat. Although open water is often required for providing access to fish and waterfowl in winter, Lerg and Pierce (1981) found that carrion (e.g., road kill) is also an important food supply in Michigan. Winter feeding areas may be miles distant from roost sites. In Michigan, these roosts are usually small (Southern 1963). The degree of wind protection is important for roost selection in sheltered timber stands (Steenhof 1978; Keister et al. 1985).

Ecology and Life History: In Michigan, eagles arrive at their nesting territories between mid-February and mid-March. Nesting pairs are usually faithful to previous nesting sites. Territorial boundaries at active nests average a distance of nearly one-half mile (0.6 km) (Mahaffy and Frenzel 1987). The massive stick nests are renovated and may be used in consecutive years. Established pairs may alternate between two or more nests. These nests may reach a depth of 12 feet (3.7 m) and measure 8 feet (19.5 m) across. Most are placed in the crotch of tall, live trees with a leaf canopy providing shade. In Ontario, nest heights range from 30 to 90 feet (9.0 to 27.5 m), averaging 50 to 70 feet (15.0 to 21.5 m) (Peck and James 1983).

An elaborate cartwheel courtship display precedes the laying of one to four (usually two) eggs. However, nesting is not initiated every year. Incubation lasts approximately 35 days. Individual eagles pair for life, but replacement of lost mates occurs between seasons as well as within the same season (Grubb et al. 1988). Both sexes care for the young until they fledge (10 to 12 weeks) and forage independently (13 to 16 weeks after hatching). By October and November, immatures move southward and adults remain in their northern Michigan breeding range.

Bald eagles prefer fish, but, depending on food availability, are opportunistic foragers and frequently feed on carrion, waterfowl, and other birds and mammals (Todd et al. 1982; Stalmaster 1987; Harper et al. 1988). Tate and Postupalsky (1965) found that northern pike (*Esox lucius*) comprised more than 50% of the prey remains at a nest in Seney National Wildlife Refuge, Schoolcraft County. Road kill, particularly white-tailed deer (*Odocoileus virginianus*), may be important during initiation of nesting in late winter.

Pairs often cooperate in flushing and killing prey. Prominent perches (e.g., snags) adjacent to water are important components of suitable foraging habitat, although eagles will also hunt from flight. Ospreys, and sometimes other raptors and fish-eating birds, may be robbed of their prey by pirating eagles (kleptoparasitism). During food scarcity, pirating increases in frequency (Jorde and Lingle 1988), but scavenging is still preferred (Prevost 1979; Knight and Knight 1983; Fischer 1985; Knight and Skagen 1988). Kleptoparasitism between bald eagles also occurs (Southern 1963; Stalmaster and Gessman 1984).

Conservation/Management: The bald eagle experienced initial declines in its breeding populations from general nesting and foraging habitat degradation as well as general persecution. Downward population trends were exacerbated in the 1950s and 1960s by toxic pollutants, coupled with mortality from shooting, incidental trapping, poisoning, and other human-related activities.

Widespread use of organochlorine pesticides (especially DDT) caused dramatic breeding population declines throughout the contiguous United States. Ingestion of DDE, a metabolite of DDT (highly concentrated in fish), had a severely detrimental effect by reducing eggshell thickness, which reduced hatching success (Cooke 1973; Hickey and Anderson 1968; Krantz et al. 1970; Wiemeyer et al. 1972; Wiemeyer et al. 1984). These and other toxic pollutants, such as dieldrin and mercury, have caused poisoning deaths of bald eagles (Mulhern et al. 1970). Some of these pesticide compounds now have been banned or significantly reduced, although long lasting residues still persist (Frenzel and Anthony 1989).

Although eagle populations are recovering and nesting pairs along the Great Lakes shorelines are increasing (34 nesting attempts in 1988), pairs associated with the Great Lakes are characterized by extremely low production of young. Recent studies show a continued and significant contamination by organochlorines and PCBs in the Great Lakes ecosystem. These chemicals are known to reduce reproductive success.

Habitat degradation continues. Direct loss of nesting, foraging, and winter habitats remain a threat to recovering populations in Michigan. Dominant trees in undisturbed wooded areas adjacent to large bodies of water are needed for optimal nesting and roosting sites. In heavily logged areas, suitable nest trees are few and winter roost site habitat is marginal. Nesting pairs occasionally accept artificial nest structures, but much less frequently than the more adaptable osprey. On riverine stretches, optimal winter roost sites are located within undisturbed woodland edges near dams (Paruk 1987). Protection of communal roosts may be crucial to the viability of local populations.

Bald eagles are sensitive to human intrusion, especially during the first 12 weeks of the breeding cycle. Activities such as commercial and residential development, logging, and motorized vehicle use (e.g., ORVs, snowmobiles) near the nest site can lead to reproductive failures. Several studies show that human activity affects the distribution and behavior of nesting and wintering eagles (Stalmaster and Newman 1978; Gerrard and Ingram 1985; Peterson 1986). Human-related activities on waterways with bald eagle winter roosts should be excluded within a minimum distance of one-quarter mile (Knight and Knight 1984).

Conservation of eagles remains crucial; potential management measures include continuing strict controls on known limiting factors, purchasing areas with active nests and wintering roosts, designating state and federal management guidelines, continuing public education, and preventing the loss of nesting and roosting trees to logging operations. Other problems also need to be monitored, such as eagle mortality from leghold traps, lead poisoning, lake acidification, rough-fish removal (i.e., nongame fish) from lakes near active nests, and protection of winter roosting and feeding sites from human disturbance. Adult eagles are prone to lead poisoning through the ingestion of lead shot in prey species (Kaiser et al. 1980; Pattee and Hennes 1983; Reichel et al. 1984). The incidence of lead shot in live eagles can be monitored from regurgitated pellets (Lingle and Krapu 1988).

Most important, existing nest sites must be protected from human intrusion during the breeding season. Depending on the tolerance of individual eagles, forest cover, and topography, little human-related activity should occur within 330 to 660 feet (100 to 200 m) of the nest tree (Mathisen et al. 1977). Limited activity in a secondary zone (660 to 1,320 feet [200 to 400 m]) is possible as long as dramatic changes (e.g., new development) are avoided. Logging within a 1,500-foot (457-m) secondary zone significantly disturbs bald eagles (Wood et al. 1989).

Areas with a potential for recovery can be assisted with translocation of eggs, fostering, and hacking (Postupalsky and Holt 1975; Engel and Isaacs 1982). In the United States, these three techniques had fledging success rates of 26%, 84%, and 100%, respectively, between 1974 and 1981. Nearly 1,000 individuals have been hacked in the last decade. These methods have not been needed in Michigan, but nestlings have been transferred to other states.

The bald eagle's success in Michigan is a tribute to a strong conservation ethic; continued efforts will further ensure healthy populations.

LITERATURE CITED

American Ornithologists' Union. 1957. *The A.O.U. check-list of North American birds.* 5th ed. Am. Ornith. Union, Baltimore.

Anonymous. 1990. Fish and Wildlife Service undertakes review of the bald eagle's status. *Endangered Species Update* 15 (2): 3.

Anthony, R. G., R. L. Knight, G. T. Allen, B. R. McClelland, and J. I. Hodges. 1982. Habitat used by nesting and roosting bald eagles in the Pacific Northwest. *Trans. N. Am. Wildl. Nat. Resour. Conf.* 47:332–42.

Barrows, W. B. 1912. *Michigan bird life.* Mich. Agric. Coll. Spec. Bull., E. Lansing.

Bird, D. M. (ed.). 1983. *Biology and management of bald eagles and ospreys.* Harpell Press, Ste. Anne de Bellevue, Quebec.

Brewer, R., G. A. McPeek, and R. J. Adams, Jr. 1991. *The atlas of breeding birds in Michigan*. Mich. State Univ. Press, E. Lansing.

Clark, W. S. 1983. The field identification of North American eagles. *Am. Birds*. 37:822–26.

Cooke, A. S. 1973. Shell thinning in avian eggs by environmental pollutants. *Environ. Pollut.* 4:85–152.

Engel, J. M. 1986. Bald Eagle production in the North Central Region, 1973–1985. *U.S. Fish Wildl. Serv.*, Twin Cities, MN, Unpubl. Rept.

Engel, J. M., and F. B. Isaacs. 1982. Bald eagle translocation techniques. *U.S. Fish Wildl. Serv.*, Region 3, Twin Cities, MN.

Evers, D. C. 1989. Michigan raptor populations may be on the upswing. *Eyas* 12:10–15.

Fischer, D. L. 1985. Piracy of wintering bald eagles. *Condor* 87:246–51.

Frenzel, R. W., and R. G. Anthony. 1989. Relationship of diets of environmental contaminants in wintering bald eagles. *J. Wildl. Mgmt.* 53:792–802.

Gerrard, J. M., P. Gerrard, W. J. Maher, and D. W. A. Whitfield. 1975. Factors influencing nest site selection of bald eagles in northern Saskatchewan and Manitoba. *Blue Jay* 33:169–76.

Gerrard, J. M., and T. N. Ingram (eds.). 1985. *The bald eagle in Canada*. White Horse Plains Publ., Headingly, Man.

Grier, J. W. 1977. Quadrat sampling of a nesting population of bald eagles. *J. Wildl. Mgmt.* 41:438–43.

Grier, J. W. 1982. Ban of DDT and subsequent recovery of reproduction in bald eagles. *Science* 218:1232–34.

Grubb, T. G., L. A. Forbis, M. McWhorter, and D. R. Sherman. 1988. Adaptive perch selection as a mechanism of adoption by a replacement bald eagle. *Wilson Bull.* 100:302–5.

Harper, R. G., D. S. Hopkins, and T. C. Dunstan. 1988. Nonfish prey of wintering bald eagles in Illinois. *Wilson Bull.* 100:688–90.

Hensel, R., and W. Troyer. 1964. Nesting studies of the bald eagle in Alaska. *Condor* 66:282–86.

Hickey, J. J., and D. W. Anderson. 1968. Chlorinated hydrocarbons and eggshell changes in raptorial and fisheating birds. *Science* 162:271–73.

Jorde, D. G., and G. R. Lingle. 1988. Kleptoparasitism by bald eagles wintering in south-central Nebraska. *J. Field Ornith.* 59:183–88.

Kaiser, T. E., W. L. Reichel, L. N. Locker, E. Cromartie, A. J. Krynitsky, T. G. Lamont, B. M. Mulhern, R. M. Prouty, C. J. Stafford, and D. M. Swineford. 1980. Organochlorine pesticides, PCB, and PBB residues and necropsy data for bald eagles from 29 states—1975–1977. *Pest. Monit. J.* 13:145–49.

Keister, G. P., R. G. Anthony, and H. R. Holbo. 1985. A model of energy consumption in bald eagles: An evaluation of night communal roosting. *Wilson Bull.* 97:148–60.

King, W. B. (comp.). 1981. *Endangered birds of the world: The ICBP bird red data book*. Smithsonian Inst. Press, Washington, D.C.

Knight, R. L., and S. K. Knight. 1983. Aspects of food finding by wintering bald eagles. *Auk* 100:477–84.

Knight, R. L., and S. K. Knight. 1984. Responses of wintering bald eagles to boating activity. *J. Wildl. Mgmt.* 48:999–1004.

Knight, R. L., and S. K. Skagen. 1988. Agonistic asymmetries and the foraging ecology of bald eagles. *Ecology* 69:1188–94.

Krantz, W. C., B. M. Mulhern, G. E. Bagley, A. Sprunt, F. J. Ligas, and W. B. Robertson. 1970. Organochlorine and heavy metal residues in bald eagle eggs. *Pest. Monit. J.* 4:136–40.

Lerg, J. M., and V. B. Pierce. 1981. The status of bald eagle wintering in Michigan. *Mich. Acad. Sci.* 14:131–40.

Lingle, G. R., and G. L. Krapu. 1988. Ingestion of lead shot and aluminum bands by Bald Eagles during winter in Nebraska. *Wilson Bull.* 100:327–28.

McCollough, M. A. 1989. Molting sequence and aging of bald eagles. *Wilson Bull.* 101:1–10.

Mahaffy, M. S., and L. D. Frenzel. 1987. Elicited territorial responses of northern bald eagles near active nests. *J. Wildl. Mgmt.* 51:551–54.

Mathisen, J. E. 1968. Identification of bald eagle and osprey nests in Minnesota. *Loon* 40:113–14.

Mathisen, J. E., D. J. Sorenson, L. D. Frenzel, and T. C. Dunstan. 1977. Management strategy for bald eagles. *Trans. N. Am. Wildl. Nat. Resour. Conf.* 42:86–92.

Mulhern, B. M., W. L. Reichel, L. N. Locke, T. G. Lamont, A. Belisle, E. Cromartie, G. E. Bagley, and R. M. Prouty. 1970. Organochlorine residues and autopsy data from bald eagles, 1966–68. *Pest. Monit. J.* 4:141–44.

Paruk, J. D. 1987. Habitat utilization by bald eagles wintering along the Mississippi River. *Trans. Ill. Acad. Sci.* 80:333–42.

Pattee, O. H., and S. K. Hennes. 1983. Bald eagles and waterfowl: The lead shot connection. *N. Am. Wildl. Conf.* 48:230–37.

Peck, G. K., and R. D. James. 1983. Breeding birds of Ontario: Nidiology and distribution (vol. 1): Nonpasserines. *Royal Ont. Mus. Life Sci. Misc. Publ.*, Toronto.

Pendleton, B. G. (ed.). 1989. Proceedings of the Northeast raptor management symposium and workshop. *Natl. Wildl. Fed., Sci. Tech. Ser.* No. 13.

Peterson, A. 1986. Habitat suitability index models: Bald eagle (breeding season). *U.S. Fish Wildl. Serv. Biol. Rept.* 82(10.126).

Postupalsky, S. 1976. Banded northern bald eagles in Florida and other southern states. *Auk* 93:835–36.

Postupalsky, S., and J. B. Holt. 1975. Adoption of nestlings by breeding bald eagles. *J. Raptor Res.* 9:18–20.

Prevost, Y. 1979. Osprey-bald eagle interactions at a common foraging site. *Auk* 96:413–14.

Reichel, W. L., S. K. Schmeling, E. Cromartie, T. E. Kaiser, A. J. Krynitsky, T. G. Lamont, B. M. Mulhern, R. M. Prouty, C. J. Stafford, and D. M. Swineford. 1984. Pesticide, PCB, and lead residues and necropsy data for bald eagles from 32 states—1978–1981. *Environ. Monit. Assess.* 4:395–403.

Southern, W. E. 1963. Winter population, behavior, and seasonal dispersal of bald eagles in northwestern Illinois. *Wilson Bull.* 75:186–87.

Southern, W. E. 1967. Further comments on subadult bald eagle plumages. *Jack-Pine Warbler* 45:70–80.

Sprunt, A., W. B. Robertson, S. Postupalsky, R. J. Hensel, C. E. Knoder, and F. J. Ligas. 1973. Comparative productivity of six bald eagle populations. *Trans. N. Am. Wildl. Nat. Resour. Conf.* 38:96–106.

Stalmaster, M. V. 1987. *The Bald Eagle.* Universe Books, New York.

Stalmaster, M. V., and J. A. Gessman. 1984. Ecological energetics and foraging behavior of overwintering bald eagles. *Ecol. Monogr.* 54:407–28.

Stalmaster, M. V., and J. R. Newman. 1978. Behavioral responses of wintering bald eagles to human activity. *J. Wildl. Mgmt.* 42:506–13.

Steenhof, K. 1978. Management of wintering bald eagles. *U.S. Fish Wildl. Serv.*, FWS/OBS-78/79.

Stocek, R. F., and P. A. Pearce. 1981. Status and breeding success of New Brunswick bald eagles. *Can. Field-Nat.* 95:428–33.

Tate, L. J., and S. Postupalsky. 1965. Food remains at a bald eagle nest. *Jack-Pine Warbler* 43:146–47.

Todd, C. S., L. S. Young, R. B. Owen, and F. J. Gramlich. 1982. Food habits of bald eagles in Maine. *J. Wildl. Mgmt.* 46:636–45.

Whitfield, D. W. A., J. M. Gerrard, W. J. Maher, and D. W. Davis. 1974. Bald eagle nesting habitat, density and reproduction in central Saskatchewan and Manitoba. *Can. Field-Nat.* 99:399–407.

Wiemeyer, S. N., T. G. Lamont, C. M. Bunck, C. R. Sindelar, F. J. Gramlich, F. D. Fraser, and M. A. Byrd. 1984. Organochlorine pesticide, PCB, and mercury residues in bald eagle eggs, 1969–79, and their relationships to shell thinning and reproduction. *Archives Environ. Contam. Tox.* 3:529–49.

Wiemeyer, S. N., B. M. Mulhern, F. J. Ligas, R. J. Hensel, J. E. Mathisen, F. C. Robards, and S. Postupalsky. 1972. Residues of organochlorine pesticides, polychlorinated biphenyls, and mercury in bald eagle eggs and changes in shell thickness—1969 and 1970. *Pest. Monit. J.* 6:50–55.

Wood, P. B., T. C. Edwards, Jr., and M. W. Collopy. 1989. Characteristics of bald eagle nesting habitat in Florida. *J. Wildl. Mgmt.* 53:441–49.

Red-shouldered Hawk

· *Buteo lineatus* (Gmelin)
· **State Threatened**

Map 12. Confirmed red-shouldered hawk nesting pairs. Historically statewide, rare in the Upper Peninsula (Barrows 1912); ● = 1983 to 1992 (Brewer et al. 1991).

Status Rationale: Red-shouldered hawks nested throughout much of the Lower Peninsula as recently as the 1940s. Since then, breeding populations have shifted to northern Michigan and nearly disappeared in the southern Lower Peninsula, primarily due to the loss of extensive, mature lowland forests. In northern Michigan, recent studies have found only two-thirds of the number of young produced per active nest needed to maintain a stable population. Fewer than 100 nesting territories are currently known.

Field Identification: Like all soaring hawks (*Buteo* spp.), the red-shouldered hawk has a wide, rounded tail and broad wings, and its wingspan ranges from 39 to 44 inches (99 to 112 cm). Adults can be recognized by their rufous underparts and wing linings, checkered upper wing pattern, five to six narrow, white tail bands, and translucent, crescent-shaped patches ("windows") at the base of the primaries. The red shoulders, for which this hawk is aptly named, vary in intensity and may be difficult to observe in the field. Immatures lack the rufous coloring of the adults and instead are heavily streaked with brown teardrop markings on the underparts. Immatures may be best recognized by many light tail bands and underwing "windows." During the breeding season, this hawk's distinctive loud screaming *kee-yer* (the second syllable drops in pitch) may be heard while it aggressively defends its territory. The subspecies found in Michigan is *B. l. lineatus*.

120   Red-shouldered hawks are most easily confused with Michigan's other two

nesting buteos, the red-tailed and broad-winged hawks (*B. jamaicensis* and *B. platypterus*, respectively). The adult red-tailed can be quickly identified by its dark belly band and distinctive pinkish red tail; the smaller adult broad-winged is distinguished by its three broad black-and-white tail bands, which are of equal width. In flight, the broad-winged hawk's creamy underwings are clearly different than the red-shouldered's dark underwings. Immature red-shouldered hawks have plumages similar to other Michigan *Buteos* but can be distinguished by the "windows" in the wing tips. Other hawks either have short, rounded wings and long tails (*Accipiter* spp.) or long, pointed wings and long tails (*Falco* spp.). The red-shouldered hawk is commonly associated with the barred owl (*Strix varia*), because they share similar habitats; antagonism between the two is rare (Stewart 1949). Care should be taken when identifying the red-shouldered by call because it is often imitated by the blue jay (*Cyanocitta cristata*).

Total Range: The red-shouldered hawk occurs in two separate geographic populations in North America. In the West, this hawk's primary breeding range is in California, extending south to Baja California, east to Arizona (Glinski 1982), and north to Oregon. Eastern populations have a larger range and may be found nesting from Minnesota and southern Quebec south to Florida, Texas, and along the Gulf of Mexico to central Mexico. Western populations are relatively stable, while most northern and eastern winter populations have declined markedly, probably reflecting declining breeding populations (Brown 1971). Northern populations typically shift to the southern part of the red-shouldered hawk's range in winter. The southeastern United States contains the greatest concentration of wintering red-shouldered hawks (Bock and Lepthien 1976) as well as high densities of resident nesting pairs (Robbins et al. 1986).

State History and Distribution: Up to the early 1900s, this species was a very common hawk in southern Lower Michigan (Barrows 1912), but, by the late 1950s, the red-shouldered began to disappear from this region, a decline from which it has never recovered. The changing red-shouldered populations in Superior Township, Washtenaw County, clearly illustrates the severity of their decline. Of the 22 nesting pairs occurring there in 1942 (Craighead and Craighead 1969), only a single pair could be located in 1966 (Wallace 1969).

The red-shouldered hawk recently has shifted its primary nesting range from southern to northern Michigan. Known concentrations now occur in the Grand Traverse area from Leelanau County south to Manistee County, and in the Straits area from Charlevoix County east to Alpena County and north to the Straits of Mackinac. However, recent studies have found that even in these strongholds it may be declining slowly (Ebbers 1986). During the 1980s, the red-shouldered hawk has been observed more frequently in the Upper Peninsula and is expanding into the area. Summer occurrences have been reported occasionally in the past, but confirmation of nesting in the Upper Peninsula was not recorded until 1978 (Postupalsky 1980). Since that discovery, breeding evidence has been found in eight Upper Peninsula counties (Brewer et al. 1991).

The red-shouldered hawk now nests only sporadically in southern lower

Michigan. Less fragmented forests, such as those found in Allegan and Barry counties, have the highest potential of retaining breeding populations. Currently, a dense breeding population occurs along the Manistee River, Manistee County. It is the last stronghold of this hawk within its primary historic range. Fewer than 100 nest sites are known in the state, with fewer than ten pairs in the southern Lower Peninsula (Evers 1989). This hawk winters in small numbers in southern Michigan (Payne 1983; Evers 1989).

Habitat: The red-shouldered hawk typically has specific habitat requirements. It needs large tracts of mature forest in which a closed canopy inhibits undergrowth, with numerous openings that are used as hunting areas. Floodplain forests are optimal, since they are commonly produced by meandering rivers or creeks, which create small marshy openings over time. Another important factor is the structure of a floodplain forest; it is situated on level ground and usually has minimal underbrush. The importance of mature floodplain forests as nesting habitat for red-shouldered hawks has been documented by several studies (Henny et al. 1973; Campbell 1975; Portnoy and Dodge 1979; Titus and Mosher 1981; Bednarz and Dinsmore 1982). Nest sites are usually near water and wetland openings. Stewart (1949) found red-shouldered hawks absent only one-quarter mile (0.4 km) from floodplain forests.

In Michigan, red-shouldered hawks appear to be slightly less selective than in other areas and are successful in nonriverine habitats, such as swamps and isolated woodlots with adjacent openings, and riverine habitat with steep banks (Craighead and Craighead 1969; Ebbers 1986). In the Upper Peninsula, nests have been found in upland mixed forests (Postupalsky 1980) and in similar upland habitat adjacent to lakes. Bednarz and Dinsmore (1981) found that large, contiguous upland forests may compensate for floodplain forest habitat, although mature forest tracts with a minimum of 300 floodplain acres (121 ha) and 170 upland acres (69 ha) are preferred (Bednarz and Dinsmore 1982).

Ecology and Life History: Although Michigan's red-shouldereds may be found in small numbers during the winter, the majority of nesting birds arrive from southern wintering areas between late February and early April. Territories are often utilized for several consecutive years, but nests are commonly rebuilt each year at different sites within the territory (Craighead and Craighead 1969). Courtship flights are usually over wetland hunting areas and have been found to be most frequent around noon (Portnoy and Dodge 1979). While defending their nesting territories, red-shouldered hawks are easily located by their aggressive vocal behavior. Their highly territorial nature also is evident by the relative uniformity of distances between nests, which is especially obvious in large tracts of suitable habitat.

Nests are typically placed in the main trunk-crotch or support branches of large deciduous (rarely coniferous) trees (Stewart 1949; Morris et al. 1982; Titus and Mosher 1987). Nest placement averages 50 feet (15 m) above ground (Henny et al. 1973; Stewart 1949), usually in or below the tree canopy (Titus and Mosher 1981; Woodrey 1986; Ebbers 1986). In early to mid-April, both parents become secretive while incubating an average of three to four eggs for nearly a month. Studies suggest that after hatching, the young are probably fed solely by the female, while the male is responsible for supplying the female with food (Portnoy and Dodge 1979). The most critical time for fledging success is during the first two weeks (Henny et al. 1973). Young fledge at about six weeks but remain dependent upon their parents for several more weeks. Individuals probably begin breeding at two years of age (Henny 1972).

Red-shouldered hawks are opportunistic hunters, usually obtaining most of their prey from such openings as wet meadows within forested areas (Bednarz and Dinsmore 1985). A typical diet consists mainly of small rodents (Ernst 1945; Portnoy and Dodge 1979), along with a variety of birds, reptiles, amphibians, and insects (Barrows 1912; Stewart 1949; Pettingill 1976). A comprehensive study in southeastern Michigan found nesting red-shouldered hawks to prey most heavily upon small rodents and birds, with less emphasis on snakes, frogs, and crayfish (Craighead and Craighead 1969).

Conservation/Management: The primary reason for the decline of the red-shouldered hawk is disturbance to its floodplain forest habitat. Although logging has the obvious effect of reducing suitable habitat, intense selective cutting can restrict breeding pairs from developing territories in more subtle ways. By opening the tree canopy and increasing sunlight, new understory growth is stimulated. This deteriorates otherwise suitable red-shouldered habitat by making it less attractive to this species and more appealing to the opportunistic red-tailed hawk (Bryant 1986). Red-tailed and red-shouldered hawks initiate nesting at approximately the same time in Michigan, but Bednarz and Dinsmore (1981) found that the number of red-shouldered hawks in a given area may be dependent on the number of red-tailed hawks already present.

Other detrimental practices include the development of buildings and roads (Bednarz and Dinsmore 1982), dam construction and channelization (Bednarz and Dinsmore 1981), forest fragmentation (Morris and Lemon 1983), and possibly pesticide contamination (Campbell 1975), as also indicated for other woodland hawks (Pattee et al. 1985). Henny et al. (1973) concluded that the levels of dieldrin and DDE in red-shouldered eggs probably did not have a detrimental effect on reproductive performance in their study area, although there was evidence of decreased eggshell thicknesses. The red-shouldered's major natural predators are the great horned owl (*Bubo virginianus*) (Portnoy and Dodge 1979; Wiley 1975; Crocoll and Parker 1989) and the raccoon (*Procyon lotor*). Severe raccoon predation of eggs and young may occur after heavy early spring rains in the northern Lower Peninsula (Ebbers 1986). Northern goshawks (*Accipiter gentilis*) contribute to chick mortality in zones of range overlap.

To moderate the decline of red-shouldered hawks, Michigan's existing floodplain forests and wet, mature wooded areas need to be preserved. Large contiguous forest tracts with abundant small openings and edge are necessary for successful breeding. Bednarz and Dinsmore (1981) found that prior breeding territory estimates of approximately 250 acres (101 ha) may be too small and suggested that each pair may require a territory as large as an estimated 615 acres (249 ha).

The same study also suggested that mature forests be maintained at 370 to 1,000 trees for every $2^1/_2$ acres (6 ha) with openings comprising around 15% of suitable habitat. If natural openings are scarce, artificial open areas (less than 10 acres) may be created.

In general, if a nest or breeding activity center is located, disturbance within an estimated one-half mile (0.8 km) radius should be kept to a minimum. The critical period is mid-March through early June, after which the young should be at least two weeks old (Bednarz and Dinsmore 1981). Poor reproductive success in northern Michigan in 1986–88 (0.80 to 1.20 fledged young per pair) was attributed partly to prolonged and heavy precipitation (Ebbers 1986). Henny (1972) and Henny et al. (1973) estimated that 1.95 to 2.10 fledged young per breeding pair is necessary for a self-sustaining population.

Michigan's red-shouldered hawk has been pushed northward to areas with greater climate extremes and marginal habitat. To avoid the eventual loss of this *Buteo* in Michigan, protected southern populations need to be established to provide a focus for a permanent statewide population. Comprehensive surveys using past successful techniques are needed (DeVaul 1988), such as the broadcast of red-shouldered hawk vocalizations (Fuller and Mosher 1981). Only then, when research, management, and education programs are established, will our formerly most common raptor be secure in Michigan.

LITERATURE CITED

Barrows, W. B. 1912. *Michigan bird life.* Mich. Agric. Coll. Spec. Bull., E. Lansing.

Bednarz, J. C., and J. J. Dinsmore. 1981. Status, habitat use, and management of red-shouldered hawks in Iowa. *J. Wildl. Mgmt.* 45:236–41.

Bednarz, J. C., and J. J. Dinsmore. 1982. Nest-site and habitat of red-shouldered and red-tailed hawks in Iowa. *Wilson Bull.* 94:31–45.

Bednarz, J. C., and J. J. Dinsmore. 1985. Flexible dietary response and feeding ecology of the red-shouldered hawk, *Buteo lineatus*, in Iowa. *Can. Field-Nat.* 99:262–64.

Bock, C. E., and L. W. Lepthien. 1976. Geographical ecology of the common species of *Buteo* and *Parabuteo* wintering in North America. *Condor* 78:554–57.

Brewer, R., G. A. McPeek, and R. J. Adams, Jr. 1991. *The atlas of breeding birds in Michigan.* Mich. State Univ. Press, E. Lansing.

Brown, W. H. 1971. Winter population trends in the red-shouldered hawk. *Am. Birds* 25:813–17.

Bryant, A. W. 1986. Influence of selective logging on red-shouldered hawks, *Buteo lineatus*, in Waterloo Region, Ontario, 1953–1978. *Can. Field-Nat.* 100:520–25.

Campbell, C. A. 1975. Ecology and reproduction of red-shouldered hawks in the Waterloo region, southern Ontario. *J. Raptor Res.* 9:12–17.

Craighead, J. J. and F. C. Craighead, Jr. 1969. *Hawks, owls and wildlife.* Dover, New York.

Crocoll, S. T., and J. W. Parker. 1989. The breeding biology of broad-winged and red-shouldered hawks in western New York. *J. Raptor Res.* 23:125–39.

DeVaul, H. 1988. Survey techniques for woodland hawks in the Northeast. Pp. 301–10 *in* B. G. Pendleton (ed.), Proc. of the Northeast raptor management symposium and workshop. *Nat. Wildl. Fed., Sci. Tech. Ser.* No. 13.

Ebbers, B. C. 1986. Distribution, status and reproductive ecology of red-shouldered hawks in the northern Lower Peninsula of Michigan. *Mich. Dept. Nat. Resour.*, Unpubl. Rept.

Ernst, S. G. 1945. The food of the red-shouldered hawk in New York State. *Auk* 62:452–53.

Evers, D. C. 1989. Michigan raptor populations may be on the upswing. *Eyas* 12:10–15.

Fuller, M. R., and J. A. Mosher. 1981. Methods of detecting and counting raptors: A review. *Stud. Avian Biol.* 6:235–46.

Glinski, R. L. 1982. The red-shouldered hawk (*Buteo lineatus*) in Arizona. *Am. Birds* 36:801–3.

Henny, C. J. 1972. An analysis of the population dynamics of selected avian species. *U.S. Fish Wildl. Serv., Wildl. Resour. Mgmt. Rept.* 1.

Henny, C. J., F. C. Schmid, E. M. Martin, and L. L. Hood. 1973. Territorial behavior, pesticides, and the population ecology of red-shouldered hawks in central Maryland, 1943–1971. *Ecology* 54:545–54.

Morris, M. M., and R. E. Lemon. 1983. Characteristics of vegetation and topography near red-shouldered hawk nests in southwestern Quebec. *J. Wildl. Mgmt.* 47:138–45.

Morris, M. M., B. L. Penak, R. E. Lemon, and D. M. Bird. 1982. Characteristics of red-shouldered hawk, *Buteo lineatus*, nest sites in southwestern Quebec. *Can. Field-Nat.* 96:139–42.

Pattee, O. H., M. R. Fuller, and T. E. Kaiser. 1985. Environmental contaminants in eastern cooper's hawk eggs. *J. Wildl. Mgmt.* 49:1040–44.

Payne, R. P. 1983. A distributional checklist of the birds of Michigan. *Univ. Mich. Mus. Zool. Misc. Publ.* No. 164.

Pettingill, O. S. 1976. The prey of six species of hawks in northern lower Michigan. *Jack-Pine Warbler* 54:70–74.

Portnoy, J. W., and W. E. Dodge. 1979. Red-shouldered hawk nesting ecology and behavior. *Wilson Bull.* 91:104–17.

Postupalsky, S. 1980. The red-shouldered hawk breeding in Michigan's Upper Peninsula. *Jack-Pine Warbler* 58:73–76.

Robbins, C. S., D. Bystrak, and P. H. Geissler. 1986. The breeding bird survey: Its first fifteen years, 1965–1979. *U.S. Fish Wildl. Serv., Resour. Publ.* 157.

Stewart, R. E. 1949. Ecology of a nesting red-shouldered hawk population. *Wilson Bull.* 61:26–35.

Titus, K., and J. A. Mosher. 1981. Nest-site habitat selected by woodland hawks in central Appalachians. *Auk* 98:270–81.

Titus, K., and J. A. Mosher. 1987. Selection of nest tree species by red-shouldered and broad-winged hawks in two temperate forest regions. *J. Field Ornith.* 58:274–83.

Wallace, G. J. 1969. Endangered and declining species of Michigan birds. *Jack-Pine Warbler* 47:70–75.

Wiley, J. W. 1975. The nesting and reproductive success of red-tailed hawks and red-shouldered hawks in Orange County, California, 1973. *Condor* 77:133–39.

Woodrey, M. S. 1986. Characteristics of red-shouldered hawk nests in southeastern Ohio. *Wilson Bull.* 98:466–69.

Merlin

· *Falco columbarius* (Linnaeus)
· **State Threatened**

Map 13. Breeding evidence of merlin nesting pairs.
▲ = Historically in the Upper Peninsula (Payne 1983);
● = 1983 to 1992 (Brewer et al. 1991).

Status Rationale: This falcon has been a rare Michigan breeder historically, largely because the state is at the southern periphery of its nesting range. Although Great Lakes' populations have recently increased and apparently have surpassed historical levels, the merlin's chemically contaminated South and Central American wintering grounds are a source of concern. An estimated 35 to 55 nesting pairs occur in Michigan.

Field Identification: The swift-flying merlin is intermediate in size between the slightly smaller American kestrel (*Falco sparverius*) and the much larger, federally endangered peregrine falcon (*Falco peregrinus*). Distinguishing characteristics are its long, pointed wings, long, heavily barred tail, and vertically streaked underparts. As with most birds of prey, the female merlin is larger than the male—an average of 1^{1}/3 times heavier (Evans 1982). This species averages 11 inches (28 cm) long with a wingspan of nearly 2 feet (0.6 m). Adult females and immatures differ from the blue-gray backed adult male with their dark brown upperparts. Like other falcons, merlins repeatedly give rapid, high-pitched calls when alarmed. The taiga merlin subspecies (*F. c. columbarius*) occurs in Michigan.

In Michigan, two groups of raptors could be confused with the merlin. Of the other falcon species, the more common kestrel lacks heavy ventral streaking and

126

has rufous tail and wings (or blue wings in the male). The much larger peregrine falcon has large and distinct black "sideburns" in contrast to the merlin's faint ones. The accipiter group also may be confused with the merlin, especially the immature sharp-shinned and Cooper's hawks (*Accipiter striatus* and *A. cooperii*, respectively). These accipiters, however, lack any evidence of "sideburns," and the rounded wings and rusty-barred breast of adults should adequately separate them from the merlin. At a distance, the flight pattern of a mourning dove (*Zenaida macroura*) may appear to be similar to that of the merlin. The merlin's barred, rounded tail and larger head size, however, should quickly distinguish it from the dove's unbarred, pointed tail and small head. In general, the merlin's heavily streaked underparts are the best field mark; its extremely quick, powerful, and agile flight, aggressive behavior toward other birds, and loud calls also serve as useful identification characteristics.

Total Range: Found throughout the northern hemisphere, North American merlins occur in the boreal forests of Alaska and Canada, south to the northern portions of the United States from the Pacific Coast, northern Rockies, and northern Great Plains through the northern Great Lakes region to New England. The paler, prairie subspecies (*F. c. richardsoni*) breeds in the prairie parklands of central Canada and the United States, the black subspecies (*F. c. suckleyi*) breeds in the Pacific coastal regions, and the more widespread taiga subspecies (*F. c. columbarius*) occupies the remaining range (primarily boreal forest); there is little breeding range overlap among the subspecies. As a species, the merlin is not in danger of local extinction in most of its range and recently has been removed from its long-standing watch status on the National Audubon Society's Blue List (Tate and Tate 1982). The taiga merlin generally winters from the Gulf states to northern South America.

State History and Distribution: In Michigan, this falcon always has been a rare breeding bird, restricted to the Upper Peninsula. The first confirmed Michigan nesting was in 1955 in the Huron Mountains, Marquette County (Dodge 1955; Gysendorfer 1955). Since this discovery, 10 of the 15 counties in the Upper Peninsula have confirmed breeding, with summer observations in each of the remaining counties. Isle Royale National Park, Keweenaw County, maintains a consistent and high density of merlin nesting pairs. In the early 1980s, only 3 territorial pairs were known annually on the island. In 1988, intensive surveys for the Michigan Breeding Bird Atlas estimated approximately 11 to 25 nesting pairs on the island. In addition to Isle Royale, Keweenaw County, merlin nesting pairs are currently concentrated in the Keweenaw Peninsula, at Porcupine Mountains State Park, Ontonagon County, and in the Huron Mountains and Huron Islands, Marquette County.

The merlin has not been confirmed as a Michigan nesting resident outside the Upper Peninsula, although there are several summer records in the jack pine (*Pinus banksiana*) plains of the Kirtland's Warbler Management Area in north-central lower Michigan. In 1986, the first nesting record of the merlin was recorded in the Lower Peninsula in Antrim County; along the shores of Lake Huron, 1–2 pairs (with breeding evidence) were present in 1988 and 1989 in the Lower Peninsula's Alpena County (Brewer et al. 1991). In 1989, the estimated population was 35 to 55 nesting pairs (Evers 1989).

During migration, merlins often are seen throughout the state. Spring migration concentrations are especially notable at Whitefish Point Bird Observatory (WPBO) in Chippewa County. Standardized counts by WPBO since 1983 have documented increases in the number of spring migrants, indicating an expanding Great Lakes breeding population. In particular, 60 individuals were officially recorded in 1986, more than double the 1983–85 WPBO average of 25 individuals (Evers et al. 1986). Merlins occasionally winter in southern Michigan (Payne 1983), but observations should be carefully documented.

Habitat: The typical nesting habitat of the merlin is the boreal forest. In Michigan, this species usually prefers to nest in spruce forests near bogs or open water (Johnson and Coble 1967). Large trees adjacent to a body of water (particularly Lake Superior) are the preferred nesting sites. Recently, merlins have nested within the city limits of Marquette, Marquette County, similar to behavior exhibited by an urban population in Thunder Bay, Ontario (Escott 1986). Lake shorelines and other open areas are used as hunting grounds, although merlins also will hunt within forest habitats (Lawrence 1949).

Ecology and Life History: Spring arrival of this falcon in Michigan usually coincides with the main migration of the small birds upon which it preys—late April to early May. Males may arrive up to one month before females to establish the breeding territory (Trimble 1975). Females may disperse more widely than males (Schempf and Titus 1989). After the mid-May courtship, an average clutch of four to five eggs is laid. Like other falcons, the merlin does not build its own nest. Instead, it commonly uses abandoned American crow (*Corvus brachyrhynchos*), common raven (*Corvus corax*), and hawk nests, as well as witches' brooms located in coniferous trees. In Doolittle's (1988) upper Great Lakes study, most merlin pairs were found using abandoned corvid nests in pine trees. The female does most of the incubating during the 28- to 32-day period and will renest if the eggs are destroyed early in incubation.

The male defends the nesting territory and hunts for the larger female until the young no longer need constant brooding. Adult merlins are aggressive toward intruders in the nesting area and defend it with repeated alarm cries (Lawrence 1949), even diving and striking the potential predator (Craighead and Craighead 1940). This agitated calling behavior often can be used to locate a nest. The young fledge approximately 25 to 30 days after hatching and are dependent for up to five weeks. By mid-October, most individuals have migrated south of Michigan. Merlins typically initiate breeding at two years of age.

Unlike the high, powerful dives of the peregrine falcon, the merlin generally hunts by ambushing potential prey from a perch or in flight. When prey is sighted, the merlin either strikes it down to the ground or kills it in midair. Captured prey usually are taken to a prominent snag. Small- to medium-sized birds are potential prey (Johnson and Coble 1967), although high-flying or open country birds may be most vulnerable. Merlins generally are not prey specific and therefore will concentrate on the species that is most prevalent or available. Shorebird concentrations, in particular, attract hunting merlins (Boyce 1985). Although usually considered to be strict bird hunters, merlins are known to take ground squirrels (Becker 1985), bats (Dekker 1972; Doolittle 1988), and dragonflies as well as other large insects (Bond 1951; Street 1960; Johnson and Coble 1967), particularly in summer when they are most abundant.

Conservation/Management: Since the taiga merlin's major breeding grounds are north of Michigan, its population in the Upper Peninsula is expected to be a low and fluctuating one. Therefore, the added protection of this peripheral nesting area, by inclusion in Michigan's Endangered Species Act, may contribute to a self-sustaining population in the upper Great Lakes region. Currently, its habitat is not threatened seriously in Michigan; and there are no known factors that are critically limiting its present population in Michigan. Reproductive inhibition due to organochlorine residues in its diet may still be occurring because the use of destructive chemicals (e.g., DDT and DDE) south of the United States continues unabated. Chemical contamination has been shown in the taiga (Schempf and Titus 1989) and *richardsonii* subspecies (Becker and Sieg 1987).

Beside pesticide effects, from which some populations have slowly rebounded, other potential limiting factors include disturbance of nesting areas by humans and lack of suitable nesting sites (since it prefers a large, prebuilt nest, in a spruce or mixed boreal forest, near an open body of water). Systematic surveys of its distribution and density in the Upper Peninsula are required to determine its current status and population trend. Broadcast merlin vocalizations have proven successful in locating territorial pairs (Doolittle 1988). Studies of pesticide levels and sources, protection of known nest sites, and a program of artificial nest building would help sustain the population increase currently exhibited by Michigan's merlin population.

LITERATURE CITED

Becker, D. M. 1985. Food habits of Richardson's merlins in southeastern Manitoba. *Wilson Bull.* 97:226–30.

Becker, D. M., and C. H. Sieg. 1987. Eggshell quality and organochlorine residues in eggs of merlins, *Falco columbarius*, in southeastern Montana. *Can. Field-Nat.* 101:369–72.

Bond, R. M. 1951. Pigeon hawk catches dragonflies. *Condor* 53:256.

Boyce, D. A. 1985. Merlins and the behavior of wintering shorebirds. *J. Raptor Res.* 19:94–96.

Brewer, R., G. A. McPeek, and R. J. Adams, Jr. 1991. *The atlas of breeding birds of Michigan*. Mich. State Univ. Press, E. Lansing.

Craighead, J. J., and F. C. Craighead. 1940. Nesting pigeon hawks. *Wilson Bull.* 52:241–48.

Dekker, D. 1972. Pigeon hawk catches bat. *Blue Jay* 30:256.

Dodge, P. 1955. Pigeon hawks nesting at Huron Mountain, Michigan. *Jack-Pine Warbler* 33:132–33.

Doolittle, T. C. J. 1988. Status, distribution, and breeding biology of the merlin (*Falco c. columbarius*) in the upper midwest region. *Sigurd Olson Environ. Inst.*, Unpubl. Rept.

Escott, N. G. 1986. Thunder Bay's nesting merlins. *Ontario Birds* 4:97–101.

Evans, D. L. 1982. Status reports on twelve raptors. *U.S. Fish Wildl. Serv., Spec. Sci. Rept.—Wildl.* No. 238.

Evers, D. C. 1989. Michigan raptor populations may be on the upswing. *Eyas* 12:10–15.

Evers, D. C., T. P. Wiens, and D. A. Cristol. 1986. Spring 1986 raptor migration study at Whitefish Point, Michigan. *Whitefish Pt. Bird Observ.*, Unpubl. Rept.

Gysendorfer, R. 1955. Pigeon hawks nesting in Marquette County, Michigan. *Jack-Pine Warbler* 33:130.

Johnson, W. J., and J. A. Coble. 1967. Notes on the food habits of pigeon hawks. *Jack-Pine Warbler* 45:97–98.

Lawrence, L. D. 1949. Notes on nesting pigeon hawks at Pimisi Bay, Ontario. *Wilson Bull.* 61:15–25.

Payne, R. P. 1983. A distributional checklist of the birds of Michigan. *Univ. Mich. Mus. Zool. Misc. Publ.* No. 164.

Schempf, P. F., and K. Titus. 1989. Status of the merlin (*Falco c. columbarius*) in interior Alaska. *U.S. Fish Wildl. Serv., Raptor Mgmt. Ser.*, Unpubl. Rept.

Street, M. G. 1960. Pigeon hawk catches dragonflies. *Blue Jay* 18:124.

Tate, J., and D. J. Tate. 1982. The blue list for 1982. *Am. Birds* 36:126–35.

Trimble, S. A. 1975. Habitat management series for unique or endangered species, merlin. *U.S. Bur. Land Mgmt., Tech. Note* 15.

Peregrine Falcon

· *Falco peregrinus* Tunstall
· **Federally and State Endangered**

Map 14. Confirmed peregrine falcon nesting pairs.
▲ = before 1957 (Payne 1983); ● = 1983 to 1992 (Michigan DNR).

Status Rationale: The peregrine falcon has experienced severe population declines throughout much of its worldwide range. Its status became critical in North America by the mid-1960s, when it disappeared from much of its breeding range, particularly east of the Mississippi River. Since then, the peregrine falcon has made a comeback with assistance from restoration programs. In Michigan, these successful efforts were realized in 1989 with the first confirmed nesting pair since 1957.

Field Identification: The peregrine falcon is crow sized with long, pointed wings and a narrow tail. Like most birds of prey, females are larger than the males—an average of 1.6 times heavier (Evans 1982). Length and wingspan measurements range from 15 to 20 inches (38 to 51 cm) and from 43 to 46 inches (109 to 117 cm), respectively. The flight is powerful with little gliding, although migrating peregrines will soar for extended periods of time. Adults have slate gray or bluish backs with lightly barred underparts and a black cap. A distinctive feature of the peregrine is its dark malar stripe ("sideburns"). Immatures are brownish with dense breast streaks. Subspecies are difficult to distinguish in the field. In Michigan, *F. p. anatum* was generally the subspecies present, although *F. p. tundrius* has been recorded as an occasional transient (Payne 1983). The smaller *tundrius* subspecies appears white to gray compared to the more brown and rufous-colored *anatum* subspecies (White 1968). When agitated, the peregrine will give a repeated *hek hek hek* call.

131

The similar coloration and shape of the state-threatened merlin (*Falco colum-barius*) may present difficulties in identification. Adult peregrines can be distinguished from the merlin by the much larger size and heavier sideburns. The identification of immatures is more difficult but can usually be made on the basis of the characteristics given for adults. Other falcons lack the distinct facial pattern of the peregrine; accipiters can be quickly identified by their short, rounded wings and their "flap-flap-glide"style of flight.

Total Range: The peregrine is nearly worldwide in distribution. The original breeding range of Michigan's subspecies (*F. p. anatum*) extended from the taiga portions of Alaska and Canada south to Baja California, the Rocky Mountains (Colorado), the upper and middle Mississippi River basins, the upper Great Lakes region, the Appalachian Mountains, and Chesapeake Bay.

Presently, peregrine populations are stable or increasing in areas where it was not completely exterminated by pesticides. In areas where it has disappeared (particularly the eastern United States), hacking programs have restored this falcon into historical nesting sites and other areas with suitable habitat. This was made possible by the Peregrine Fund, supported by Cornell University, New York (now based at the World Center for Birds of Prey in Boise, Idaho). This project established a captive breeding program in 1970 to provide a source of falcons for restocking purposes. More than 1,000 young peregrines have been successfully released in the eastern United States; in 1988, 258 young survived to release age.

The peregrine restoration programs have been so successful that the Eastern Peregrine Recovery Plan's goal of 175 nesting pairs with at least 20 pairs in four separate regions (Barclay and Cade 1983; Bollengier et al. 1979; USF&WS 1987) soon may be met. In 1988, 67 confirmed pairs were on territories in the East; a substantial increase since the first returning nesting pair in 1980. Midwest peregrine falcon releases began in 1981 as an effort separate from the work of the Peregrine Fund, spearheaded by the University of Minnesota Raptor Center. The source of the falcons were birds domestically produced by falconers in North America. The first release occurred in Minnesota; subsequent releases were in Illinois, Michigan, Wisconsin, Iowa, Ohio, and Ontario. A total of 370 falcons have been released in the first eight years of the restoration effort. By 1989, 23 territories were occupied (Redig and Tordoff 1990), and 16 nesting pairs produced 35 young. The Midwest region's recovery goal is to maintain at least a three-year average of 30 territorial pairs.

After the nesting season, the peregrine winters in the southern part of its breeding range, south into Central and South America. Most individuals overwintering in the United States are associated with the Atlantic and Gulf coasts (Enderson 1965). Recently, hacking programs have been responsible for a wider wintering range; individuals occasionally remain in the northern United States.

State History and Distribution: Historically, the peregrine was limited to nesting on cliff faces of the Upper Peninsula. Only ten historical nest site locations are recorded in the state: Goose Lake Escarpments, Huron Islands, Huron Mountains, and Lake Michigamme in Marquette County; Grand Island and Pictured Rocks in

Alger County; Garden Peninsula in Delta County; Isle Royale in Keweenaw County; Mackinac Island in Mackinac County; and South Fox Island in Leelanau County (Issacs 1976).

In 1986, the DNR began restoration of the peregrine falcon with the release of five peregrines in the city of Grand Rapids, Kent County. Although this urban area did not have historical peregrine nesting records, it provided suitable hacking sites and an abundant prey source. The success of this project spurred releases in 1987 in Detroit (five individuals) and Grand Rapids (six individuals). The first restoration project near a historical nesting site also occurred that year on Isle Royale, Keweenaw County, with five released birds. In 1988, five more birds were released at this Lake Superior island site and in Grand Rapids, the city's final designated year. Eight individuals also were released at a natural site in the Ottawa National Forest, Ontonagon County, in 1988.

In 1989 releases of 10 individuals each were made at Isle Royale National Park, Pictured Rocks National Lakeshore, and the Ottawa National Forest. An additional 86 peregrines were released throughout the U.S. upper Great Lakes region and other Midwest locations for a total of 116 "hacked" peregrines in 1989 (Fettig 1989). By 1992, 125 peregrines were released in Michigan and three nesting pairs were known: one in Detroit and two in the western Upper Peninsula. Michigan's recovery goal by the year 2000 is eight nesting pairs in the Upper Peninsula and two in the Lower Peninsula.

During migration, peregrine falcons are seen annually throughout Michigan, and there has been a noticeable increase of state observations since 1983. Most migrants are observed along the Great Lakes shorelines. Occasionally, individuals are found in winter (Payne 1983). Recently, winter sightings have become more frequent in southern Michigan, probably related to the increased number of released birds thoughout the Great Lakes region.

Habitat: Peregrines prefer areas with high cliffs overlooking water and other expansive openings. Nesting sites are usually on horizontal ledges or caves in vertical, isolated cliff faces. Artificial structures in cities (e.g., bridges and buildings of at least 10 stories, preferably 30 stories) are also attractive sites for nesting pairs; strategically placed nest boxes enhance nesting success. Occasionally, pairs use sand dunes, abandoned large-tree stick nests, and large tree cavities. In Michigan, the original nesting population depended upon sandstone or granite cliffs overlooking the Great Lakes shoreline.

Prey availability is crucial to nesting success, and habitats including undisturbed lake shorelines should be protected (Ratcliffe 1980; Hunter et al. 1988). Although foraging flights may extend more than 8 miles (12.9 km) from the nest site (Hunter et al. 1988), the majority of excursions are less than 2 miles (3.2 km) (Bird and Aubry 1987; Enderson and Kirven 1983). Along the Great Lakes, concentrations of waterfowl and colonies of gulls (*Larus* spp.) and terns (*Sterna* spp.) are important for the successful production of young (Ludwig 1982). In urban settings, rock doves (*Columbia livia*) and European starlings (*Sturnus vulgaris*) are preferred prey, as they are most accessible. The large, strong talons enable the peregrine to capture relatively large prey species.

Life History and Ecology: Peregrine falcons migrate north in early spring following their prey (e.g., waterfowl). Males typically arrive before females, although females may initially establish a territory. Cooperative hunting and quiet perching by a pair of birds are early indications of a successful pairing. The courtship flight is a spectacular, high-diving aerial display.

Like other falcons, the peregrine does not construct a nest but utilizes sites with "prebuilt" eyries, such as cliff and building ledges. Adult nesting pairs are faithful to old sites; a behavior that can be used to estimate population stability. Prior to its extirpation, at least 14 eastern North American eyries were occupied by peregrine pairs for more than 50 years (Hickey 1942). Release programs in the upper Midwest indicate that most human-raised individuals return to within 25 miles (40.2 km) of the original site; however, dispersal distances of up to 350 miles (56.3 km) are known for territorial birds (Tordoff and Redig 1988).

The three to four eggs are laid between late March and late May. Incubation and brooding is mainly performed by the female, lasting approximately one month. The tiercel (male) hunts and vigorously defends the nesting territory. If the first clutch is destroyed, a second clutch is usually laid within a few weeks. Generally one to three falcons fledge five to six weeks after hatching. Juveniles are dependent for another month. Stable populations produce two to three young per successful pair (Hickey and Anderson in Hickey 1969). Young peregrines have high survival in the nest, but, after fledging, the mortality rate increases dramatically to three times that of adults: 50% to 75% of the young probably do not survive the first year (Enderson in Hickey 1969). Peregrines typically begin to breed at 2 to 3 years of age and may live 18 to 20 years. One-year-olds, particularly females, are known to establish territories and nest (Tordoff and Redig 1988).

The peregrine falcon hunts in flight, stooping and striking its prey in midair. Its major prey items are birds, especially waterfowl, gulls, shorebirds, and high-flying passerines. Species that exhibit contrasting color patterns (e.g., light-colored wing patches) often are selected. Several studies indicate a strong preference for northern flickers (*Colaptes auratus*) and mourning doves (*Zenaida macroura*) during the breeding season (Bent 1938; Cade et al. 1968; Cade 1982; Enderson et al. 1982; Beaver in Erickson et al. 1988) and during migration (Errington 1933; Hunt et al. 1975). Peregrine dives have been estimated nearing 200 mph (322 kmph).

Conservation/Management: The unequivocal factor for the peregrine falcon's catastrophic decline was the proliferation of organochlorine pesticides in the environment (Cade et al. 1968; Enderson and Berger 1968; Hickey and Anderson 1968; Peakall and Kiff 1979; Ratcliffe 1980). Because high levels of these organic residues concentrate in prey, reproductive failures due to eggshell thinning have decimated peregrine populations throughout their worldwide range. According to Snow (1972), the peregrine retains one of the highest levels of DDE residues of all vertebrates.

North America's peregrines remain susceptible to harmful pesticides in the food chain because they and their neotropical prey share similar wintering grounds in Central and South America (Peakall 1976; Henny et al. 1982)—a region that continues the use of DDT and other environmentally unsound organochlorine pesticides.

Breeding depression due to chemical pressures was so influential that the peregrine was extirpated east of the Mississippi River. It was not until 1980, after massive research and stocking efforts, that peregrines were once again nesting in the eastern United States. Since these populations are now self-sustaining, the southern Appalachians and Great Lakes region are being emphasized (USF&WS 1987). This success, both in urban and wild environments, is reflected by an average annual breeding population increase of nearly 30% since 1984. In particular, urban areas contained 30 to 32 territorial pairs in 1988 across 22 cities in North America, producing more than two young per successful nest.

While some bioconcentrated toxin levels are falling in North America (particularly DDE, a metabolite of DDT) (Peakall 1976), other chemicals, such as PCBs, are causing concern for the eventual recovery of the peregrine. Other current major limiting factors are locating appropriate hacking sites and identifying and protecting suitable natural nesting sites. Although marginal nesting habitats may be enhanced by artificially enlarged ledges (Boyce et al. 1982) and artificial towers (Barclay 1980), human disturbance levels and great horned owl (*Bubo virginianus*) predation can severely prohibit peregrine nesting success and hacking projects (Sherrod et al. 1981; Ausloos and Lien 1988; Cade et al. 1989).

Recent studies have examined known limiting factors for nesting peregrines in Michigan in order to evaluate potential release sites (Christopher 1980; Ludwig 1982; LaValley 1987). These included the physical structure of predicted nesting sites, level of human disturbance, density of great horned owls, prey availability, and level of bioaccumulated toxins. From these criteria, 34 sites were evaluated and several were chosen as having high potential. One of the most favorable natural sites identified was Feldtman Ridge in Isle Royale National Park. Subsequent releases beginning in 1988 have been successful at this and other Isle Royale sites (Fettig 1989).

LaValley's (1987) study identified six mainland regions with high potential for successfully releasing young peregrines: Trap Hills in Ontonagon County, Bear Bluff in Keweenaw County, Ives Mountain in Marquette County, Grand Portal and Grand Island in Alger County, and Burnt Bluff in Delta County. Three of these areas now are used as successful release sites in Ontonagon (since 1988) and Alger (since 1989) counties.

LITERATURE CITED

Ausloos, B. J., and R. Lien. 1988. The 1988 peregrine falcon release at Madison, Wisconsin. *Passenger Pigeon* 50:305–9.

Barclay, J. H. 1980. *Release of captive-produced peregrine falcons in the eastern United States, 1975–1979.* M.S. thesis, Mich. Tech. Univ., Houghton.

Barclay, J. H., and T. J. Cade. 1983. Restoration of the peregrine falcon in the eastern United States. Pp 3–40 *in* S. A. Temple (ed.), *Bird Conservation.* Univ. Wisc. Press, Madison.

Bent, A. C. 1938. Life histories of North American birds of prey (part 2). *Bull. U.S. Nat. Mus.*, No. 170.

Bird, D. M., and Y. Aubry. 1987. Reproductive and hunting behavior in falcons, *Falco peregrinus*, in southern Quebec. *Can. Field-Nat.* 96:167–71.

Bollengier, R. M., J. Baird, L. P. Brown, T. J. Cade, M. G. Edwards, D. C. Hagar, B. Halla, and E. McCaffrey. 1979. Eastern peregrine falcon recovery plan. *U.S. Fish Wildl. Serv.*, Washington, D.C.

Boyce, D. A., C. M. White, R. E. Escano, and W. E. Lehman. 1982. Enhancement of cliffs for nesting peregrine falcons. *Wildl. Soc. Bull.* 10:380–81.

Cade, T. J. 1982. *Falcons of the world.* Cornell Univ. Press, Ithaca, NY.

Cade, T. J., P. T. Redig, and H. B. Tordoff. 1989. Peregrine falcon restoration: Expectation vs. reality. *Loon* 61:160–62.

Cade, T. J., C. M. White, and J. R. Hough. 1968. Peregrines and pesticides in Alaska. *Condor* 70:170–78.

Christopher, K. L. 1980. *A survey of peregrine falcon habitat in Upper Michigan with emphasis on reintroduction potential.* M.S. thesis, Mich. Tech. Univ., Houghton.

Enderson, J. H. 1965. A breeding and migration survey of the peregrine falcon. *Wilson Bull.* 77:327–39.

Enderson, J. H., and D. D. Berger. 1968. Chlorinated hydrocarbon residues in peregrines and their prey species from northern Canada. *Condor* 70:149–53.

Enderson, J. H., and M. N. Kirven. 1983. Flights of nesting peregrine falcons recorded by telemetry. *J. Raptor Res.* 17:33–37.

Enderson, J. H., G. R. Craig, W. A. Burnham, and D. D. Berger. 1982. Eggshell thinning and organochlorine residues in Rocky Mountain peregrines, *Falco peregrinus*, and their prey. *Can. Field-Nat.* 96:255–64.

Erickson, G., R. Fyfe, R. Bromley, G. L. Holroyd, D. Mossop, B. Munro, R. Nero, C. Shank, and T. Wiens. 1988. *Anatum* peregrine falcon recovery plan. Western Raptor Tech. Committee, *Can. Wildl. Serv.*

Errington, P. S. 1933. Food habits of southern Wisconsin raptors. *Condor* 35:19–29.

Evans, D. L. 1982. Status reports on twelve raptors. *U.S. Fish and Wildl. Serv.*, Washington, D.C.

Fettig, S. 1989. Raptor inventory: 1989 update and analysis. *Natl. Park Serv. Res. Mgmt. Rept.* 89-2.

Henny, C. J., F. P. Ward, K. E. Riddle, and R. M. Prouty. 1982. Migratory peregrine falcons *Falco peregrinus* accumulate pesticides in Latin America during winter. *Can. Field-Nat.* 96:333–38.

Hickey, J. J. 1942. Eastern populations of the duck hawk. *Auk* 59:176–204.

Hickey, J. J. (ed.). 1969. *Peregrine falcon populations: Their biology and decline.* Univ. Wisc. Press, Madison.

Hickey, J. J., and D. W. Anderson. 1968. Chlorinated hydrocarbons and eggshell changes in raptorial and fish-eating birds. *Science* 162:271–83.

Hunt, G. W., R. R. Rogers, and D. J. Slowe. 1975. Migratory and foraging behaviour of peregrine falcons on the Texas Coast. *Can. Field-Nat.* 89:111–23.

Hunter, R. E., J. A. Crawford, and R. E. Ambrose. 1988. Prey selection by peregrine falcons during the nestling stage. *J. Wildl. Mgmt.* 52:730–36.

Issacs, F. B. 1976. *Historical survey of peregrine falcon eyries in National Park Service lands bordering Lake Superior.* M.S. thesis, Mich. Tech. Univ., Houghton.

LaValley, S. 1987. Site evaluation for reintroducing peregrine falcons in Michigan's Upper Peninsula. *Mich. Dept. Nat. Resour.*, Unpubl. Rept.

Ludwig, J. P. 1982. Report on spring-early summer, 1982 survey for nesting and individual peregrine falcons in Michigan at historical sites and areas believed to be acceptable peregrine habitat. *Ecol. Res. Serv.*, Iron River, MI, Unpubl. Rept.

Payne, R. P. 1983. A distributional checklist of the birds of Michigan. *Univ. Mich. Mus. Zool. Misc. Publ.* No. 164.

Peakall, D. B. 1976. The peregrine falcon (*Falco peregrinus*) and pesticides. *Can. Field-Nat.* 90:301–7.

Peakall, D. B., and L. B. Kiff. 1979. Eggshell thinning and DDE residue levels among peregrine falcons (*Falco peregrinus*): A global perspective. *Ibis* 121:200–204.

Ratcliffe, D. A. 1980. *The peregrine falcon*. Buteo Books, Vermillion, SD.

Redig, P. T., and H. B. Tordoff. 1990. Midwest peregrine falcon restoration, 1989 report. *Minn. Dept. Nat. Resour.*, Unpubl. Rept.

Sherrod, S. K., W. R. Heinrich, W. A. Burnham, J. H. Barclay, and T. J. Cade. 1981. Hacking: A method for releasing peregrine falcons and other birds of prey. *The Peregrine Fund*, Ithaca, NY, Unpubl. Rept.

Snow, C. 1972. American peregrine falcon, *Falco peregrinus anatum* and Arctic peregrine falcon, *Falco peregrinus tundrius*. Habitat management series for endangered species. *U.S. Dept. Int., Bur. Land Mgmt. Rept.* No. 1.

Tordoff, H. B., and P. T. Redig. 1988. Dispersal, nest site selection, and age of first breeding in peregrine falcons released in the Upper Midwest, 1982–1988. *Loon* 60:148–51.

United States Fish and Wildlife Service. 1987. Revised peregrine falcon, eastern population recovery plan. *U.S. Fish Wildl. Serv.*, Newton Corner, MA.

White, C. M. 1968. Diagnosis and relationships of North American tundra-inhabiting peregrine falcons. *Auk* 85:179–91.

Yellow Rail

· *Coturnicops noveboracensis* (Gmelin)

· **State Threatened**

Map 15. Breeding evidence of yellow rails. ▲ = before 1983 (Zimmerman and Van Tyne 1959); ● = 1983 to 1992 (Brewer et al. 1991).

Status Rationale: Recent surveys indicate Michigan's yellow rail is restricted to scattered populations in the eastern Upper Peninsula. It generally occupies only one habitat type, open sedge meadows, that is naturally transitory and requires active management. Although current sites are relatively protected, potential sites for populations displaced by habitat succession and changes are declining.

Field Identification: The diminutive yellow rail is golden buff to yellow with a checkered pattern of dark and light patches on its back. The dark back feathers are edged with frosted tips. Its length and wingspan range from 6 to 7^1/$_2$ inches (15 to 19 cm) and 10 to 13 inches (25 to 33 cm), respectively. Both sexes have similar coloration; males are larger and have a yellow bill, whereas the female's bill is olive-green. The precocial chicks are downy black, changing into plumage similar to adults except for duskier markings and fine barring on the upper breast. Most rails are larger; the immature sora (*Porzana carolina*) is nearly twice as large as the yellow rail and has a solid brown back, white undertail coverts, and lacks the white wing patch.

This rail is rarely flushed unless provoked at night. In flight, the broad white band on the secondaries is diagnostic, along with its short wings, stocky body, and dangling legs. The yellow rail is most commonly found by its rhythmic, ticking calls, often given in a long series that resemble the sound of a typewriter at a distance. A close approach will reveal a distinct series of four or five alternating notes, such as *tic-tic, tic-tic-tic*. Calling is most pronounced at night, although it can be heard intermittently during the day.

Total Range: The total range of this species is unknown. It is primarily found in central and southern Canada, from northwestern Alberta and the southern Mackenzie District east to the Maritime provinces. In the United States, regular breeding populations are currently known from the northern Great Lakes region, and the extreme Northeast. Disjunct, more southern populations may be gone from Illinois and Ohio, but are still reported in scattered locales in the West and Mexico.

Wintering sites are primarily associated with dry to moist grasslands and marshes along the ocean coast, from North Carolina west through Texas into Mexico. It also winters in central California.

State History and Distribution: Historically, the yellow rail has had a widely scattered and poorly known distribution. Breeding records are known for the eastern Upper Peninsula since the 1930s. Today, Upper Peninsula summering populations are known in Seney National Wildlife Refuge (NWR) and its surrounding environs in Schoolcraft County, near Shingleton in Alger County, and Sleeper Lake and associated sedge meadows in Luce County. These eastern Upper Peninsula populations are isolated; the closest regularly occurring breeding populations are in northwestern Minnesota (Janssen 1987) and in the sedge associated with Hudson Bay in Ontario (Prescott in Cadman et al. 1987).

The largest breeding population in Michigan occurs within the Great Manistique Swamp, much of which is protected by the Seney NWR and the Manistique River State Forest. In 1982, a total of 52 calling males was counted (Bart et al. 1984), primarily south of Marsh Creek Pool. This count represented the majority of the local population, although isolated calling males frequently occur at nearby sites in and outside Seney NWR.

Calling males regularly occupy wetlands associated with Houghton Lake in Roscommon County (Brewer et al. 1991), and recent observations in the northern Lower Peninsula indicate other possible summering populations.

Habitat: Emergent wetlands are the predominant breeding habitat for the yellow rail. In the Great Lakes region, it is nearly exclusively associated with expansive, wet meadows dominated by mat-forming sedge (*Carex* spp.) (Devitt 1939; Walkinshaw 1939; Stahlheim 1974; Stenzel 1983). In Seney NWR, optimal breeding habitat is characterized by extensive homogeneous areas of sedge (*Carex lasiocarpa*), interspersed with islands of scrub-shrub wetlands and sand ridges with young to mature woody growth. In large openings, winds may align dead and dying sedge, forming a permanent microhabitat canopy. Areas with sedge densities averaging 700 to 800 stems per square meter are preferred, but, the rail's tolerance to shrubs is unknown (Stenzel 1983). Territorial males in optimal sedge habitat will permit some willow (*Salix* spp.) growth. Cattail (*Typha* spp.) stands are occasionally used by calling males.

Standing water is often present in the beginning of the breeding season, but slowly retreats through the summer. At Seney NWR, calling males were never found in habitat with water depths greater than 18 inches (46 cm). Nests are placed over water depths of 1 to 4 inches (2 to 10 cm)(Elliot and Morrison 1979; Stenzel 1983), or moist soil (Devitt 1939; Walkinshaw 1939; Terrill 1943; Stenzel 1983). Nests may be found by following the volelike runways made by foraging females (Stenzel 1983).

Ecology and Life History: Although this secretive rail is rarely discovered during migration, males do call while moving north (Easterla 1962; Reynard 1974; Stahlheim 1974), and possibly move in small flocks (Stenzel 1983). At Seney NWR, males may arrive in late April, but usually first appear between early to mid-May, depending on weather and local water depths. Females do not call.

Males establish territories (average of 19.0 acres [7.8 ha]) within one week of their arrival, giving their clicking calls nightly during the preincubation period, which lasts for about one month (Stenzel 1983, Bookhout and Stenzel 1987). Male territories may encompass multiple female territories. Areas used by females average 3.0 acres (1.2 ha) during preincubation, decreasing to 0.7 acre (0.3 ha) during incubation (Bookhout and Stenzel 1987). Although calling is most incessant and pronounced at night, the yellow rail is not actively nocturnal. At night, it is sedentary (Bookhout and Stenzel 1987), exhibiting much of its feeding behavior during the day (Stahlheim 1974).

In Seney NWR, the first nests of six to ten eggs are laid between the last few days of May and in early June (Stenzel 1983). Nests are placed over shallow water under herbaceous vegetation that forms a canopy (Peck and James 1983; Stenzel 1983). Unlike other rails, only the female incubates, with the incubation period ranging from 13 days (Lane 1962) to 18 days (Stahlheim 1974; Elliot and Morrison 1979). Renesting may occur if initial nests are unsuccessful. Hatching is nearly synchronous, and within 1 day the young vacate the nest site. They feed independently at 11 days on such aquatic invertebrates as snails (Stenzel 1983). Young yellow rails attain juvenile plumage within three weeks and fledge within five weeks (Stahlheim 1974). Calling generally ends in mid-August. Autumn migration begins in late August and continues through early October in southern Michigan.

Conservation/Management: Few limiting factors are readily apparent for the yellow rail in the Great Lakes region. Breeding habitat destruction and alteration formerly played the primary role in its disappearance in Munuscong Bay and probably for local summering individuals in southeastern Michigan. Today, the few known sites are relatively protected by private, state, and federal agencies. Management of these ephemeral wetland habitats is crucial, especially in habitats dominated by *Carex lasiocarpa* stands. Vegetation succession, changes in hydrology, and human disturbance are three factors to investigate while forming a long-term management plan.

Impeding succession in known occupied sites is one method of maintaining suitable habitat. Controlled burns are the optimal management tool, serving to rejuvenate sedge growth, limit woody growth, and impede the establishment of boreal flora such as sphagnum moss. Areas should be periodically burned to avoid the buildup of dead vegetation. Excessive dead plant material produces hot fires that will destroy the roost structure of the sedge mat. Breeding habitat also may be created by burning tamarack-spruce-sedge bogs to reestablish monotypic stands of sedge (e.g., within parts of the Lake Superior State Forest). Understanding water level fluctuations are crucial for supporting long-term yellow rail populations. Marginal changes in seasonal and annual water depths are natural, but by ditching or altering water flows, habitats become drier; diking and flooding areas have similar negative effects on preferred microhabitat structure.

Special surveys are needed to document the distribution and abundance of the yellow rail in Michigan. Although highly secretive and nocturnal, this rail is relatively easy to locate with proper census methods. First, aerial infrared photographs can be used to identify suitable habitat (sedge meadows are a greenish color) (Bart et al. 1984). Rail presence is then determined by visiting open sedge meadows after sunset between late May and mid-June. At this time, calling intensity is at its peak, generally occurring nightly. Once found, densities can be estimated with a strip-transect approach and individuals can be easily captured and monitored (Stenzel 1983; Bart et al. 1984). Capture methodology has been refined by Bart et al. (1984) and includes imitating the clicking calls of the male, luring it to a central spot, and netting it.

Direct human disturbance and mortality probably are not limiting. The U.S. hunting season on yellow rails has been closed since 1968 (Eddleman et al. 1988). However, continual visits by large groups may damage the rail's microhabitat and disturb breeding success. This rail's dependency on transitory habitats suggests that it can colonize new sites as they become available. Therefore, protecting occupied as well as unoccupied, potentially suitable habitat throughout the state is an important issue for long-term survival of the yellow rail in Michigan.

LITERATURE CITED

Bart, J., R. A. Stehn, J. A. Herrick, N. A. Heaslip, T. A. Bookhout, and J. R. Stenzel. 1984. Survey methods for breeding yellow rails. *J. Wildl. Mgmt.* 48:1382–86.

Bookhout, T. A., and J. R. Stenzel. 1987. Habitat and movements of breeding yellow rails. *Wilson Bull.* 99:441–47.

Brewer, R., G. McPeek, and R. J. Adams, Jr. 1991. *The atlas of breeding birds in Michigan.* Mich. State Univ. Press, E. Lansing.

Cadman, M. D., P. F. J. Eagles, and F. M. Helleiner. 1987. *Atlas of the breeding birds of Ontario.* Univ. Waterloo Press. Waterloo, Ont.

Devitt, D. E. 1939. The yellow rail breeding in Ontario. *Auk* 56:238–43.

Easterla, D. 1962. Some foods of the yellow rail in Missouri. *Wilson Bull.* 74:94–95.

Eddleman, W. R., F. L. Knopf, B. Meanley, R. A. Reid, and R. Zembal. 1988. Conservation of North American rallids. *Wilson Bull.* 100:458–75.

Elliot, R. D., and R. I. G. Morrison. 1979. The incubation period of the yellow rail. *Auk* 96:422–23.

Janssen, R. B. 1987. *Birds in Minnesota.* Univ. Minn. Press, Minneapolis.

Lane, J. 1962. Nesting of the yellow rail in southwestern Manitoba. *Can. Field-Nat.* 76:189–91.

Peck, G. K., and R. D. James. 1983. *Breeding birds of Ontario: nidiology and distribution (vol. 1).* Life Sci. Misc. Publ., Royal Ontario Mus., Toronto.

Reynard, G. B. 1974. Some vocalizations of the black, yellow and Virginia rails. *Auk* 91:747–56.

Stahlheim, P. S. 1974. *Behavior and ecology of the yellow rail (Coturnicops noveboracensis).* M.S. thesis, Univ. Minn., Minneapolis.

Stenzel, J. R. 1983. *Ecology of breeding yellow rails at Seney National Wildlife Refuge.* M.S. thesis, Ohio State Univ., Columbus.

Terrill, L. M. 1943. Nesting habits of the yellow rail in Gaspe County, Quebec. *Auk* 60:171–80.

Walkinshaw, L. H. 1939. The yellow rail in Michigan. *Auk* 56:227–37.

Zimmerman, D. A., and J. Van Tyne. 1959. A distributional checklist of the birds of Michigan. Univ. Mich., Mus. Zool. Occ. Papers. No. 608.

King Rail

· **Rallus elegans** Audubon
· **State Endangered**

Map 16. Breeding evidence of king rails. ▲ = before 1983 (Barrows 1912; Michigan DNR); ● = 1983 to 1992 (Brewer et al. 1991).

Status Rationale: Although the king rail remains common in the southern part of its U.S. range, it has experienced a long-term population decline in Michigan. Habitat loss is one major limiting factor; however, recent surveys indicate additional problems because apparently suitable wetland habitat is unoccupied. An estimated 20 to 35 nesting pairs occur in Michigan.

Field Identification: The king rail is a rust-colored, large and slender marsh bird with a long bill. Its laterally compressed body is well suited for moving through thick marsh vegetation. Although seldom flushed, its flight is short. This species is the largest North American rail; males average slightly larger than females, but neither sex exhibits obvious differences in plumage. Average length and wingspan of adult birds are 15 to 19 inches (38.1 to 48.3 cm) and 21 to 25 inches (53.3 to 63.5 cm), respectively. Because of their secretive behavior, rails are more often heard than seen.

King rails respond to playing recordings of their loud and diagnostic calls, a reliable method to detect their presence (Marion et al. 1981). They are most responsive to playing recordings within a 300-foot (91.4-m) radius, during evening hours, and before the young have hatched (Rabe 1986). The call most commonly used is gruntlike, and may be described as *jupe-jupe-jupe*. The mating call is basically a series of staccato notes that may end in a slur (e.g., *kik-kik-kik kurrrr*).

The similar Virginia rail (*Rallus limicola*) occurs in the same habitats, but is a gray-cheeked, smaller version of its larger relative and lacks the king rail's extensive barring on the sides and undertail coverts.

Total Range and Taxonomic Status: In general, the king rail breeds from the Great Plains through southern Ontario and New England's Atlantic coast to the Gulf coast and Cuba. It is absent throughout the Appalachian Mountains and is only a local breeder north of the marsh-rice belts of the southern states and tidal marshes of the Atlantic coast. An isolated population was reported in central Mexico (Warner and Dickerman 1959), but this population's current status is not known.

The clapper rail (*Rallus longirostrus*), which inhabits saltwater marshes along ocean coasts, and the king rail are considered by some authorities to be the same species (Ripley 1977). Hybridizing populations of the two species exist in brackish marshes in Delaware and possibly in other areas along the Atlantic coast (Meanley and Wetherbee 1962). The king rail is a migratory species throughout most of its northern range (Bateman 1980). Autumn migrations are usually heaviest in late August to mid-September. It winters primarily along the tidal marshes and rice fields of the Gulf and South Atlantic coasts, areas where it is most abundant in summer.

State History and Distribution: The primary breeding range of this elusive rail formerly extended from Saginaw Bay west to Muskegon County south; although there are several summer and nesting records in the northern Lower Peninsula and wanderers have been reported as far north as the Upper Peninsula (Ilnicky 1969). In the early 1900s, king rails were frequently reported during the breeding season along Saginaw Bay and in the southern four-county tier of the Lower Peninsula. Since the mid-1900s, king rail populations have not recovered in Michigan and are currently confined to large marshes along Lake Erie, Lake St. Clair, and Saginaw Bay.

In 1984, surveys documented king rails in the St. Clair delta, St. Clair County, as well as the Erie and Pt. Mouilee State Game areas (Evers 1984). King rails were also found in Sterling State Park, Monroe County, that year; although disturbance of the shrinking cattail marsh already may have made this area unsuitable for nesting king rail pairs. This survey correlates with Meanley's (1969) research, which determined that the western Lake Erie marshes and the St. Clair delta contained the highest breeding population in the North Central states. A subsequent survey in 1986 also found these two regions to comprise the major breeding populations in the state, particularly on Harsen's and Dickinson islands within the St. Clair delta (Rabe 1986). Adjacent Walpole Island in Canada also harbors king rails (Herdendorf and Raphael 1986; Cadman et al. 1987).

Although Saginaw Bay has historical records (Mershon 1923; Kenaga 1983), recent observations have been limited in this area. However, several pairs were found in 1986 at Nayanquing Point Wildlife Area, Bay County (Rabe 1986). Since then, king rails also have been found along the Saginaw River in the Crow Island State Game Area; at least four were found in 1987 (Brewer et al. 1991). This is a region where no birds were located during the intensive 1984 (Evers 1984) and 1986 (Rabe 1986) surveys. These recent records may indicate shifting populations or even recovery in some areas. Scattered pairs remain inland throughout the lower third of Michigan but are widely scattered. Recent reports of this secretive species have been made in Ingham County, and, in 1988, breeding was confirmed (Brewer et al. 1991). According to Barrows (1912) they were formerly "fairly abundant" in that area.

Habitat: The king rail prefers permanent freshwater marshes and, occasionally, open young swamps. Studies in Michigan found king rails in monotypic cattail (*Typha* spp.) stands, cattail-sedge-shrub mixtures, and tussock-forming, sedge-grass wetlands (Evers 1984, Rabe 1986). Although expansive stands of marshy herbaceous vegetation are typically considered preferred habitats, Rabe (1986) found king rails occupying marsh habitats interspersed with willow (*Salix* spp.) and dogwood (*Cornus* spp.). Meanley (1969) claims this species has the widest variety of habitats of any rail. In the Great Lakes region, the king rail requires large, undisturbed marshes. Nest sites are usually in shallow water depths of less than 10 inches (25.4 cm) (Meanley 1969). The nest is placed in a clump of grass above water level and generally has a canopy and an entrance ramp. In uniform stands of vegetation, this canopy (which is formed by bending over the nearby plant stalks) may be very noticeable.

Ecology and Life History: King rails arrive at Michigan marshes in late April, and pairs often return to the same marsh in consecutive years. Territories are aggressively defended against rival king rails as well as soras (*Porzana carolina*) and virginia rails (Meanley 1969). During courtship, the male attracts a female by strutting with its tail held vertically, exposing the white undertail coverts (Meanley 1957). The nest site is generally chosen by the male, which also assumes most of the nest building responsibilities. Egg laying may begin in early May, with an average clutch of 10 to 11 eggs. Both parents incubate the eggs for 21 to 23 days. After hatching, the black, precocial young quickly vacate the nest. Drying natural swales are crucial foraging sites during this time (Eddleman et al. 1988). Meanley (1969) estimated that the survival rate of the young until two weeks of age was approximately 50%. By mid-August, most have fledged.

King rails typically feed among dense vegetation and in shallow water approximately 2 to 3 inches (5.1 to 7.6 cm) in depth. Occasionally, individuals forage in open water and cultivated fields adjacent to suitable wetland habitats. The diet of this species consists of small crustaceans and aquatic insects. Fish, frogs, terrestrial insects (e.g., grasshoppers), and aquatic plant seeds are also eaten when available.

Conservation/Management: The king rail's Midwest breeding population is declining due to wetland destruction and degradation as well as high pesticide residues. The king rail's decline in Michigan is not fully understood, but these factors further stress a population already constrained by its peripheral status. Migratory northern king rail populations may be geographically isolated from the more abundant and sedentary southern populations. Therefore, even though self-sustaining rail populations are present in the southern United States, northern populations in Michigan cannot rely on recruitment from southern populations.

Lead poisoning may also be a significant factor in declining rail populations. Soras have been found with ingested lead in their gizzards, indicating lead as a major contaminant among rails (Artman and Martin 1975; Stendell et al. 1980). Lead-poisoned birds are readily identified by their reluctance or inability to fly, drooped wings, and green-colored feces (Friend 1987). Meanley (1969) considered predation, road kills, and hitting barbed wire fences as crucial mortality factors. Traps set for furbearers may also incidentally take rails (Meanley 1969; Stocek and Cartwright 1985), and there are annual harvests of king rails on their wintering grounds (Eddleman et al. 1988).

Since the availability of suitable habitat is a major limiting factor, protection of occupied habitats is needed as well as artificial manipulation to enhance areas for migrating and nesting rails. Suitable rail habitat can be created by flooding impoundments in spring to permit shallow water depths (less than 10 inches [25.4 cm]), followed by drawdowns in late summer to maintain vegetation density and coverage. Water depth and vegetation structure are probably more important than plant species composition. Brood foraging sites with open mud flats adjacent to dense vegetation are also crucial.

Marsh habitat is declining, but many historically occupied areas are still available and lack summering king rails. This suggests the decline is caused by more than the loss of nesting habitat, although the effects of wetland degradation and fragmentation have not been addressed. More research needs to be completed (e.g., evaluating the relationship between chemical use and the levels ingested from aquatic insects). Floating nesting platforms are a potential management tool and have increased breeding success with clapper rails in areas with limited nesting sites (Wiley and Zembal 1989).

LITERATURE CITED

Artman, J. W., and E. W. Martin. 1975. Incidence of ingested lead shot in sora rails. *J. Wildl. Mgmt.* 39: 514–19.
Barrows, W. B. 1912. *Michigan bird life.* Mich. Agric. Coll. Spec. Bull., E. Lansing.
Bateman, H. A. 1980. Rails and gallinules. Pp. 93–104 in G. C. Sanderson (ed.), *Management of migratory shore and upland game birds in North America.* Int. Assoc. Fish Wildl. Agen., Washington, D.C.
Brewer, R., G. A. McPeek, and R. J. Adams, Jr. 1991. *An atlas of breeding birds in Michigan.* Mich. State Univ. Press, E. Lansing.
Cadman, M. D., P. F. Eagles, and F. M. Helleiner. 1987. *Atlas of the breeding birds of Ontario.* Univ. Waterloo Press, Waterloo, Ont.
Eddleman, W. R., F. L. Knopf, B. Meanley, F. A. Reid, and R. Zembal. 1988. Conservation of North American rallids. *Wilson Bull.* 100:458–75.
Evers, D. C. 1984. King rail survey—1984. *Mich. Nat. Feat. Invent.,* Unpubl. Rept.
Friend, M. (ed.). 1987. Field guide to wildlife diseases. *U.S. Fish Wildl. Serv., Resour. Publ.* 167.
Herdendorf, C. E., and C. N. Raphael. 1986. The ecology of Lake St. Clair wetlands: A community profile. *U.S. Fish Wildl. Serv., Biol. Rept.* 85(7.7).
Ilnicky, N. J. 1969. Sighting of a king rail in the Upper Peninsula. *Jack-Pine Warbler* 47:104
Kenaga, E. E. 1983. *Bird, birders, and birding in the Saginaw Bay Area.* Chippewa Nature Center, Midland, MI.

Marion, W. R., T. E. O'Meara, and D. S. Maehr. 1981. Use of playback recordings in sampling elusive or secretive birds. *Stud. Avian Biol.* 6:81–85.

Meanley, B. 1957. Notes on the courtship behavior of the king rail. *Auk* 74:433–40.

Meanley, B. 1969. Natural history of the king rail. *U.S. Bur. Sport Fish. Wildl., N. Am. Fauna* No. 67.

Meanley, B., and D. K. Wetherbee. 1962. Ecological notes on mixed populations of king rails and clapper rails in Delaware Bay marshes. *Auk* 79:453–57.

Mershon, W. B. 1923. *Recollections of my fifty years hunting and fishing.* Stratford Co., Boston.

Rabe, M. L. 1986. King rail census—1986: Population status and habitat utilization. *Mich. Dept. Nat. Resour.*, Unpubl. Rept.

Ripley, S. D. 1977. *Rails of the world: A monograph of the family Rallidae.* David R. Godine, Boston.

Stendell, R. C., J. W. Artman, and E. Martin. 1980. Lead residues in sora rails from Maryland. *J. Wildl. Mgmt.* 44:525–27.

Stocek, R. F., and D. J. Cartwright. 1985. Birds as nontarget catches in the New Brunswick furbearer harvest. *Wildl. Soc. Bull.* 13:314–17.

Warner, D. W., and R. W. Dickerman. 1959. The status of *Rallus eleganstenuirostris* in Mexico. *Condor* 61:49–51.

Wiley, J. W., and R. Zembal. 1989. Concern grows for light-footed clapper rail. *Endangered Species Tech. Bull.* 14:6–7.

Wood, N. A. 1951. The birds of Michigan. *Univ. Mich. Mus. Zool. Misc. Publ.* No. 75.

Piping Plover

- *Charadrius melodus* Ord
- **Federally and State Endangered**

Map 17. Confirmed piping plover nesting pairs. ▲ = historically locally distributed along the Great Lakes shoreline (Lambert and Ratcliff 1981); ● = 1983 to 1992 (Michigan DNR).

Status Rationale: The piping plover is declining throughout its range, particularly in the Great Lakes region. Disturbances associated with human activity are the primary cause for this decline in Michigan. Only 16 nesting pairs are known along the Great Lakes shoreline, each confined to limited areas in northern Michigan.

Field Identification: This diminutive plover is characterized by its pale, sand-colored back and head, white rump, orange legs, and a single black band on the breast and on the forehead. This dark breast band may completely encircle the neck (particularly in the Great Lakes population) or form an incomplete or very thin collar (typical in other populations). Adults average 7 inches (18 cm) long with a 15-inch (38-cm) wingspan. As its name implies, this plover's call is a melodic piping sound (a clear *peep* or *peep-lo*), frequently made during courtship. Its cryptic coloration still makes it difficult to locate, however. There is minor sexual dimorphism; dark markings tend to be less pronounced in females. Juveniles and wintering adults lack the characteristic breast and head bands, and the orange-colored, black-tipped bill of the adults turns solid black.

Few other Michigan shorebirds are confused with the piping plover. Shorebirds, such as the spotted sandpiper (*Actitis macularia*) and killdeer (*Charadrius vociferus*) are frequent associates of the piping plover in summer (Brown 1987a). Both are easily distinguishable, the spotted sandpiper by its spotted breast and continuous bobbing behavior, and the larger killdeer by its 2 breast bands. During

147

migration, piping plovers occasionally flock and forage with other shorebirds and are often confused with the semipalmated plover (*Charadrius semipalmatus*), which has a distinctly darker back, lacks a white rump, and has two black head bands.

Total Range: This plover is typically divided into three geographic breeding populations: (1) the Atlantic coast from Newfoundland to North Carolina; (2) the silty flats, saline wetlands, and river and lake sand flats of the northern Great Plains region, and (3) the Great Lakes shoreline. However, there is no genetic evidence supporting the division of these geographic populations (Haig and Oring 1988c). Severe population declines led to the addition of the piping plover to the federal endangered and threatened species list in the United States and Canada. The northern Great Plains and Atlantic coast populations are classified as threatened; populations in Canada (Haig 1985) and along the Great Lakes shoreline are endangered. Individuals on the wintering grounds have a threatened status.

The total number of piping plover nesting pairs is around 2,000. The Atlantic coast population has experienced precipitous declines (Master and French 1984). In the northern Great Plains, the piping plover reaches its greatest abundance and widest distribution. In 1989, the number of nesting pairs was 1,007 to 1,064 (729 in the United States and 278 to 335 in Canada). This is down considerably from a 1985 estimated population of 1,700 to 2,000 nesting pairs (Haig 1985). The Great Lakes shoreline population has experienced the most dramatic decline, and currently faces immediate threats to its continued existence. Although no historical counts are known, Russell (1983) estimated a former population of 500 to 680 nesting pairs; 125 to 130 nesting pairs were concentrated along the limited shoreline of Illinois.

In winter, most individuals remain on the Atlantic and Gulf coasts, from North Carolina to Texas. Occasionally, birds enter Mexico and the West Indies (Haig and Oring 1988a). Major Gulf coast wintering areas are along the Texas coast, such as Galveston Island and Port Aransas, on Dauphin Island in Alabama, and the Chandeleur islands in Louisiana. Breeding populations from the northern Great Plains mix with those of the Great Lakes region on the Gulf of Mexico in winter; the Atlantic coast population moves south in winter, remaining on the ocean shoreline (Haig and Oring 1988c). Winter sightings of piping plovers banded in Michigan are known from the Gulf coast near Galveston, Texas, and Naples, Florida (Haig and Oring 1988a).

State History and Distribution: Historically, the piping plover resided on the Great Lakes shoreline wherever conditions were favorable. Lambert and Ratcliff (1981) noted 18 coastal counties with records of nesting pairs and the following 7 counties with adults observed during the breeding season: Alpena, Iosco, Leelanau, Mackinac, Mason, Presque Isle, and Wayne. An example of the species' early abundance was along Lake Michigan in Illinois, where a reported 30 pairs nested on a two-mile stretch of beach (Nelson in Barrows 1912).

Human-plover conflicts began in the mid-1800s when market hunting was at its peak and all shorebirds were commonly taken. Around the turn of the century, when piping plovers numbers were extremely low, hunting ceased due to the Migratory Bird Treaty Act. By the 1930s, their populations had fully recovered and remained relatively stable until the 1950s (Cottrille 1957). At this time, the plover began its second decline, due to increased human development of coastal areas, and was eventually eliminated from Lakes Ontario, Erie, and Huron.

By 1981, the Michigan population of the piping plover comprised 17 known nesting pairs, restricted to a handful of sites along Lakes Michigan and Superior. This was nearly a 50% decrease from the 33 known nesting pairs in 1979 (Lambert and Ratcliff 1981). Between 1982 and 1988, annual surveys sponsored by the Michigan Department of Natural Resources indicated 13 to 20 nesting pairs in Michigan. Populations fluctuated greatly, sometimes as much as a third between surveys.

In 1989, the first real increase of nesting pairs was realized since the 20 nesting pairs in 1985. A total of 19 nesting pairs were confirmed. This followed a highly productive 1988 nesting season, where nearly three-quarters of the hatched young survived through the crucial three-week period. This also was the first year in which selective nests were covered with predator exclosures. Assuming this is 75% to 80% of the total adult population, nearly 50 adults were probably present in 1989.

The last Michigan strongholds of this plover are in Wilderness State Park (Emmet County) and along the Lake Superior shoreline between Whitefish Point (Chippewa County) west through Luce County to Grand Marais (Alger County). Even in these strongholds there have been major declines.

Other areas usually harboring pairs are near Sturgeon Bay in Emmet County, Cathead Bay and North Manitou Island, Leelanau County, and the Beaver Island chain (e.g., High Island), Charlevoix County. In 1989, two new sites were discovered, the first indication of range expansion in many years. One pair nested within the Hiawatha National Forest, Mackinac County, along Lake Michigan and another in Cheboygan State Park, Cheboygan County. This latter location is the first Lake Huron active nesting site in decades. In 1992, 16 nesting pairs produced 13 young (4 pairs in the Upper Peninsula, the rest in the northern Lower Peninsula).

Habitat: In the Great Lakes region, the piping plover prefers to nest and forage on sparse or nonvegetated sand-pebble beaches, averaging 100 feet (30.5 m) in width (Lambert and Ratcliff 1981; Nordstrum 1990). Vegetative cover is usually less than 5% (Niemi and Davis 1979). Associated bodies of water (e.g., beach pools and stream mouths) and interdunal wetlands enhance these areas by increasing food availability. Optimal foraging areas are especially crucial along Lake Superior, where shoreline and benthic invertebrate communities are known to be naturally sparse (Mason et al. 1985), possibly limiting plover nesting success (Nordstrum 1990). Nests are generally placed in level areas between the water's edge and the first dune (upper beach). Lambert and Ratcliff (1981) reported that, in Michigan, the average nest was 82 feet (25 m) from the water's edge, 52 feet (16 m) from the first dune, and approximately 360 feet (110 m) from the nearest tree line.

Specific winter habitat requirements have been studied on the Atlantic and Gulf coasts (Johnson 1987; Nicholls 1989). Plovers typically use ocean lagoons and sand flats near beaches and coastal inlets (Haig and Oring 1988a); unfortunately, these areas presently are being quickly altered by development (Haig 1985).

Ecology and Life History: Piping plovers arrive in Michigan in mid- to late April. Males soon establish territories and attract mates with flight displays. The male aggressively defends its territory against intruding male piping plovers and other birds. Allan (1985) found Michigan males defending large territories, averaging 25,000 square meters along Lake Superior.

Evidence suggests relatively high site faithfulness (regardless of breeding success in the previous year), and that adult males tend to return to the same beach site as the previous year, while adult females tend to choose new beach sites in the general geographic area as the previous year. Adults typically pair with new mates each year and frequently during the breeding season (Wilcox 1959; Haig and Oring 1988b). Piping plover nesting pairs will associate with common tern (*Sterna hirundo*) colonies, benefiting from the tern's aggressive behavior toward intruders (Niemi and Davis 1979; Burger 1987).

Following an average two-week courtship, a set of four eggs is typically laid. Scrapes (nests) are made primarily by the male during courtship, although females will help. First nests with complete clutches occur from late May to early June. In Michigan, the scrape is located within a substrate of sand and pebbles. Both sexes share in the 28-day incubation period (ranging from 22 to 31 days). Pairs will frequently renest if the first attempt is unsuccessful. Pairs may change mates within the breeding season when renesting (Haig and Oring 1988b), although no more than one brood is raised annually (Cairns 1982; Haig and Oring 1988b). Like the killdeer, piping plovers protect their eggs and young with a broken-wing act to distract potential predators.

Chicks from first nests hatch in late June and early July. The precocial young are protected and brooded by the adults but not fed by them; females may desert the brood and leave chick rearing to the male (Haig and Oring 1988b). Adults will at least temporarily accept chicks from other family units (Flemming 1987). Chick survival increases after the first 17 to 20 days (Allan 1986; Flemming 1987). Young generally remain within 400 to 500 feet of the nest site until fledging four weeks after hatching (Wilcox 1959).

Adults typically depart before young-of-the-year, beginning in July; by late August, Michigan piping plovers have left the breeding grounds. Most return to the same wintering areas (Johnson and Baldassarre 1988). First-year individuals rarely return to their natal breeding grounds (Haig and Oring 1988c), although they may be seen at nearby sites. Adult mortality is low and piping plovers live a relatively long time compared to other shorebirds (Wilcox 1959).

While feeding, open shoreline is preferred to vegetated beach areas. Females apparently move shorter distances, but forage more intensively than do males (Allan 1986). The piping plover frequently forages by running short distances, stopping, and then staring at the ground with its head tilted to the side looking for small insects, worms, crustaceans, and other invertebrates.

Conservation/Management: Major threats to this species in the Great Lakes region are associated with the levels of human activity in the nesting area (e.g., continual visitation and vehicles) and low food availability. For example, on remote and undisturbed beaches, reproductive success averages 1.3 to 2.1 fledged young per pair (Cairns 1982; Wiens and Cuthbert 1984; Prindiville-Gaines and Ryan 1988). At frequently disturbed beaches, reproductive success is considerably lower, generally less than 1.0 fledged young per pair. Success rates are dependent on the parent's ability to effectively keep the eggs or young protected from predators and overexposure. Areas with high human use (e.g., ORV traffic) cause a greater frequency of exposed nests and young. Additionally, chick mortality increases when behavior must shift from freely feeding to increased sitting and danger avoidance (Flemming et al. 1988).

Even though human presence during the plover's courtship, incubation, and brooding is detrimental, free-running dogs pose an even greater threat (Debinsk in Pike et al. 1987). Any disturbance interrupts incubation (which increases the chance of egg failure), disrupts family units, and inhibits juvenile survival; however, canid visitations cause a greater degree of disturbance for a longer period of time. In areas with piping plovers, dogs should always be on a leash.

Other threats include loss of nesting habitat (through beach erosion, rising lake levels, inclement weather, and succession) and predation by unnaturally high densities of corvids, gulls, and mammals (e.g., raccoon [*Procyon lotor*]). Predation is particularly imposing in many breeding populations (Prindiville-Gaines and Ryan 1988). In Michigan, the American crow (*Corvus brachyrhynchos*) is a major predator of local breeding populations (Brown 1987b), as well as ring-billed (*Larus delawarensis*) and herring (*Larus argentatus*) gulls (Pike and Wolinski 1987). Possible losses from pesticides, migration hazards, and limitations on its wintering grounds also may severely limit the viability of breeding populations. Preliminary results have shown that wintering areas are subject to water level manipulations, oil spills, and recreational activities due to increased coastline development (Haig 1985).

An important first step in this plover's protection is to reduce human activity between mid-May and early August, primarily by restricting vehicles, people, and dogs at designated beaches. By simply avoiding the upper beach zones, many human-plover conflicts could be avoided. Posting informative signs (Flemming et al. 1988) and placing psychological or physical fencing around nests in conjunction with public programs are helpful, and can be effective if augmented with daily patrols (Master and French 1984). A fence of twine and wooden posts works well for rerouting people away from nest sites. If predators appear to be a problem, wire mesh fencing is an effective method of deterrence, as long as the adults can freely pass and fencing can be checked daily. Michigan studies have shown some success with specially constructed mesh wire boxes placed over the nest.

Piping plovers have utilized artificial sites, including industrial disposal sites (Switzer 1979), vegetated areas cleared by sand excavation (Niemi and Davis 1979), disturbed inland lake beaches (Russell 1973), gravel dike roads (Haig 1985), and artificial islands (Lakela 1946). These modified habitats may be useful when determining management plans and habitat manipulation options.

The federal (Haig et al. 1988) and state recovery (Pike et al. 1987) plans address crucial conservation issues and provide specific measures for recovery of the Great Lakes population. According to the plan, several criteria must be met before reclassifying the piping plover to a threatened status, including a population increase to 150 nesting pairs (100 in Michigan).

The five objectives outlined to meet this recovery goal are (1) management of nesting areas, (2) provision for maximum protection, (3) development and implementation of a standardized ecological monitoring program during the breeding season, (4) management and protection during migration and winter, and (5) artificially increasing its population (if the species fails to respond to protective measures and conservation strategies)(Pike et al. 1987).

Ideally, all human use of piping plover breeding areas from mid-May until early August should be eliminated. Beach areas containing the few traditional breeding sites should be closed annually during this part of the year. Currently, many areas are closed, including North Manitou Island in Leelanau County, state lands, and the Lake Superior Vermillion Station in Chippewa County. If these relatively clear-cut and simple solutions are implemented, the Michigan piping plover's future can be secured.

LITERATURE CITED

Allan, T. A. 1985. Nesting biology, habitat use, and reproductive success of piping plovers on the Lake Superior shoreline in the eastern Upper Peninsula of Michigan. *Mich. Dept. Nat. Resour.*, Unpubl. Rept.

Allan, T. A. 1986. Nesting biology, behavior, and reproductive success of piping plovers on the Lake Superior shoreline in the eastern Upper Peninsula of Michigan. *Mich. Dept. Nat. Resour.*, Unpubl. Rept.

Barrows, W. B. 1912. *Michigan bird life.* Mich. Agric. Coll. Spec. Bull., E. Lansing.

Brown, S. C. 1987a. *Comparative breeding biology of the piping plover and spotted sandpiper.* M.S. thesis, Univ. Mich., Ann Arbor.

Brown, S. C. 1987b. Field tests to control crow predation on piping plover eggs. *Mich. Dept. Nat. Resour.*, Unpubl. Rept.

Burger, J. 1987. Physical and social determinants of nest-site selection in piping plover in New Jersey. *Condor* 89:811–18.

Cairns, W. E. 1982. Biology and behavior of breeding piping plovers. *Wilson Bull.* 94:531–45.

Cottrille, B. C. 1957. Summer distribution of the piping plover in Michigan. *Jack-Pine Warbler* 35:26–33.

Flemming, S. P. 1987. Natural and experimental adaption of piping plover chicks. *J. Field Ornith.* 58:270–73.

Flemming, S. P., R. D. Chiasson, P. C. Smith, P. J. Austin-Smith, and R. P. Bancroft. 1988. Piping plover status in Nova Scotia related to its reproductive and behavioral responses to human disturbance. *J. Field Ornith.* 59:321–30.

Haig, S. M. 1985. The status of the piping plover in Canada. *Natl. Mus. Can.*, Ottawa.

Haig, S. M., and L. W. Oring. 1988a. Distribution and dispersal in the piping plover. *Auk* 105:630–38.

Haig, S. M., and L. W. Oring. 1988b. Mate, site, and territory fidelity in piping plovers. *Auk* 105:268–77.

Haig, S. M., and L. W. Oring. 1988c. Genetic differentiation of piping plovers across North America. *Auk* 105:260–67.

Haig, S. M., W. Harrison, R. Lock, L. Pfannmuller, E. Pike, M. Ryan, and J. Sidle. 1988. Piping plover recovery plan for the Great Lakes and northern Great Plains. *U.S. Fish Wildl. Serv.*, Washington, D.C.

Johnson, C. M. 1987. *Aspects of the wintering ecology of the piping plover in coastal Alabama*. M.S. thesis, Auburn Univ., Auburn, AL.

Johnson, C. M., and G. A. Baldassarre. 1988. Aspects of the wintering ecology of piping plovers in coastal Alabama. *Wilson Bull.* 100:214–23.

Lakela, O. 1946. Additional observations on the nesting of piping plovers in Duluth. *Flicker* 18:11.

Lambert, A., and B. Ratcliff. 1981. Present status of the piping plover in Michigan. *Jack-Pine Warbler* 59:44–52.

Mason, J. W., M. H. Albers, and E. M. Brick. 1985. An evaluation of beach nourishment on the Lake Superior shore. *Wisc. Dept. Nat. Resour., Tech. Bull.* No. 157.

Master, L., and T. French. 1984. Notes from the piping plover and least tern inventory/protection/management workshop for northeastern states. *The Nature Conservancy*, Unpubl. Rept.

Nicholls, J. L. 1989. *Distribution and other ecological aspects of piping plovers wintering along the Atlantic and gulf coasts*. M.S. thesis, Auburn Univ., Auburn, AL.

Niemi G. J., and T. E. Davis. 1979. Notes on the nesting ecology of the piping plover. *Loon* 51:74–79.

Nordstrum, L. H. 1990. *Assessment of habitat suitability for reestablishment of piping plovers in the Great Lakes National Lakeshores*. M.S. thesis, Univ. Missouri, Columbia.

Pike, E. A., and R. A. Wolinski. 1987. A survey of the piping plover in Michigan: 1987. *Mich. Dept. Nat. Resour.*, Unpubl. Rept.

Pike, E., T. Allan, D. Ewert, M. Holden, and J. Weinrich. 1987. Michigan piping plover recovery plan. *Mich. Dept. Nat. Resour.*, Unpubl. Rept.

Prindiville-Gaines, E. P., and M. R. Ryan. 1988. Piping plover habitat use and reproductive success in North Dakota. *J. Wildl. Mgmt.* 52:266–73.

Russell, R. 1973. The extirpation of the piping plover as a breeding species in Illinois and Indiana. *Ill. Aud. Bull.* 165:46–48.

Russell, R. P., Jr. 1983. The piping plover in the Great Lakes Region. *Am. Birds* 37:951–55.

Switzer, F. A. 1979. Piping plovers use man-made habitat. *Blue Jay* 37:116

Wiens, T. P., and F. J. Cuthbert. 1984. Status and reproductive success of the piping plover in Lake of the Woods. *Loon* 56:106–9.

Wilcox, L. 1959. A twenty year banding study of the piping plover. *Auk* 76:129–52.

Caspian Tern

· *Sterna caspia* Pallas
· State Threatened

Map 18. Confirmed caspian tern nesting pairs. ▲ = before 1983 (Wood 1951; Ludwig 1965); ● = 1983 to 1992 (Michigan DNR).

Status Rationale: The Caspian tern nests at only a few sites, primarily in northern Lake Michigan. With increasing human disturbance, limited suitable habitat, high subadult mortality, and recent severe effects from pollutants, a self-sustaining Caspian tern population depends on protection and management.

Field Identification: The Caspian tern is the largest tern in the world. Adults vary from 19 to 23 inches (48 to 58 cm) in length and have a 50- to 55-inch (127 to 140 cm) wingspan. The large size combined with a short, wedge-shaped tail and the habit of soaring and circling gives this species a gull-like appearance. However, the large, stout red bill, conspicuous black cap, ternlike habit of flying slowly with its bill pointed downward, and hovering fishing behavior distinguishes it from gulls. Immatures and adults in winter plumage have black caps streaked with white. The orange-footed immatures can be distinguished from fall-plumaged adults, which have black feet. The Caspian tern commonly gives a low, harsh call while in flight, similar to a *karrr* or *kraa-ah*. The subspecies *S. c. caspia* occurs in Michigan. The large stocky build, stout red bill, and lack of a deeply forked tail immediately distinguishes the Caspian from other white Michigan terns.

Total Range: This cosmopolitan species breeds in at least six disjunct populations on the coasts and inland waters of North America. On the Pacific coast, Caspian terns breed locally in Washington and California into Baja California; on the Atlan-

154

tic coast they breed locally in Newfoundland and Quebec and from Virginia to northern Florida. Nesting colonies also occur along the Gulf coast from Florida to Mexico. Inland populations are located in the Great Lakes northwest to central Manitoba and in central Mackenzie (Great Slave Lake); colonies also occur locally in the Great Salt Lake and nearby regions. Most of the six populations are stable or increasing in size; the U.S. breeding population estimate for 1976 through 1982 was nearly 19,000 individuals in 48 colonies (Spendelow and Patton 1988).

The Great Lakes contain nearly one-third of the continental population (Martin in Ludwig 1979; Spendelow and Patton 1988) and is isolated from other Caspian tern breeding areas. Because little mixing occurs among colonies of different geographic areas, regionally created protection measures are needed to prevent local extinction.

The Caspian tern winters on the southern coast of the United States south, including the West Indies, to northern South America. The Great Lakes population migrates through the Mississippi Flyway and along the Atlantic coast to reach these wintering grounds (Ludwig 1942; Ludwig 1965).

State History and Distribution: Historically, the Caspian tern was an uncommon breeding species in Michigan. The first population estimates for Caspian terns breeding in the northern Great Lakes region (made by individuals who visited the colonies on banding expeditions) suggest that approximately 2,400 pairs nested in this region in the 1920s; by 1960 the population had declined to 1,400 nesting pairs (Ludwig 1965). Ludwig attributed this decline to low fledging success caused by intensive commercial fishing activities in the early to mid-1900s (Ludwig 1965).

In the mid-1950s, the alewife (*Alosa pseudoharengus*) rapidly expanded into the upper Great Lakes after entering the Welland Canal in the 1870s. This range expansion provided Caspian terns with an easily available prey alternative to over-exploited native fishes (Ludwig 1965). By 1963, an estimated 2,000 nesting pairs were present in Lakes Huron, Michigan, and Ontario; the number increased to 3,600 nesting pairs in 1978 (Ludwig 1979) and approximately 4,300 nesting pairs in the early 1980s (Kress et al. 1983; Spendelow and Patton 1988). Throughout the recent history of the Great Lakes Caspian tern population, Michigan's colonies increased from at least 525 nesting pairs in 1962 (Ludwig 1962) and 925 nesting pairs in 1963, to over 1,400 nesting pairs in 1967 (Ludwig 1979). Between 1975 and 1982, the Michigan population averaged 1,800 nesting pairs.

Although the stability and apparent viability of the Caspian tern colonies is reassuring, the low number of nesting sites and human-related pressures warrant continued close monitoring. Since the early 1980s, new sites have been colonized, including an artificial disposal dike in Saginaw Bay, Bay County, in 1982 (Scharf and Shugart 1983). These new sites were partly responsible for an increase in the population from 1,615 nesting pairs in 1978 to 2,241 in 1985 (Scharf and Shugart 1985) and 2,720 in 1987. Censuses in 1989 indicate an additional increase in population.

Historically, Caspian terns nested on the following islands: Gravelly and Snake in Delta County; Hat, High, and Shoe in Charlevoix County; Ile aux Galets in Emmet County; Goose in Mackinac County; and Black River Island in Alcona

County (Scharf and Shugart 1985). In recent years, this species has continued to nest regularly on Gravelly, Hat, High, and Ile aux Galets islands and occurs at three new sites: Gull Island, Delta County; Thunder Bay Island, Alpena County; and the Saginaw Bay Confined Disposal Facility, Bay County.

During the breeding season, nonbreeding individuals or unsuccessful pairs move among colony sites throughout Michigan, searching for suitable breeding sites in northern Lakes Michigan and Huron. Although no records of inland or Lake Superior nesting sites are known for this species, the increase in population, the annual presence of individuals in some areas, and the recent development of new colony sites suggest that biologists need to be alert to the establishment of additional colonies.

Habitat: Like many terns, Caspians prefer to nest on islands in large bodies of water; nest sites are characterized by sparse vegetation and sand-gravel substrate. In the Great Lakes region, this species nests primarily in large colonies. Great Lakes water levels, larid competition, and vegetative succession all impact selection by nesting colonies. Terns occasionally use peninsula tips for nesting (Pettingill 1957), but these sites are often vulnerable to mammalian predation. Recently, artificial dikes, such as the Saginaw Bay Confined Disposal Facility, have been colonized. However, toxins may enter the surrounding ecosystem at these sites and may negatively impact the long-term success of Michigan's Caspian tern population. Creation and management of noncontaminated habitat is needed to more evenly distribute nesting colonies.

Ecology and Life History: Arriving in mid-April, pairs begin scraping a depression in the substrate. Most individuals return to the same island for more than one breeding season, and virtually all return to the general breeding area, particularly nesting pairs that were successful the previous season (Cuthbert 1988). Mate retention is low, possibly due to the instability of insular habitat (Cuthbert 1985). Caspian terns are highly colonial, often nesting within 3 feet (.9 m) of one another (Ludwig 1965) and within several feet of other larid species. In Michigan, the largest colonies on Hat and Gravelly islands in 1987 reached 955 and 584 nesting pairs, respectively (Kurita et al. 1987). Solitary nesting is rare but recorded in Michigan (Pettingill 1958). Nests with eggs are known from mid-May to mid-July. The average clutch is two to three eggs. About half of all unsuccessful breeding pairs will renest, typically relocating at a different colony (Cuthbert 1988).

Incubation is shared by both sexes for approximately 26 days. Young fledge in July and August (six to eight weeks after hatching), although parental care may continue away from the natal colony (Ludwig 1968). Subadult mortality is relatively high; Ludwig (1965) found 62% died before breeding at age 3 to 4 years in the northern Great Lakes. As in other bird species, band records demonstrate that the first six months are particularly hazardous (Ludwig 1942). Young-of-the-year generally remain on the wintering grounds for the first summer, but return to the Great Lakes region to breed (Ludwig 1968; Cuthbert 1988). Adults are relatively long lived, with life spans occasionally exceeding 20 years.

During the breeding season, Caspian terns feed almost exclusively on fish

(Ludwig 1965; Vermeer 1973); occasionally crayfish and insects are eaten. Two techniques are used for finding and catching fish; one is hovering at approximately 10 to 30 feet (approximately 3 to 10 m) above the water with the bill pointed downward and plunging underwater. The other, less common, technique is swimming on the water surface and dipping the bill underwater. It also will rob other larids of freshly caught fish.

Caspian terns typically catch small fish between 2 to 6 inches (5 to 15 cm) long. In Michigan, alewife, American smelt (*Osmerus mordax*), and yellow perch (*Perca flavescens*) are common prey. Today, alewives comprise much of the Caspian tern diet, usually one-half to three-quarters (Ludwig 1965; Shugart et al. 1978).

Conservation/Management: The Caspian tern apparently was never common or widespread in the Great Lakes region. Limited suitable insular habitat, interspecific competition, human disturbance, and environmental contaminants currently are believed to affect population viability. Suitable insular habitat appears to be limited in many areas of the northern Great Lakes, especially during years of higher than average water levels (Cuthbert 1981) and gull densities. Washouts from wind-driven waves can destroy entire nesting colonies. Regional studies show that over half of the nest failures experienced in Caspian tern colonies are caused by washouts, and up to 12% of the nests may be inundated annually (Shugart et al. 1978; Cuthbert 1988).

Often, Caspian terns are forced to use marginal nesting habitats due to the crowding out of prime nesting habitat by earlier nesting herring and ring-billed gulls (*Larus argentatus* and *L. delawarensis*, respectively) and double-crested cormorants (*Phalacrocorax aurus*). Although Caspian terns can displace ring-billed gulls, encounters with cormorants and herring gulls are less decisive. Caspian tern colonies on islands with low numbers of herring gulls may still be severely impacted by their predation (Kurita et al. 1987). Caspian tern colonies with normal nesting densities can successfully drive away herring gulls by a combined mobbing action.

Because all Michigan nesting colonies occur on islands, human disturbance is generally minimal, particularly in outer island areas. Isolated cases of high-impact boating and island visitation, however, are growing and are potentially harmful to entire colonies. Posting of signs in areas such as High Island have reduced confrontations.

If there is access to nesting colonies, foxes (*Vulpes vulpes*) and coyotes (*Canis latrans*) may significantly disrupt reproductive success. On High Island, eggs and young frequently are eaten, sometimes resulting in nearly total reproductive failure for all individuals in the colony (Shugart 1977). Other mammals, such as raccoons (*Procyon lotor*) and skunks (*Mephitis mephitis*), are also potential predators (Fetterolf and Blokpoel 1983).

Toxic chemical loads in the Great Lakes have severe implications for the long-term survival of Caspian tern populations because of direct poisoning and bioaccumulation of environmental contaminants in their prey. Colonial waterbirds studied by Nisbet (1980) showed that toxins can reduce productivity or cause eggshell thinning, but do so without a noticeable decline in nesting success. Michigan Caspian tern colonies are currently exposed to many toxins, including PCBs, dioxins, and other organochlorines in the Great Lakes. Although a 1980–81 study indicated that chemical contamination in Lakes Huron, Michigan, and Ontario did not adversely affect reproduction (Struger and Weseloh 1985), recent studies in the northern Great Lakes found lower productivity as well as the presence of defective embryos, whose occurrence appears to be associated with widely distributed toxins (Kurita et al. 1987).

Primary conservation efforts need to focus on the protection of traditional sites from all forms of human disturbance. Additionally, terns currently using contaminated sites for nesting need to be provided with alternative breeding sites constructed from uncontaminated substrate. Colonies should be monitored on a regular basis (e.g., alternate years) to estimate the number of breeding pairs, hatching, and fledging success. Individual colonies characterized by lower-than-expected size or productivity should be carefully monitored to determine factors effecting the observed changes.

Protection and control of competitors and mammalian predators is crucial at some existing colonies. All banding operations should strive for minimal nest disturbance. Such innovative techniques as plastic tunnel systems used as blinds (Shugart et al. 1981) and nest covers (Quinn 1984) are important steps. Other human-related impacts, such as mercury contamination of eggs (Vermeer 1973) and adult mortality in nylon monofilament fishline (Dunstan 1969), should be monitored.

LITERATURE CITED

Cuthbert, F. J. 1981. *Caspian tern colonies of the Great Lakes: Responses to an unpredictable environment.* Ph.D. diss., Univ. Minn., Minneapolis.

Cuthbert, F. J. 1985. Mate retention in caspian terns. *Condor* 87:74–78.

Cuthbert, F. J. 1988. Reproductive success and colony-site tenacity in Caspian terns. *Auk* 105:339–44.

Dunstan, T. C. 1969. Tern mortality due to entanglement in nylon monofilament fishline. *Loon* 41:50–51.

Fetterolf, P. M., and H. Blokpoel. 1983. Reproductive performance of Caspian terns at a new colony in Lake Ontario, 1979–1981. *J. Field Ornith.* 54:170–86.

Kress, S. W., E. H. Weinstein, and I. C. T. Nisbet. 1983. The status of tern populations in the northeastern United States and adjacent Canada. *Colonial Waterbirds* 6:84–106.

Kurita, H., J. P. Ludwig, and M. E. Ludwig. 1987. Results of the 1987 Michigan colonial waterbird monitoring project on Caspian terns and double-crested cormorants: Egg incubation and field studies of colony productivity, embryologic mortality and deformities. *Ecological Res. Serv.*, Bay City, MI. Unpubl. Rept.

Ludwig, F. E. 1942. Migration of caspian terns banded in the Great Lakes Area. *Bird Banding* 13:1–9.

Ludwig, J. P. 1962. A survey of the gull and tern populations of Lakes Huron, Michigan, and Superior. *Jack-Pine Warbler* 40:104–19.

Ludwig, J. P. 1965. Biology and structure of the caspian tern (*Hydroprogne caspia*) population of the Great Lakes from 1896–1964. *Bird Banding* 36:217–33.

Ludwig, J. P. 1968. *Dynamics of ring-billed gull and caspian tern populations of the Great Lakes.* Ph.D. diss., Univ. Mich., Ann Arbor.

Ludwig, J. P. 1979. Present status of the caspian tern population of the Great Lakes. *Mich. Acad.* 12:64–77.

Nisbet, I. C. T. 1980. Effects of toxic pollutants on productivity in colonial waterbirds. *Trans. Linnaean Soc. of New York, Colonial Waterbird Group* 9:103–13.

Pettingill, O. S., Jr. 1958. Unusual nesting of the Caspian tern. *Jack-Pine Warbler* 36:183–84.

Quinn, J. S. 1984. Egg predation reduced by nest covers during research activities in a Caspian tern colony. *Colonial Waterbirds* 7:149–51.

Scharf, W. C., and G. W. Shugart. 1983. New Caspian tern colonies in Lake Huron. *Jack-Pine Warbler* 61:13–15.

Scharf, W. C., and G. W. Shugart. 1985. Population sizes and status recommendation for double-crested cormorant, black-crowned night-heron, caspian tern, common tern, and Forster's tern in the Michigan Great Lakes in 1985. *Mich. Dept. Nat. Resour.*, Unpubl. Rept.

Shugart, G. W. 1977. Resident red fox predation upon an island gull colony. *Jack-Pine Warbler* 55:199–205.

Shugart, G. W., M. A. Fitch, and V. M. Shugart. 1981. Minimizing investigator disturbance in observational studies of colonial birds: Access to blind through tunnels. *Wilson Bull.* 93:565–69.

Shugart, G. W., W. C. Scharf, and F. J. Cuthbert. 1978. Status and reproductive success of the caspian tern (*Sterna caspia*) in the U.S. Great Lakes. *Proc. Colonial Waterbird Group,* 1978:146–56.

Spendelow, J. A., and S. R. Patton. 1988. National atlas of coastal waterbird colonies of the contiguous United States. *U.S. Fish Wildl. Serv. Biol. Rept.* 88 (5).

Struger, J., and D. V. Weseloh. 1985. Great Lakes Caspian terns: Egg contaminants and biological implications. *Colonial Waterbirds* 8:142–49.

Vermeer, K. 1973. Comparison of food habits and mercury residues in Caspian and common terns. *Can. Field-Nat.* 87:305.

Wood, N. A. 1951. The birds of Michigan. Univ. Mich. Mus. Zool. Misc. Publ. 75.

Common Tern

· *Sterna hirundo* (Linnaeus)
· **State Threatened**

Map 19. Confirmed common tern nesting pairs. ▲ = before 1983 (Ludwig 1962; Michigan DNR); ● = 1983 to 1992 (Michigan DNR).

Status Rationale: The population of this formerly abundant tern has declined greatly from historic levels in the Great Lakes region. Because of the common tern's naturally fluctuating population and its vulnerability to human-related and natural factors, its protection is crucial, even though total numbers may indicate otherwise.

Field Identification: This tern is identified by its slender body, long, pointed wings, and deeply forked tail. Adults are 13 to 16 inches (33 to 41 cm) long with an average wingspan of 31 inches (79 cm). During the breeding season, adult common terns have a red bill with a black tip, black crown, and red legs. Sexes are similar in plumage and size, although males have longer, deeper, and wider bills than females (Coulter 1986). Immatures and wintering adults have a black nape and dark bill. The most frequent call given is a drawled *kee-ar.* The subspecies *S. h. hirundo* occurs in Michigan.

The species most easily confused with the common tern are the uncommon Forster's tern (*Sterna forsteri*) and the extremely rare and transient arctic tern (*Sterna paradisaea*). Adult Forster's terns have a diagnostic nasal, low-pitched, slurred call compared to the common tern's distinctively higher pitched call. Recently fledged young of each species, however, have similar vocalizations. While in flight, the arctic tern has a shorter neck profile, a lighter build, and longer tail streamers than the common tern. In addition, during the nesting season, the arctic tern's red bill typically has no black tip. Experienced observers can also distinguish these three terns by the trailing edge pattern of the underwing; the arctic tern has a translucence effect, while others have various shades of gray, being darkest in the common tern.

Total Range: The common tern nests throughout much of the temperate zone of the Northern Hemisphere. In North America, its primary breeding range includes the Atlantic coast (from Nova Scotia to North Carolina), the Great Lakes region (extending east along the St. Lawrence River to the Gulf of St. Lawrence), and the northern Great Plains. In this latter region, it occurs in disjunct colonies from central Minnesota, northeastern Montana, and eastern Washington north to south-central Mackenzie, northern Saskatchewan, and southern James Bay in Ontario.

Common tern populations are generally stable in much of their range. Downward population trends are known for southeastern Ontario, the Niagara-Champlain region, Ohio, various locales on the Atlantic coast (Tate 1981), and Michigan. Great Lakes reproductive success is lower than that of Atlantic coast colonies (Morris et al. 1980) and numbers have declined since the mid-1960s (Courtney and Blokpoel 1983). Colonies within the Great Lakes are generally smaller and more transitory than Atlantic coast populations (Ludwig 1962; Spendelow and Patton 1988).

From a peak population in the entire Great Lakes region of approximately 32,000 individuals in the 1960s, the common tern declined to an estimated 10,000 individuals in the late 1970s (Courtney and Blokpoel 1983). Since the late 1970s, populations have stabilized, although at a much reduced number of individuals and colonies (Shugart and Scharf 1983; Scharf and Shugart 1985). In 1980, Canadian Great Lakes colonies contained three times as many individuals as U.S. Great Lakes colonies (Kress et al. 1983).

The U.S. Great Lakes region currently contains approximately 7% (5,000 individuals at 37 colonies) of the U.S. population of over 72,000 individuals in nearly 300 colonies (Spendelow and Patton 1988). Concentrations in the Great Lakes region include the St. Marys River, northern Lake Michigan (e.g., Beaver Island chain), and northern Lake Huron (e.g., Thunder Bay); these areas contained more than 50% of the region's population.

Great Lakes common terns migrate along the Atlantic coast (Austin 1953; Haymes and Blokpoel 1978) and winter on the U.S. Gulf coast, the Atlantic coast along Florida, the Caribbean, and north and west coasts of South America. Adults have a wider winter distribution than juveniles (Blokpoel et al. 1987).

State History and Distribution: Once abundant and widespread on the Great Lakes and inland waters (Barrows 1912), the common tern was first threatened with extinction at the turn of the century due to the plume and feather trade. In 1916, it was given federal protection under the Migratory Bird Treaty and soon recovered. During the mid-1900s, Great Lakes common terns maintained relatively healthy populations. The first comprehensive census estimates were between 1960 and 1962 (Ludwig 1962). During that time, populations fluctuated widely. Since the mid-1970s, an average of 1,800 nesting pairs occurred in Michigan, with a known low of 1,400 nesting pairs in 1977 (Shugart and Scharf 1983) to a known high of 2,100 nesting pairs in 1982 (Shugart and Scharf 1982). By 1985, declining nesting habitat quality on artificial and natural sites affected reproductive success and the total Michigan population fell to 1,500 nesting pairs (Scharf and Shugart 1985). A similar number of nesting pairs was counted in 1989.

In Michigan, the common tern currently nests nearly exclusively on the Great Lakes. Formerly, however, it frequently occurred inland on lakes, rivers, and other bodies of water (Barrows 1912). Zimmerman and Van Tyne (1959) reported common terns nesting inland in Saginaw County, South Manistique Lake in Mackinac County, and in Schoolcraft County. Summering and nonbreeding adults may still be seen in inland areas such as in the Manistique Lake system in the eastern Upper Peninsula. Inland nesting colonies also remain widespread in Ontario (Cadman et al. 1987).

During the 1980s, islands in the Great Lakes with colonies of 100 or more nesting pairs included areas near Lime Island in St. Marys River, Snake (Grassy) Island and Sand Products Harbor in northern Lake Michigan, the Saginaw Bay Confined Disposal Facility in Lake Huron, near the mouth of the Clinton River in Lake St. Clair, and at Pt. Mouilee State Game Area in Lake Erie (Scharf and Shugart 1985). The stability of these and other sites are crucial for self-sustaining common tern populations; many pairs nest individually or in small colonies that are subject to higher competition and predator pressures. Only one site (Snake or Grassy Island) is natural, the other major nesting colonies are dependent on artificially created sites. These island habitats are now the most important breeding areas. During much of the 1980s, the Saginaw Bay Confined Disposal Facility comprised one-quarter of the state's nesting population. By the late 1980s, however, it was frequently deserted.

Habitat: Michigan's common tern typically nests in colonies on sand-gravel, sparsely vegetated beaches on islands, coastal points, and isolated artificial sites on the Great Lakes. Several natural and human-related factors, such as island and beach development and off-road vehicles, now limit this optimal habitat. In many areas, marginal sites are now used for nesting, such as sandbars that are frequently flooded and beaches with relatively heavy growths of vegetation (Harris and Matteson 1975a; Morris et al. 1980; Scharf 1981; Kress et al. 1983).

Artificial and human-modified sites currently provide the most favorable habitat. Sites formed from dredged material are used extensively; colonies on this habitat occur along the St. Marys River in Chippewa County, the Saginaw Bay Confined Disposal Facility in Bay County, and Pt. Mouilee State Game Area in Monroe County. Occasionally other associated artificial nesting habitats are used, including abandoned wooden piers (Harris and Matteson 1975b), flat roofs (MacFarland 1977), and ash pilings from mining operations (Harris and Matteson 1975a; Courtney and Blokpoel 1983).

In winter, ocean shoreline habitats are used for roosting and foraging. Recently, sand beaches along the Peruvian coast in western South America were shown to provide habitat for important staging areas of Great Lakes adult common terns (Blokpoel et al. 1987).

Ecology and Life History: Common terns arrive in Michigan beginning in mid-April. After reaching the nesting territory, adults pair according to bill size (Coulter 1986) and age (Nisbet et al. 1984). Pairs cooperate in nest building, which ranges from a minimal scratching of sand and gravel to the construction of slightly raised mounds with linings of fine grasses and other nearby materials. Nest placement usually is associated with low, herbaceous vegetation and driftwood (Blokpoel et al. 1987). Like most larids, common terns prefer to nest in relatively large colonies. Nest and fledgling success is generally higher in large colonies because of cooperative defense against competitors and predators. Adult terns aggressively protect eggs and young by repeatedly diving and striking at intruders; males vigorously defend their territory.

In Michigan, the nesting season ranges from late May to mid-July (including renesting). Food availability and precipitation determine the timing of initial nesting (Becker et al. 1985; Safina and Burger 1988). An average clutch of two to three eggs is laid; both adults share the 22- to 25-day incubation period. Initial nest loss is frequent but is usually compensated by second nestings. Typically single brooded, common tern pairs occasionally attempt to raise two broods (Hay 1984; Wiggins et al. 1984). Chick rearing is also shared by the pair; males generally deliver larger fish (Wagner and Safina 1989). Young are capable of flight approximately four weeks after hatching. Reproductive maturity is attained at three years of age. The average annual mortality rate for adults is approximately 8% (Di-Costanzo 1980).

Postbreeding dispersal (beginning in July) of adults and juveniles is common. Individuals typically move along the Great Lakes coast but may move inland; dispersal distances of nearly 300 miles (550 km) from the breeding colony are known (Blokpoel et al. 1987). Autumn migration begins in late August and extends through October.

The common tern feeds mainly on small fish, from 1 to 3 inches (2.5 to 7.6 cm) in length. They are opportunistic, foraging on small fish species that are most available (Courtney and Blokpoel 1980). Fish are captured by hovering 15 to 30 feet (4.6 to 9.1 m) over the water and then diving and seizing them with the bill. In late summer and autumn, insects are frequently caught while flying; Vermeer (1973) found that insects dominated the diet of common terns in certain Canadian locales.

Conservation/Management: It is difficult to assess the true population status of the common tern due to the scattered and ephemeral nature of its colonies. Many limiting factors naturally pose restrictions, but when combined with human-related influences, self-sustaining populations are severely impacted. Limited nesting habitat, competition for resources, predation, chemical contaminants, and human-related disturbance all contribute to today's low and struggling breeding population.

Currently dependent on nesting habitat in the Great Lakes, the common tern is regularly affected by fluctuating water levels, which sometimes vary several feet. Vegetation succession and erosion continually reduce current and potential sites suitable for nesting.

Resource competition, both nesting and foraging, can be intense and overwhelming. Ring-billed gulls (*Larus delawarensis*) and herring gulls (*L. argentatus*) are dominant, aggressive birds. Both species have large and increasing populations in the Great Lakes (Spendelow and Patton 1988). Since the extent of common tern displacement varies with the amount of suitable nesting habitat available, the intensity of gull competition and predation increasingly becomes one of the more limiting factors (Ludwig 1962; Morris and Hunter 1976; Shugart and Scharf 1983).

Many common tern colonies are required to compete for nesting space as well as fend off predators. Many times, insular habitat lacks mammalian predators, although Norway rats (*Rattus norvegicus*) and native canids (Shugart and Scharf 1983) do limit reproductive success. Other species known to prey on common terns and their nest contents include the black-crowned night-heron (*Nycticorax nycticorax*) (Collins 1970; Hunter and Morris 1976; Nisbet and Welton 1984), Canada goose (*Branta canadensis)* (Courtney and Blokpoel 1981), eastern garter snake (*Thamnophis sirtalis*) (Cuthbert 1980), and great horned owl (*Bubo virginianus*) (Nisbet and Welton 1984). Nocturnal disturbance by great horned owls, early in the breeding season, gives species like the ruddy turnstone (*Arenaria interpres*), a known egg predator on terns (Parkes et al. 1971; Lofton and Sutton 1979), an opportunity to eat the eggs (Morris and Wiggins 1986).

Direct human-related interference includes the release of chemical contaminants into the environment. Organochlorine contaminants have been found in adult and young common terns along the Atlantic coast (Custer et al. 1985). Recently, Kurita et al. (1987) found severe chick deformation and mortality in Caspian tern colonies throughout Michigan's Great Lakes, and attribute it to PCBs, dioxins, and other toxins; a forewarning that waterbirds, particularly terns, are exposed and affected by chemical buildups.

Nesting habitat may be increased in several ways. In areas impacted by vegetative succession, the best strategy is burning or cutting to expose the ground surface. Willows (*Salix* spp.) and other brush removed from West Sugar Island, Chippewa County, in 1986 resulted in more than doubling the existing colony (Scharf 1986). Since common terns prefer to nest near low vegetation, which provides shade for chicks, a sparse herbaceous layer less than 4 to 6 inches (10.0 to 15.2 cm) high is important (Morris et al. 1980). In contrast, barren areas can be enhanced by introducing small rocks, logs, and vegetation.

Artificial island habitats, such as dikes and spoils, have a high potential for maintaining and increasing common tern populations. Terns can be attracted to new sites with decoys and sound recordings (Kress 1983).

Control of competitors and predators increasingly becomes more crucial each year. Gull populations occur in high densities throughout much of the Great Lakes region. Nesting gull colonies frequently become established two to four weeks before common tern colonies. Various methods of scare tactics during the common tern prenesting periods should be used. For example, heavy black plastic temporarily spread on known sites will delay gull establishment. The restriction of one competitor or predator, however, is usually not sufficient to increase fledging success. Rather, intense programs are required to control all species (Cuthbert 1980).

Despite extensive research, more is needed regarding habitat availability, relationships with gulls and other competitors, and food requirements (Polosky et al. 1981), before understanding the population dynamics and long-term preservation of colonies of the common tern. Reproductive success and toxin relationships are not clear, but measures need to be taken immediately, such as habitat manipulation, if the common tern is to survive in healthy numbers within the Great Lakes ecosystem.

LITERATURE CITED

Austin, O. L., Sr. 1953. The migration of the common tern (*Sterna hirundo*) in the Western Hemisphere. *Bird Banding* 24:39–55.

Barrows, W. B. 1912. *Michigan bird life*. Mich. Agric. Coll. Spec. Bull., E. Lansing.

Becker, P. H., P. Finck, and A. Anlauf. 1985. Rainfall preceeding egg-laying—a factor of breeding success in common terns (*Sterna hirundo*). *Oecologia* 65:431–36.

Blokpoel, H., G. D. Tessier, and A. Harfenist. 1987. Distribution during post-breeding dispersal, migration, and overwintering of common terns color-marked on the lower Great Lakes. *J. Field Ornith.* 58:206–17.

Cadman, M. D., P. F. J. Eagles, and F. M. Helleiner. 1987. *Atlas of the breeding birds of Ontario.* Univ. Waterloo Press, Waterloo, Ont.

Collins, C. T. 1970. The black-crowned night heron as a predator of tern chicks. *Auk* 87:584–86.

Coulter, M. C. 1986. Assortative mating and sexual dimorphism in the common tern. *Wilson Bull.* 98:93–100.

Courtney, P. A., and H. Blokpoel. 1980. Food and indicators of food availability for common terns on the lower Great Lakes. *Can. J. Zool.* 58:1318–23.

Courtney, P. A., and H. Blokpoel. 1981. Canada goose predation on eggs of common terns. *Ont. Field Biol.* 34:40–42.

Courtney, P. A., and H. Blokpoel. 1983. Distribution and numbers of common terns on the lower Great Lakes during 1900–1980: A review. *Colonial Waterbirds* 6:107–20.

Custer, T. W., C. M. Bunck, and C. J. Stafford. 1985. Organochlorine concentrations in prefledging common terns at three Rhode Island colonies. *Colonial Waterbirds* 8:150–54.

Cuthbert, F. J. 1980. An evaluation of the effectiveness of fence enclosures in reducing predation of common tern nests by snakes. *Mich. Dept. Nat. Resour.*, Unpubl. Rept.

DiCostanzo, J. 1980. Population dynamics of a common tern colony. *J. Field Ornith.* 51:229–43.

Harris, J. J., and S. W. Matteson. 1975a. Gulls and terns nesting at Duluth. *Loon* 47 69–77.

Harris, J. J., and S. W. Matteson. 1975b. Gulls and terns on Wisconsin's Lake Superior shore. *Passenger Pigeon* 37:99–110.

Hay, H. 1984. Common terns raise young from successive broods. *Auk* 101:274–80.

Haymes, G. T., and H. Blokpoel. 1978. Seasonal distribution and site tenacity of the Great Lakes common tern. *Bird Banding* 49:142–51.

Hunter, R. A., and R. D. Morris. 1976. Nocturnal predation by a black-crowned night heron at a common tern colony. *Auk* 93:629–33.

Kress, S. W. 1983. The use of decoys, sound recordings, and gull control for re-establishing a tern colony in Maine. *Colonial Waterbirds* 6:185–96.

Kress, S. W., E. H. Weinstein, and I. C. T. Nisbet (eds.). 1983. The status of tern populations in northeastern United States and adjacent Canada. *Colonial Waterbirds* 6:84–106.

Kurita, H., J. P. Ludwig, and M. E. Ludwig. 1987. Results of the 1987 Michigan colonial waterbird monitoring project on Caspian terns and double-crested cormorants: Egg incubation and field studies of colony productivity, embryologic mortality and deformities. *Ecological Res. Serv.*, Bay City, MI, Unpubl. Rept.

Lofton, R. W., and S. Sutton. 1979. Ruddy turnstones destroy royal tern colony. *Wilson Bull.* 91:133–35.

Ludwig, J. P. 1962. A survey of the gull and tern populations of Lake Huron, Michigan, and Superior. *Jack-Pine Warbler* 40:104–19.

MacFarland, A. 1977. Roof-nesting by common terns. *Wilson Bull.* 89:475–76.

Morris, R. D., and R. A. Hunter. 1976. Factors influencing desertion of colony sites by common terns (*Sterna hirundo*). *Can. Field-Nat.* 90:137–43.

Morris, R. D., J. R. Kirkham, and J. W. Chardine. 1980. Management of a declining common tern colony. *J. Wildl. Mgmt.* 44:241–45.

Morris, R. D., and D. A. Wiggins. 1986. Ruddy turnstones, great-horned owls, and egg loss from common tern clutches. *Wilson Bull.* 98:101–9.

Nisbet, I. C. T., J. M. Winchell, and A. E. Heise. 1984. Influence of age on the breeding biology of common terns. *Colonial Waterbirds* 7:117–26.

Nisbet, J. C., and M. J. Welton. 1984. Seasonal variations in breeding success of common terns: Consequences of predation. *Condor* 86:53–60.

Parkes, K. C., A. Poole, and H. Lapham. 1971. The ruddy turnstone as an egg predator. *Wilson Bull.* 83:306–8.

Polosky, R., K. Standen, W. Miller, W. Stoeffler, and J. Noftz. 1981. Michigan common tern—a recovery plan. *Mich. Dept. Nat. Resour.*, Unpubl. Rept.

Safina, C., and J. Burger. 1988. Prey dynamics and the breeding phenology of common terns (*Sterna hirundo*). *Auk* 105:720–26.

Scharf, W. C. 1981. The significance of deteriorating man-made island habitats to common terns and ring-billed gulls in the St. Marys River. *Colonial Waterbirds* 4:61–67.

Scharf, W. C. 1986. Habitat improvement for common terns. *Mich. Dept. Nat. Resour.*, Unpubl. Rept.

Scharf, W. C., and G. W. Shugart. 1985. Population sizes and status recommendation for double-crested cormorants, black-crowned night-herons, Caspian terns, common terns, and Forster's terns in the Michigan Great Lakes in 1985. *Mich. Dept. Nat. Resour.*, Unpubl. Rept.

Shugart, G. W., and W. C. Scharf. 1982. Reproductive assessment of Michigan Great Lakes common terns (*Sterna hirundo*) in 1982. *Mich. Dept. Nat. Resour.*, Unpubl. Rept.

Shugart, G. W., and W. C. Scharf. 1983. Common terns in the northern Great Lakes: Current status and population trends. *J. Field Ornith.* 54: 160–69.

Spendelow, J. A., and S. R. Patton. 1988. National atlas of coastal waterbird colonies in the contiguous United States: 1976–82. *U.S. Fish Wildl. Serv. Biol. Rept.* 88 (5).

Tate, J. T. 1981. The blue list for 1981: The first decade. *Am. Birds* 35:3–10.

Vermeer, K. 1973. Comparison of food habits and mercury residues in Caspian and common terns. *Can. Field-Nat.* 87:305.

Wagner, R. H., and C. Safina. 1989. Relative contribution of the sexes to chick feeding in Roseate and common terns. *Wilson Bull.* 101:497–500.

Wiggins, D. A., R. D. Morris, I. C. T. Nisbet, and T. W. Custer. 1984. Occurrence and timing of second clutches in common terns. *Auk* 101:281–87.

Zimmerman, D. A., and J. Van Tyne. 1959. A distributional check-list of the birds of Michigan. *Univ. Mich. Mus. Zool. Occ. Pap.* No. 608.

Barn Owl

· *Tyto alba* (Scopoli)
· **State Endangered**

Map 20. Confirmed barn owl nesting records. ▲ = before 1983 (Wallace 1948; Payne 1983; Lerg 1984); ● = 1983 to 1992 (Brewer et al. 1991).

Status Rationale: Although historical records are few, original habitats suggest that the barn owl was restricted to Michigan's southernmost counties. Early in this century, it adapted to the expansive openings created by European settlement. Recently, however, barn owl numbers have declined noticeably , and it now may be extirpated in Michigan. Other Midwest states have also experienced similar downward trends.

Field Identification: This medium-sized owl has a diagnostic white, heart-shaped facial disk, long legs, white underparts and wing linings, and dark eyes. The golden back is sprinkled with white and black spots. Females tend to be slightly darker and larger than males. By about eight to ten weeks of age, the plumage of an immature bird resembles that of an adult. Mature barn owls are 13 to 15 inches (33 to 38 cm) long and have a wingspan of 42 to 47 inches (107 to 119 cm). A standing barn owl is approximately 15 inches (38 cm) tall. Barn owls do not hoot, but they can give a variety of screeching calls, including alarm shrieks. Young give a raspy, begging call throughout most of the night and a defensive hissing call if the nest is disturbed. Barn owls are highly nocturnal. The subspecies that occurs in Michigan is *T. a. pratincola*.

A barn owl's flight appears light, often reeling from side to side, similar to a moth. Its flight profile may be confused with the state-threatened short-eared owl (*Asio flammeus*), which has similar foraging behaviors. The brown-bodied, short-

167

eared owl is distinguished by its streaked breast, darker face and underparts, yellow eyes, and distinct, dark "wrist" markings. The larger snowy owl (*Nyctea scandiaca*), which winters in Michigan, has white underparts and upperparts but lacks the inset facial disk and has yellow eyes.

Total Range: This nearly cosmopolitan species ranges in North America from the Atlantic to Pacific coasts and from southern British Columbia, Colorado, North Dakota, and southern New England south through Central and South America. Although northern and Midwest barn owl populations have experienced dramatic declines, other populations remain stable in many parts of the United States (Tate and Tate 1982). Midwest population estimates range from 45 to 65 breeding pairs (Pendleton and Krahe 1991). Local populations have responded favorably to recent nest-box management programs (Marti et al. 1979; Colvin 1986). Populations from the northern part of its range may shift southward in the winter, but, throughout most of its distribution, the barn owl is a permanent resident. Winter observations are particularly numerous in areas bordering the Atlantic and Pacific coasts (Stewart 1980).

State History and Distribution: Historically, the barn owl was a rare Michigan bird (Barrows 1912), probably restricted to the prairies, wet meadows, marshes, and oak savannas extending across much of the three lower county tiers (Chapman 1984), including the western Lake Erie region (Mayfield 1988). By the turn of the century, however, increased observations occurred in Ohio and Indiana as grassland-associated agriculture expanded. Similarly, by the early 1920s, dozens of pairs nested regularly in scattered locales throughout southern lower Michigan (Lerg 1984). By the late 1940s, however, their numbers decreased, and, by the 1970s, nesting was known only in Monroe County in extreme southeastern lower Michigan and in Berrien, St. Clair, and Tuscola counties.

In 1977, a three-year program was initiated to locate existing barn owls, identify limiting factors, and implement management actions (Lerg 1984). Survey results were discouragingly low, with few confirmed observations. During the survey period, only two to three barn owl pairs were known to have nested in Michigan. Recently, the Michigan Breeding Bird Atlas surveys failed to locate breeding individuals (Brewer et al. 1991).

Individuals are still occasionally seen throughout Michigan; in the winter of 1986–87, barn owls were found in Midland and Calhoun counties; and a Monroe County bird was present in the autumn of 1989. This indicates that barn owls may still breed within the state. Immigrants from southern populations may be providing recruits. Records are few, but barn owls do wander into the northern Lower Peninsula (Payne 1983; Smith 1984).

Habitat: The barn owl is primarily a bird of open fields and other clearings. Before European settlement, barn owls probably depended upon the historic prairies, wet meadows, marshes, and oak savannas as foraging areas, and large tree cavities for nest sites. As human encroachment advanced, expanding agriculture provided abundant foraging areas. Barn owls adapted to newly available habitat nesting sites, including barn lofts, mine shafts, silos, church steeples, and other minimally dis-turbed structures. These locations provided additional nesting sites in areas where an abundant supply of the barn owl's primary prey (rodents) could be found.

Foraging areas utilized included wet meadows, lightly grazed pastures, infre-quently mowed hayfields, marshes, and other farmlands. The quality of this owl's hunting grounds is crucial and is the determining factor for a self-sustaining popu-lation (Colvin et al. 1984). Barn owls can regularly forage up to 2 miles (3.2 km) or more from their nest site (Colvin 1984; Hegdal and Blaskiewicz 1984). Woodlots and artificial structures are used as daytime roosting sites (Hegdal and Blaskiewicz 1984). In winter, roosting barn owls are frequently observed in medium-sized pines and spruces (Wilson 1938).

Ecology and Life History: During winter, barn owls may shift southward or re-main in their general nesting area. This owl has a flexible reproductive cycle and may begin breeding before warm weather returns. Some pairs may nest year round if rodent populations are abundant. Prior to egg laying, courtship includes paired flights and screeching and ticking calls given in series. In Michigan, egg laying typically occurs in April or early May, although egg dates are known from February to August (Wallace 1948). The barn owl does not build a true nest. Instead, it commonly lays the eggs, cradled only in shredded owl pellets, on a hard surface (e.g., wood). Typical Michigan clutches average five to seven eggs. Brood size is dependent upon rodent abundance, and the numbers of chicks within a brood declines when prey availability is limited. During the early part of the breeding cycle, barn owls are sensitive to human-related disturbances. The female is respon-sible for the month-long incubation period. During courtship and incubation, male barn owls bring prey to the female at the nest; occasionally, several rodents can be found cached at the nest site (Wallace 1948).

Approximately 7¹/₂ to 8 weeks after hatching, most of the young will have fledged. Fledglings often remain at or near the nest site for 2 to 3 weeks, during which time they frequently beg for food from the adults. An average of about 4 fledglings (Reese 1972; Henny 1969) were found in self-sustaining populations. Adults commonly roost away from the nest during the day, especially as nesting progresses. Adults do not aggressively defend the nest, but young can be aggressive toward intruders. After a few weeks, a successful pair may attempt to produce a second brood (around mid-July), especially when there is abundant prey (Wallace 1948). Juveniles typically disperse widely from the nest sites, although Stewart (1952a) suggested that adults typically nest within 200 miles (322 km) of their hatching site. Barn owls breed in their first year, as early as seven months of age. The mortality rate for the first year is approximately 60% and declines to approxi-mately 38% for subsequent years (Henny 1969).

Barn owls are relatively short lived; however, they have high reproductive rates and exceptional colonizing abilities (Keith 1964; Colvin 1985). Like most birds of prey, barn owls have large brood sizes, possible multiple broods per year, early sexual maturation, and may practice polygyny (Colvin 1984). They com-monly change nest sites and mates; selection of breeding areas primarily depend on foraging habitat and prey availability (Keith 1964; Colvin 1985).

The barn owl typically leaves its roost approximately one hour after sunset and returns to roost before sunrise. The barn owl is able to locate and capture prey with remarkable efficiency with its nearly silent flight, large talons, and superb hearing and vision. Prey can be captured in complete darkness, solely with the aid of its acute and multidimensional sense of hearing (Payne 1971). Its eyes can refract 35 times lower light intensity than the human's eye.

Barn owls are beneficial predators, since their diet is largely comprised of rodents. Estimates of 600 to 1,000 meadow voles (*Microtus pennsylvanicus*) consumed per nesting (chicks and adults) have been recorded (Colvin et al. 1984). Wallace (1948) found that 85% of the prey taken in Michigan were meadow voles; other small rodents and shrews were also found. Norway rats (*Rattus norvegicus*) and house mice (*Mus musculus*) were infrequently taken, while birds (predominately house sparrows [*Passer domesticus*]) comprised 1% of the total diet.

Conservation/Management: The barn owl is relatively common in the southern part of its range, but, in Michigan, at the northern edge of its distribution, it historically has been rare. Its recent dramatic and prolonged decline in Michigan and in other peripherally occupied states has caused concern. While clearing and early farming practices in Michigan proved to be beneficial for barn owl colonization in the early part of this century, crucial natural openings (e.g., wet meadows and oak savannas) have been changed by succession and modern agriculture (e.g., small, mosaic, grass-associated farm fields converted into large monocultures of row crops). Along with increasingly intensive farming practices was the gradual decline of pastures needed for livestock (replaced by feedlots) and work animals. These two changes correlate with this owl's decline (as well as other grassland birds), and appear to be the principal reason for severe population reductions in the Midwest (Colvin et al. 1984; Colvin 1985). The significance of grasslands lies within their ability to provide an abundant supply of preferred prey.

Because barn owls are more of a southern species in the United States, Michigan winters may be the most crucial limiting factor. Barn owls are not well insulated by their feathers, and they have comparatively low fat reserves (Johnson 1974). They can become easily emaciated during harsh winters, and this can lead to a significant decline in breeding attempts and high mortality rates (Errington 1931; Speirs 1940; Bunn et al. 1982; Marti and Wagner 1985). The amount of snow cover may be the most critical factor (Stewart 1952b) in winter survival for barn owls, because deep snow limits the accessibility of voles. Stewart (1952b) suggested that barn owls will perish if deprived of food for more than three or four days.

Mortality factors can include vehicle collisions, predation by raccoon (*Procyon lotor*), and great horned owl (*Bubo virginianus*) densities. There is little range overlap between the two owl species on quality, open foraging areas (Rudolph 1978). Secondary poisoning from farm use of rodenticide does not pose a hazard to barn owl populations (Colvin 1984; Hegdal and Blaskiewicz 1984), although ingestion of certain organic chemicals is toxic (Mendelssohn 1977; Hill and Mendenhall 1980) and can reduce nesting success (Klass et al. 1978). Indiscriminate killing by misinformed people was formerly a problem but has diminished considerably.

Suitable nesting sites and foraging areas are widely available, but many regions do not provide both. Therefore, because suitable breeding conditions and barn owl populations are so widely scattered, the probability of natural recovery in Michigan is low. Since the barn owl historically occurred in Michigan and was once well established, its restoration in the state is important. Intensive reestablishment efforts in other Midwest states have not been successful, probably because of limited foraging habitat.

One important management technique that has a twofold value and thus a high potential for maintaining barn owl populations is a barn owl nest box program. This species readily accepts properly constructed boxes mounted in artificial structures and trees for nesting and as winter roosting sites (Marti et al. 1979). Nest boxes also produce significantly more young by minimizing exposure to weather and predators (Colvin et al. 1984). Protected winter roosting sites may be crucial for surviving cold spells.

The most crucial component for a self-sustaining barn owl population, however, is to manage for (or create) large, quality grasslands and wet meadows. Management of these habitat types is severely limited in Michigan in size and distribution. This needs to be acted on in order to restore Michigan's native grassland communities that once extended across much of the southern Lower Peninsula.

LITERATURE CITED

Barrows, W. B. 1912. *Michigan bird life.* Mich. Agric. Coll. Spec. Bull., E. Lansing.

Brewer, R., G. A. McPeek, R. J. Adams, Jr. 1991. *The atlas of breeding birds in Michigan.* Mich. State Univ. Press, E. Lansing.

Bunn, D. S., A. B. Warburten, and R. D. Wilson. 1982. *The barn owl.* Buteo Books, Vermillion, SD.

Chapman, K. A. 1984. *An ecological investigation of native grassland in southern lower Michigan.* M.S. thesis, W. Mich. Univ., Kalamazoo.

Colvin, B. A. 1984. *Barn owl foraging behavior and secondary poisoning hazard from rodenticide use on farms.* Ph.D. diss., Bowling Green State Univ., Bowling Green, OH.

Colvin, B. A. 1985. Common barn-owl population decline in Ohio and the relationship to agricultural trends. *J. Field Ornith.* 56:224–35.

Colvin, B. A. 1986. Barn owls: Their secrets and habits. *Illinois Audubon* 216:9–13.

Colvin, B. A., P. L. Hegdal, and W. B. Jackson. 1984. A comprehensive approach to research and management of common barn-owl populations. Pp. 270–82 *in* Proc. of the Workshop on management of nongame species and ecological communities, Univ. Ky., Lexington.

Errington, P. L. 1931. Winter killing of barn owls in Wisconsin. *Wilson Bull.* 43:60.

Hegdal, P. L., and R. W. Blaskiewicz. 1984. Evaluation of the potential hazard to barn owls of Talon (Brodifacoum bait) used to control rats and house mice. *Environ. Toxicol. Chem.* 3:167–79.

Henny, C. J. 1969. Geographic variation in mortality rates and production requirements of the barn owl (*Tyto alba* spp.). *Bird Banding* 40:277–90.

Hill, E. F., and V. M. Mendenhall. 1980. Secondary poisoning of barn owls with Famphur, an organophosphate insecticide. *J. Wildl. Mgmt.* 44:676–81.

Johnson, W. D. 1974. *Bioenergetics of the barn owl, Tyto alba.* M.S. thesis, Calif. State Univ., Long Beach.

Keith, A. R. 1964. A thirty-year summary of the nesting of the barn owl on Martha's Vineyard, Massachusetts. *Bird Banding* 35:22–31.

Klass, F. E., S. N. Wiemeyer, H. M. Ohlendorf, and D. M. Swineford. 1978. Organochlorine residues, eggshell thickness, and nest success in barn owls from the Chesapeake Bay. *Estuaries* 1:46–53.

Lerg, J. M. 1984. Status of the common barn owl in Michigan. *Jack-Pine Warbler* 62:38–48.

Marti, C. D., and P. W. Wagner. 1985. Winter mortality in common barn owls and its effect on population density and reproduction. *Condor* 87:111–15.

Marti, C. D., P. W. Wagner, and K. W. Denne. 1979. Nest boxes for the management of barn owls. *Wildl. Soc. Bull.* 7:145–48.

Mayfield, H. F. 1988. Changes in bird life at the western end of Lake Erie (p. I). *Am. Birds* 42:393–98.

Mendelssohn, M. 1977. Mass mortality of birds of prey caused by Azodrin, an organophosphorus insecticide. *Biol. Conserv.* 11:163–69.

Payne, R. P. 1983. A distributional checklist of the birds of Michigan. *Univ. Mich. Mus. Zool. Misc. Publ.* No. 164.

Payne, R. S. 1971. Acoustic location of prey by barn owls (Tyto alba). *J. Exp. Biol.* 54:535–73.

Pendleton, B. G., and D. L. Krahe (eds.). 1991. Proceedings of the Midwest raptor management symposium and workshop. *Natl. Wildl. Fed., Sci. Tech. Ser.,* No. 15.

Reese, J. G. 1972. A Chesapeake barn owl population. *Auk* 89:106–14.

Rudolph, S. G. 1978. Predation ecology of coexisting great-horned and barn owls. *Wilson Bull.* 90:134–37.

Smith, R. 1984. Common barn owl in northern Lower Peninsula. *Jack-Pine Warbler* 62:53.

Speirs, J. M. 1940. Mortality of barn owls at Champaign, Illinois. *Auk* 57:571.

Stewart, P. A. 1952a. Dispersal, breeding behavior, and longevity of banded barn owls in North America. *Auk* 69:227–45.

Stewart, P. A. 1952b. Winter mortality of barn owls in central Ohio. *Wilson Bull.* 63:164–66.

Stewart, P. A. 1980. Population trends of barn owls in North America. *Am. Birds* 34:698–700.

Tate, J., and D. J. Tate. 1982. The blue list for 1982. *Am. Birds* 36:126–35.

Wallace, G. J. 1948. The barn owl in Michigan. *Mich. Agric. Exp. Sta. Tech. Bull.* No. 208.

Wilson, K. A. 1938. Owl studies at Ann Arbor, Michigan. *Auk* 55:187–97.

Long-eared Owl

· *Asio otus* (Linnaeus)
· **State Threatened**

Map 21. Breeding evidence of long-eared owls. ▲ = before 1983 (Barrows 1912); ● = 1983 to 1992 (Brewer et al. 1991).

Status Rationale: The long-eared owl is a secretive and cryptic raptor, dependent to a large extent on openings that provide access to an abundance of nocturnal mammals. Recent studies indicate this owl is a rare nesting species and is confined to scattered areas in the southern Lower Peninsula. Its apparent limited distribution, restricted prey requirements, and dependence on relatively undisturbed foraging habitat adjacent to suitable nesting sites qualifies it for special protection.

Field Identification: This medium-sized, brownish colored owl is characterized by underparts that are streaked lengthwise, mottled upperparts, an orange facial disk, and two closely set ear tufts. The ear tufts typically are hidden unless an individual is threatened or agitated. The long-eared owl may assume three postures while perched: (1) a normal or resting posture; (2) a camouflage pose that it assumes by elongating its slender body, narrowing its eyes, and erecting its ear tufts to mimic a dead tree stub; and (3) a threat display that it assumes by puffing its body feathers and arching its bowed wings forward to appear large and formidable. In flight, its 36- to 43-inch (91- to 109-cm) wingspan is surprisingly long considering its body length of 13 to 16 inches (33 to 41 cm). This high wing–body size ratio produces a wingbeat similar to a large moth in flight—buoyant, with quick but deep strokes. Sexual dimorphism is minor, although females are generally heavier and have longer wingspans than males (Earhart and Johnson 1970).

The long-eared owl has a varied repertoire of calls. Although generally silent,

173

its hoots and caterwauling can be heard during the breeding season. Its calls include (1) soft hoots or low moans randomly given, (2) a *wuk-wuk-wuk* call uttered when disturbed, and (3) an assortment of catlike wailing sounds.

Similar owl species include the short-eared (*Asio flammeus*) and great horned owls (*Bubo virginianus*). The similar size and form of the short-eared owl can be most confusing, particularly at dusk. Short-eared owls are tawny, lack mottled underparts and an orange facial disk, have remnant ear tufts, and, in flight, exhibit a buffy patch on the upper wing and a more prominent black wrist mark on the underwing. The size difference between the much larger great horned owl and the crow-sized long-eared owl separate these two similarly appearing species. The great horned owl also has underparts with horizontal streaking, widely spaced ear tufts, a conspicuous white throat bib, and a more powerful and direct flight silhouette. The eastern screech owl (*Otus asio*) bears little resemblance to the long-eared owl but is the only other Michigan owl with ear tufts.

Total Range: This North American owl has a Holarctic distribution, occurring throughout much of Eurasia and northwestern Africa. In North America, its breeding range extends from southern California, northern Texas, central Indiana, and the Appalachian Mountains (south to Virginia), north throughout much of Canada to the treeline. Its northern Canadian abundance and distribution is relatively uncertain; for example, few nest records are known in northern Ontario (Peck and James 1983). It is absent from Alaska.

Its breeding status in the Northeast (Melvin et al. 1988), East coast (Bosakowski et al. 1989) and the Midwest, however, is relatively unknown. Much of the Canadian breeding population migrates south into the United States. Wintering individuals may move as far south of the breeding range as Baja California, central Mexico, and the southeastern United States (except Florida).

State History and Distribution: The first nests in Michigan were discovered near the turn of the century in Ingham, Kalamazoo, Monroe, Washtenaw (Barrows 1912), and Jackson counties (Wood 1951). At that time, Barrows (1912) stated it "does not seem to be an abundant owl in Michigan." By the late 1950s, confirmed breeding records were known from ten southern Lower Peninsula counties, two northern Lower Peninsula counties, and one county in the Upper Peninsula (Parmelee and Johnson 1955; Zimmerman and Van Tyne 1959).

By the early 1980s, the long-eared owl probably had a distribution and status similar to that of former years: an uncommon and local summer resident in the southern Lower Peninsula and rare in northern Michigan. However, surveys during the 1980s produced fewer breeding locations than expected (Brewer et al. 1991). Breeding was confirmed only in four counties of the southern Lower Peninsula. Individuals on territory were recorded in Lake and Montcalm counties. Most observations were in various state game areas in southern Michigan.

Habitat: Two habitat components are crucial during the breeding season: wooded areas for nesting sites and openings for foraging. The long-eared owl depends exclusively on abandoned nests or dense vegetation for egg-laying sites. Abandoned

American crow (*Corvus brachyrhynchos*), hawk, and heron nests, witches' brooms, and squirrel nests are frequently used. In Michigan, confirmed nesting sites are from tamarack (*Larix laricina*), white pine (*Pinus strobus*), Scotch pine (*Pinus sylvestris*), red maple (*Acer rubrum*), and elm (*Ulmus* spp.); nest trees were within coniferous and mixed forests, pine plantations, and woodlots (Armstrong 1958). All recent nest records were from pine plantations (Brewer et al. 1991).

Optimal foraging habitat juxtaposed to nesting areas is a major requirement for long-eared owl nesting success. Relatively undisturbed openings with high densities of small rodents are preferred foraging habitats; these include wet meadows, short grass fields, and abandoned farmland (Getz 1961; Bosakowski et al. 1989). The decline of the barn owl (*Tyto alba*) has been closely associated with degradation of this type of foraging habitat (Colvin 1985). Nesting pairs in contiguous forest regions rely on scattered natural openings (Armstrong 1958; Bull et al. 1989).

Winter habitat requirements are similar to those of the breeding season; recent evidence suggests that some nesting pairs remain on territory year-round in Michigan, depending on prey availability. Preferred roost sites are in conifer stands (Wilson 1938; Armstrong 1958; Craighead and Craighead 1969) as well as in tangled thickets of grape vines (Craighead and Craighead 1969) and dense shrubbery. Preferred roost sites in mixed conifer stands are characterized by closely spaced trees with extensive branching to within 6$^1/2$ feet (2 m) of the ground (Smith 1981; Bosakowski 1984).

Ecology and Life History: Most winter roosts in Michigan are accompanied by long-eared owl migrants; northward migration probably begins in late March (Smith 1981; Slack et al. 1987). At Whitefish Point in the Upper Peninsula, migration peaks in late April (Grigg 1989) and extends through early June (Carpenter 1987; Evers 1989). Although some nesting pairs may remain on territory throughout the year, most probably return from more southern environs in early March.

Courtship includes sequential competitive calling, aerial performances, and noncompetitive calling (Armstrong 1958). In Michigan, egg dates range from mid-March to late May (Wood 1951; Armstrong 1958). Nest records in May are probably cases of renesting. Egg clutch size ranges from three to eight, averaging four to five (Bent 1938). Incubation lasts 21 to 26 days. Only females incubate and brood (for the first two weeks); males provide most of the food (Wijnandts 1984; Craig et al. 1988).

Because eggs are laid at intervals of one to five days (Armstrong 1958), hatching and fledging occur on different dates. Young leave the nest between two to three weeks of age, generally perching on trees within the vicinity of the nest (Craig and Trost 1979; Hilliard et al. 1982; Marks 1986; Craig et al. 1988). Young fledge at four to six weeks but remain dependent on the adults for food for several more weeks (Whitman 1924; Weir in Cadman et al. 1987). Care for the young, by both sexes, may continue for several weeks after fledging. Communal roost sites are frequently formed in winter (Randle and Austing 1952; Armstrong 1958; Craighead and Craighead 1969), and occasionally in summer (Craig et al. 1985). Winter roosts may reach 50 to 75 individuals in a single woodlot (Weir in Cadman et al. 1987).

Long-eared owls are usually nocturnal foragers (Marti 1974; Craig et al. 1988), although they may make crepuscular flights during times of low prey abundance. Most hunting forays are made while on the wing. The long-eared owl is a prey specialist. Small mammals (less than 2.1 ounces [60 g]) comprise over 98% of this owl's diet in North America (Marti 1976). In the eastern part of its range, meadow voles (*Microtus pennsylvanicus*) are the primary prey during the winter and nesting season (Warthin and Van Tyne 1922; Spiker 1933; Geis 1952; Armstrong 1958; Getz 1961; Craighead and Craighead 1969). In Michigan, other small rodents, such as *Peromyscus* spp., shrews, and various species of passerines, also serve as prey (Armstrong 1958; Craighead and Craighead 1969). Reptiles and amphibians are rarely taken (Ross 1989).

Conservation/Management: The breeding status of the long-eared owl is difficult to assess because of its secretive behavior. Standard methods of population monitoring are initially required to determine distribution, abundance, population trends, and breeding status. Suspected limiting factors include habitat loss and degradation, predation (Melvin et al. 1988) and competition (Bosakowski et al. 1989) by great horned owls, and environmental contaminants.

Because long-eared owls maintain regular communal winter roosts (Smith 1981; Bosakowski 1984), there is ample time to conduct surveys to assess population trends and regional abundance. Christmas bird counts can assist in determining the distribution of known winter roost sites, but offer little information about abundance and population trends (Evers 1989; Melvin et al. 1988). Another survey technique includes capture and release using standardized mist netting during migration; data gathered with this method from Whitefish Point Bird Observatory, Chippewa County, suggest widely fluctuating numbers from year to year (Carpenter 1987; Evers 1989).

Playing recordings before and during nesting (primarily March and April in Michigan) allows identification of territories and potential nest sites (Voronetsky 1987; Clark 1989). Young-of-the-year can be found by following begging calls given at night; the calls are similar to a squeaky wheel (Weir in Cadman et al. 1987).

In Michigan, active winter roosts should be monitored for potential nesting pairs. The extent to which individuals in winter roosts are actually permanent residents is unknown. Armstrong (1958) found breeding areas adjacent to winter roosts in Michigan and, in the West, long-eared owl populations were permanent residents (Craig and Trost 1979). In Michigan, a few winter roosts were later occupied by nesting pairs (Brewer et al. 1991). In addition, little is known regarding annual site faithfulness of breeding territories; Marti (1974) suggested that local populations may change yearly.

In areas with suitable foraging habitat, the past establishment of pine plantations has enhanced certain sites for nesting pairs. Nest platforms placed in these areas increase nesting use and success. Human disturbance should be minimized in occupied pine groves.

Influxes of environmental contaminants (e.g., pesticides and herbicides) in open country habitats are probably playing an increasing role in limiting self-sustaining raptor populations (Fimreite et al. 1970; Mendelssohn 1977). Hegdal and

Colvin (1988) evaluated the biomagnification hazards of a rodenticide (Brodifacorum) used to control voles in orchards and found that several owl species, including the long-eared owl, may die from its effects

Standard surveys specifically geared for locating long-eared owl territories and nest sites are needed in Michigan. Only then can site faithfulness be studied. Habitat management techniques can then be used to ensure the existence of this owl in Michigan.

LITERATURE CITED

Armstrong, W. H. 1958. Nesting and food habits of the long-eared owl in Michigan. *Mich. State Univ., Publ. Mus. Biol. Ser.* 1:63–96.

Barrows, W. B. 1912. *Michigan bird life.* Mich. Agric. College Spec. Bull., E. Lansing.

Bent, A. C. 1938. Life histories of North American birds of prey. *U.S. Natl. Mus., Bull.* 170.

Bosakowski, T. 1984. Roost selection behavior of the long-eared owl (*Asio otus*) wintering in New Jersey. *J. Raptor Res.* 18:137–42.

Bosakowski, T., R. Kane, and D. G. Smith. 1989. Decline of the long-eared owl in New Jersey. *Wilson Bull.* 101:481–85.

Brewer, R., G. A. McPeek, and R. J. Adams, Jr. 1991. *The atlas of breeding birds of Michigan.* Mich. State Univ. Press, E. Lansing.

Bull, E. L., A. L. Wright, and M. G. Henjum. 1989. Nesting and diet of long-eared owls in conifer forests, Oregon. *Condor* 91:908–12.

Cadman, M. D., P. F. J. Eagles, and F. M. Helleiner. 1987. *Atlas of the breeding birds of Ontario.* Univ. Waterloo Press. Waterloo, Ont.

Carpenter, T. W. 1987. The role of the Whitefish Point Bird Observatory in studying spring movements of northern forest owls. Pp. 71–74 *in* R. W. Nero, R. J. Clark, R. J. Knapton, and R. H. Hamre (eds.), Biology and conservation of northern forest owls: Symposium proceedings. *U.S. Dept. Agric., For. Serv. Gen. Tech. Rept.* RM-142.

Clark, R. J. 1989. Survey techniques for owl species in the Northeast. Pp. 318–27 *in* B. G. Pendleton (ed.), Proc. of the Northeast raptor management symposium and workshop. *Inst. Wildl. Res., Natl. Wildl. Fed., Sci. Tech. Ser.* No. 13.

Colvin, B. A. 1985. Common barn-owl population decline in Ohio and the relationship to agricultural trends. *J. Field Ornith.* 56:224–35.

Craig, T. H., and C. H. Trost. 1979. The biology and nesting density of breeding American kestrels and long-eared owls on the Big Lost River, southeastern Idaho. *Wilson Bull.* 91:50–61.

Craig, T. H., E. H. Craig, and L. R. Powers. 1985. Food habits of long-eared owls (*Asio otus*) at a communal roost site during the nesting season. *Auk* 102:193–95.

Craig, T. H., E. H. Craig, and L. R. Powers. 1988. Activity patterns and home-range use of nesting long-eared owls. *Wilson Bull.* 100:204–13.

Craighead, J. J., and F. C. Craighead, Jr. 1969. *Hawks, owls and wildlife.* Dover, NY.

Earhart, C. M., and N. K. Johnson. 1970. Size dimorphism and food habits of North American owls. *Condor* 72:251–64.

Evers, D. C. 1989. State report: Michigan raptor populations may be on the upswing. *Eyas* 12:10–15.

Fimreite, N., R. W. Fyfe, and J. A. Keith. 1970. Mercury contamination of Canadian prairie seed eaters and their avian predators. *Can. Field-Nat.* 84:269–76.

Geis, A. D. 1952. Winter food habits of a pair of long-eared owls. *Jack-Pine Warbler* 30:93.

Getz, L. L. 1961. Hunting areas of the long-eared owl. *Wilson Bull.* 73:79–82.

Grigg, W. N. 1989. Owl banding at Whitefish Point, Michigan—Spring 1988. *N. Am. Bird Bander* 14:120–22.

Hegdal, P. H., and B. A. Colvin. 1988. Potential hazard to eastern screech-owls and other raptors of Brodifacoum bait used for vole control in orchards. *Environ. Toxicol. Chem.* 7:245–60.

Hilliard, B. L., J. C. Smith, M. J. Smith, and L. R. Powers. 1982. Nocturnal activity of long-eared owls in southwestern Idaho. *J. Idaho Acad. Sci.* 18:47–53.

Marks, J. S. 1986. Nest-site characteristics and reproductive success of long-eared owls in southwestern Idaho. *Wilson Bull.* 98:547–60.

Marti, C. D. 1974. Feeding ecology of four sympatric owls. *Condor* 76:45–61.

Marti, C. D. 1976. A review of prey selection by the long-eared owl. *Condor* 78:331–36.

Melvin, S. M., D. G. Smith, D. W. Holt, and G. R. Tate. 1988. Small owls. Pp. 88–96 *in* B. G. Pendleton (ed.), Proc. of the Northeast raptor management symposium and workshop. *Inst. Wildl. Res. Natl. Wildl. Fed., Sci. Tech. Ser.* No. 13.

Mendelssohn, H. 1977. Mass mortality of birds of prey caused by Azodrin, an organophosphorus insecticide. *Biol. Conserv.* 11:163–69.

Parmelee, D. F., and J. A. Johnson. 1955. Nesting of the saw-whet owl in Dickinson County, Michigan. *Jack-Pine Warbler* 33:75–76.

Peck, G. K., and R. D. James. 1983. *Breeding birds of Ontario: Nidiology and distribution,* vol. 2. Royal Ontario Mus., Publ. Life Sci., Toronto.

Randle, W., and R. Austing. 1952. Ecological notes on long-eared and saw-whet owls in southwestern Ohio. *Ecology* 33:422–26.

Ross, D. A. 1989. Amphibians and reptiles in the diets of North American raptors. *Wisc. Dept. Nat. Resour., Wisc. Endangered Resour. Rept.* No. 59.

Slack, R. S., C. B. Slack, R. N. Roberts, and D. E. Emord. 1987. Spring migration of long-eared owls and northern saw-whet owls at Nine Mile Point, New York. *Wilson Bull.* 99:480–85.

Smith, D. G. 1981. Winter roost site fidelity by long-eared owl in central Pennsylvania. *Am. Birds* 35:339.

Spiker, C. J. 1933. Analysis of 200 long-eared owl pellets. *Wilson Bull.* 45:198.

Voronetsky, V. I. 1987. Some features of long-eared owl ecology and behavior: Mechanism maintaining territoriality. Pp. 229–30 *in* R. W. Nero, R. J. Clark, R. J. Knapton, and R. H. Hamre (eds.), *Biology and conservation of northern forest owls: Symposium proceedings.* U.S. Dept. Agric., For. Serv. Gen. Tech. Rept. RM-142.

Warthin, A. S., Jr., and J. Van Tyne. 1922. The food of long-eared owls. *Auk* 39:417.

Whitman, F. N. 1924. Nesting habits of the long-eared owl. *Auk* 41:479–80.

Wijnandts, H. 1984. Yearling male long-eared owls breed near natal nest. *J. Field Ornith.* 56:181–82.

Wilson, K. A. 1938. Owl studies at Ann Arbor, Michigan. *Auk* 55:187–97.

Wood, N. A. 1951. The birds of Michigan. *Univ. Mich. Mus. Zool. Misc. Publ.* 75.

Zimmerman, D. A., and J. Van Tyne. 1959. A distributional check-list of the birds of Michigan. *Univ. Mich. Mus. Zool. Occ. Pap.* No. 608.

Short-eared Owl

· *Asio flammeus* (Pontoppidan)
· **State Endangered**

Map 22. Confirmed short-eared owl nesting pairs. Historically locally scattered statewide (Barrows 1912); ● = 1983 to 1992 (Brewer et al. 1991); ▲ = winter observations, 1983 to 1989 (Jack-Pine Warbler seasonal surveys).

Status Rationale: Although this wide-ranging owl still commonly occurs in much of its North American range, its numbers have declined in the Northeast and Midwest, where it is an uncommon and local nester. Reasons for this decline are not fully known but are likely related to the loss of large, quality grasslands and marshes. Only a few nesting pairs are known in Michigan.

Field Identification: The short-eared owl is a medium-sized owl with long wings and a short tail, a tawny-streaked body, and a round face with small ear tufts that are displayed when agitated. Adults are typically 13 to 17 inches (33 to 43 cm) long and have a wingspan of approximately 42 inches (107 cm). In flight, the short-eared owl looks like a large moth with its floppy and irregular wingbeat. While flying, the pale buff area at the base of the primaries is conspicuous, even at a distance. The short-eared owl is generally silent, although in the spring it may give a sharp barking call similar to that of a small dog. Indicators of this owl's presence are the 1½ to 2½ inch (3.8 to 6.4 cm) long, dark cylindrical pellets regurgitated daily at a roost or hunting perch. The pellets consist of undigestible material such as feather, hair, and bone, and they may vary in size and shape. Short-eared owl pellets are usually firm (high bone content), while the pellets of the commonly associated northern harrier (*Circus cyaneus*) tend to be spongy (low bone content) (Clark 1972; Holt et al. 1987). The subspecies recognized in Michigan is *A. f. flammeus*.

179

When gliding low over the ground, this species can be confused with the more common northern harrier. The harrier prefers similar habitats, and often coexists with the short-eared owl (Beske and Champion 1971). Although interspecific competition may be intense, different prey selection (Phelan and Robertson 1978), behavior, activity periods, and morphological features allow both species to coexist (Clark and Ward 1974). Four identifying features distinguish the harrier from the short-eared owl. The harrier has a longer tail, a distinctive white rump, and a slight wing dihedral that can be seen while gliding, and it lacks a dark "wrist" patch (carpal joint) underneath the wings.

The state-endangered barn owl (*Tyto alba*), which has similar foraging techniques and habitat preferences, can be distinguished by its mostly white underparts. Short-eareds may be confused with the long-eared owl (*Asio otus*), although the long-eared is primarily a nocturnal, forest inhabitant. However, long-eared owls will use communal winter roost sites with short-eared owls and may be seen foraging in fields together during this season. In flight, long-eared owls can be separated from short-eared owls by their darker color, dark trailing edge in the wing, and the absence of a prominent buffy patch near the tip of the upper wing.

Total Range: This widespread owl occurs as a visitor or resident on every continent except Australia. In North America, the short-eared owl breeds primarily from northern Alaska and Canada south to central California, Nevada, Missouri, and northern Ohio. It also regularly nests eastward in disjunct populations along Lakes Erie, Ontario, and Champlain, the St. Lawrence River, and the coastal marshes of the Northeast. New Jersey's large populations (Urner 1923 and 1925) have disappeared, but nesting pairs remain on Long Island and on the islands of Monomoy National Wildlife Refuge, Nantucket, and Tuckenuck in Massachusetts (Melvin et al. 1988). Western populations are more stable than eastern ones, although there are recent indications of declining populations in the northern and southern Great Plains region (Tate 1986). The species winters from southern Canada south to Central America (Guatemala).

State History and Distribution: Although its breeding range includes Michigan, the short-eared owl was never a common nester, and it can be expected only in scattered locales. Historical nesting records are few, mostly in southeastern Michigan (Barrows 1912; Wood 1951), and probably reflect the density of observers and not the true distribution of the owls. Changes in land use eventually altered the distribution of Michigan's short-eared owls. Scattered breeding records were documented in several counties (Calhoun, Clinton, Lapeer, Saginaw, Washtenaw) from the 1930s to 1950s (Zimmerman and Van Tyne 1959). However, since the mid-1900s, few nesting records have been documented.

Presently, the only known regular summering area is on the grasslands of the Rudyard Flats, Chippewa County. A 1984 survey of the open jack pine (*Pinus banksiana*) plains in the Hiawatha National Forest (Chippewa and Mackinac counties), adjacent to the Rudyard Flats, failed to find nesting short-eared owls (Petersen et al. 1984). Sandy conifer plains have suitable open nesting habitat but generally do not support a sufficient prey base.

The larger and more preferred Rudyard Flats appear to support a small but persistent population of short-eared owls. The area's large size enables nesting pairs to move annually to follow local rodent outbreaks. In 1987, one individual was observed in this region; in 1988, surveys located four owls, one which was seen carrying food in June (Brewer et al. 1991). Other summer reports during the 1980s are along the Lake Erie marshes (e.g., Pt. Mouillee State Game Area, Monroe County), in the St. Clair Flats region, in the thumb area of the Lower Peninsula (e.g., Tuscola County), and within peatlands of the Upper Peninsula.

In Michigan, the short-eared owl is seen most frequently during migration and winter (when individuals gather in roosts). Current sightings are now limited to an average of three to six birds per autumn. Migrants are regularly sighted along the Great Lakes shorelines, particularly in the southern half of Lake Michigan in autumn and at Whitefish Point, Chippewa County, in spring. Short-eared owls gather in relatively large winter roosts in southern Michigan, apparently every three to four years (indicative of the irruptive population cycles of their prey). Recent examples include 32 individuals near Juddville, Shiawassee County, in 1957 (Reed 1959); 16 owls near Williamston, Ingham County, in 1960 (Short and Drew 1962); 25 owls near Hudsonville, Ottawa County, in 1964 (Ponshair 1976); 16 owls in Macomb County in 1973; 28 in Macomb County in 1976; 10 in Lenawee County in 1980; 7 each in Macomb and Monroe counties in 1983; and 6 in Kalamazoo County in 1987. Short-eared owls typically change their wintering grounds each year, due to fluctuating rodent abundance and the succession or alteration of habitat used for foraging and roosting. Since 1970, regular winter concentrations have been established in Berrien, Kalamazoo, Macomb, Monroe, and St. Clair counties. The marshes of Lakes Erie and St. Clair annually provide stable wintering sites for paired birds.

Habitat: This species is associated exclusively with open areas, frequenting marshes, grasslands, pastures, and occasionally peatlands. Nests are generally placed in a clump of herbaceous vegetation or at the base of a low shrub. Dry sites are preferred due to the flimsy structure of the nest. Open areas with low rodent densities, such as pesticide-sprayed croplands, fields with row crops, and jack pine plains, cannot support breeding populations.

In winter, congregations of short-eared owls generally roost and hunt in open areas that offer an abundant supply of prey, such as abandoned pastures, hayfields, airports, marshes, and old fields interspersed with trees and shrubs. Short and Drew (1962) found no evidence of wintering short-eared owls hunting in farmlands containing corn stubble. For winter roosting, they will also use young pine plantations (preferably with an average height of 5 feet [1.5 m]) adjacent to clearings or marshes (Ponshair 1976) and individual conifers near large openings (Munyer 1966). Conifer roosting may provide a survival advantage over typical ground roosting sites (Bosakowski 1986), following a snowfall of 2 inches (5.1 cm) or more (Bosakowski 1986).

Ecology and Life History: Northbound migrants begin to disperse throughout Michigan in March and April. Upon its arrival on the nesting grounds, the short-eared owl performs a spectacular courtship display that may be seen during the day. This display is usually performed by the male and may occur in several forms. The most common includes the owl climbing and descending from high altitudes, during which it produces a courtship "song"(whoos) and a clapping sound produced by the wings hitting each other on the downstroke (Townsend 1938). Breeding territory size may vary according to food availability. Clark (1975) found a mean territory size (for five pairs) of 183 acres (74 ha) per pair during high prey densities, 300 acres (121 ha) per pair during low prey densities.

By mid-May, egg laying is complete. In years with high rodent concentrations, clutches of up to nine eggs are common, but the typical clutch size is five to seven eggs. If the nest is destroyed, the pair may renest. The female incubates the eggs for 24 to 28 days. Approximately two weeks after hatching, the young leave the nest and hide in the surrounding vegetation. Adults will aggressively defend the wandering juvenile owls by diving at an intruder or by attempting to lure predators away with a broken-wing act. After four to five weeks, the young are able to fly. Family groups may remain together through the first winter, which aids the survival of inexperienced young (Marr and McWhirter 1982). Short-eared owls commonly establish and maintain winter foraging territories (Clark 1975).

Although the short-eared owl prefers hunting during crepuscular and night-time hours, it is more diurnally active than other Michigan owls. This diurnal behavior may be the result of low nocturnal hunting success when prey is scarce or adverse weather conditions prevail. While hunting, the short-eared usually quarters 5 to 15 feet (1.5 to 4.6 m) above ground, hovers, and quickly descends when prey is sighted. Hunting from posts and other perches is also common (Bosakowski 1989).

From the results of several North American dietary studies, the short-eared owl is unequivocally dependent on a diet of meadow voles (*Microtus pennsylvanicus*) and secondarily on *Peromyscus* species (Errington 1932; Hendrickson and Swan 1938; Snyder and Hope 1938; Banfield 1947; Reed 1959; Short and Drew 1962; Craighead and Craighead 1969). Clark (1975) estimated that the mean daily intake of a young owl was $2^1/2$ mature voles per day. Individuals commonly invade an area and will leave only when the food supply is exhausted. Short-eared owls will occasionally take birds and other small vertebrates. They have been known to cause nest desertions of the state-threatened common terns (*Sterna hirundo*) in Michigan and along the East coast.

Conservation/Management: In Michigan, the decline of the short-eared owl probably is due to the loss of native grasslands and marshes (clearly the main reason in the Northeast). It is able to cope with some habitat changes, however, and seemingly suitable nesting habitat is presently available. The regional success of this species may be related to its dependence on high rodent concentrations. During local rodent depressions, only quality native grasslands or areas such as pastures and hayfields can regularly provide abundant prey, whereas row crops (mainly grain related) are marginal habitats for prey and, thus, owl occupation (Colvin et al. 1984).

The short-eared owl's dynamic and unpredictable breeding requirements make its true status difficult to assess. Many areas with suitable habitat, which previously contained at least periodic short-eared nesting, no longer harbor these owls. This suggests that limiting factors other than habitat alteration may be responsible for its disappearance in much of its Midwest range. Contamination from heavy pesticide and herbicide use in agricultural areas may be a major contributor to this decline (Fimreite et al. 1970; Mendelssohn 1977; Hegdal and Colvin 1988).

Protection and enhancement of this species' breeding and wintering habitats are important for landowners and managers in Michigan, particularly since the short-eared owl effectively reduces rodent outbreaks. This ecological solution is, in the long term, better for the environment and the people dependent upon it, as well as for the short-eared owl and associated species. To encourage short-eared owl occupation, the use of environmentally unsound rodenticides should be curtailed to provide the species with an extra edge for survival.

LITERATURE CITED

Banfield, A. W. 1947. A study of the winter feeding habits of the short-eared owl (*Asio flammeus*) in the Toronto region. *Can. J. Res.* 25:45–65.

Barrows, W. B. 1912. *Michigan bird life.* Mich. Agric. Coll. Spec. Bull., E. Lansing.

Beske, A., and J. Champion. 1971. Prolific nesting of short-eared owls on Buena Vista Marsh. *Passenger Pigeon* 33:99–103.

Bosakowski, T. 1986. Short-eared owl winter roosting strategies. *Am. Birds* 40:237–40.

Bosakowski, T. 1989. Observations on the evening departure and activity of wintering short-eared owls in New Jersey. *J. Raptor Res.* 23:162–66.

Brewer, R., G. A. McPeek, R. J. Adams. 1991. *The atlas of breeding birds in Michigan.* Mich. State Univ. Press, E. Lansing.

Clark, R. J. 1972. Pellets of the short-eared owl and marsh hawk compared. *J. Wildl. Mgmt.* 36:962–64.

Clark, R. J. 1975. A field study of the short-eared owl, *Asio flammeus* (Pontoppidan) in North America. *Wildl. Monogr.* No. 47.

Clark, R. J., and J. G. Ward. 1974. Interspecific competition in two species of open country raptors, *Circus cyaneus* and *Asio flammeus*. *Proc. Pap. Acad. Sci.* 48:79–87.

Colvin, B. A., P. L. Hegdal, and W. B. Jackson. 1984. A comprehensive approach to research and management of common barn-owl populations. Pp. 270–82 *in Proc. of the Workshop on Management of Nongame Species and Ecological Communities*. Univ. Ky., Lexington.

Craighead, J. J., and F. C. Craighead, Jr. 1969. *Hawks, owls and wildlife.* Dover, NY.

Errington, P. L. 1932. Food habits of southern Wisconsin raptors. *Condor* 34:176–86.

Fimreite, N., R. W. Fyfe, and J. A. Keith. 1970. Mercury contamination of Canadian prairie seed eaters and their avian predators. *Can. Field-Nat.* 84:269–76.

Hegdal, P. H., and B. A. Colvin. 1988. Potential hazard to eastern screech-owls and other raptors of Brodifacoum bait used for vole control in orchards. *Environ. Toxicol. Chem.* 7:245–60.

Hendrickson, G. O., and C. Swan. 1938. Winter notes of the short-eared owl. *Ecology* 19:584–88.

Holt, D. W., L. J. Lyon, and R. Hale. 1987. Techniques for differentiating pellets of short-eared owls and northern harriers. *Condor* 89:929–31.

Marr, T. G., and D. W. McWhirter. 1982. Differential hunting success in a group of short-eared owls. *Wilson Bull.* 94:82–83.

Melvin, S. M., D. G. Smith, D. W. Holt, and G. R. Tate. 1988. Small owls. Pp. 88–96 *in*

B. G. Pendleton (ed.), Proceedings of the Northeast raptor management symposium and workshop. *Inst. Wildl. Res. Nat. Wildl. Fed., Sci. Tech. Ser. No. 13.*

Mendelssohn, H. 1977. Mass mortality of birds of prey caused by Azodrin, an organophosphorus insecticide. *Biol. Conserv.* 11:163–69.

Munyer, E. A. 1966. Winter food of the short-eared owl (*Asio flammeus*) in Illinois. *Trans. Ill. Acad. Sci.* 59:174–80.

Petersen, G., D. C. Evers, and J. D. Paruk. 1984. Survey of the short-eared owl in the Hiawatha National Forest, Michigan. *Mich. Nat. Feat. Invent.,* Unpubl. Rept.

Phelan, F. J., and R. J. Robertson. 1978. Predatory responses of a raptor guild to changes in prey density. *Can. J. Zool.* 56:2565–72.

Ponshair, J. F. 1976. Short-eared owls roosting in pine plantations. *Jack-Pine Warbler* 54:130–31.

Reed, S. A. 1959. An analysis of 111 pellets from the short-eared owl. *Jack-Pine Warbler* 37:19–23.

Short, H. L., and L. C. Drew. 1962. Observations concerning behavior, feeding, and pellets of short-eared owls. *Am. Midl. Nat.* 67:424–33.

Snyder, L. L., and C. E. Hope. 1938. A predator-prey relationship between the short-eared owl and the meadow mouse. *Wilson Bull.* 50:110–12.

Tate, J. 1986. The blue list for 1986. *Am. Birds* 40:227–36.

Townsend, C. W. 1938. Short-eared owl. Pp. 169–82 *in* Life histories of North American birds of prey. A. C. Bent (ed.), *U.S. Natl. Mus. Bull.* No. 167, pt. 2. Washington, D.C.

Urner, C. A. 1923. Notes on the short-eared owl. *Auk* 40:30–36.

Urner, C. A. 1925. Notes on two ground-nesting birds of prey. *Auk* 42:31–41.

Wood, N. A. 1951. The birds of Michigan. *Univ. Mich. Mus. Zool. Misc. Publ.* 75.

Zimmerman, D. A., and J. Van Tyne. 1959. A distributional check-list of the birds of Michigan. *Univ. Mich. Mus. Zool. Occ. Pap.* No. 608.

Loggerhead Shrike

· *Lanius ludovicianus* Linnaeus

· **State Endangered**

Map 23. Confirmed loggerhead shrike nesting pairs. Historically Lower Peninsula and eastern Upper Peninsula (Barrows 1912); ● = 1983 to 1992 (Brewer et al. 1991); ▲ = summer observations, 1983 to 1989 (Jack-Pine Warbler seasonal surveys).

Status Rationale: The loggerhead shrike has experienced a serious population decline in Michigan and throughout the Midwest and Northeast. This downward trend is especially alarming because the limiting factors for the decline remain undefined. Recent intensive statewide surveys indicate only four to six active nesting pairs.

Field Identification: A grayish, robin-sized bird, the loggerhead shrike is usually found perched on wires, posts, or treetops. This predatory songbird has a slightly hooked bill, black mask, and large white wing patches contrasting with dark wings. It averages 8 to 10 inches (20 to 25 cm) in length with a wingspan of around 13 inches (33 cm). Although usually silent, this songbird may be heard repeating short harsh phrases similar to *queedle-queedle.* The subspecies occurring in Michigan is *L. l. migrans.*

The loggerhead shrike may be confused with the northern mockingbird (*Mimus polyglottus*), which lacks the face mask and strongly hooked bill, and its Canadian-breeding counterpart, the northern shrike (*Lanius excubitor*). The differences between the two shrikes are subtle and change with age and season. The northern shrike does not nest in Michigan, and there are no summer records; however, both shrikes are present during migration. Northern shrikes may arrive in mid-September and leave by late April (Janssen 1987). Therefore, a shrike observed from May through August is most likely a loggerhead.

185

In spring and early autumn, the loggerhead can be distinguished from the northern shrike by its solid black bill, lack of a faintly barred breast, and the continuation of the wider, blacker mask over the base of the upper mandible. In late autumn and winter, these distinguishing characteristics change and become less distinct. Zimmerman (1955) has meticulously studied the differentiating details of these two shrike's plumage and behavior. A distinctive identification feature (in any season) that can be used with practice is the loggerhead's bill; it is shorter and without an obvious hook and merges with the face mask to form a continuous black area. Juvenile loggerheads are grayer than the young, brownish northern shrikes. Clues given by the shrike's behavior (e.g., the loggerhead seldom perches more than 25 feet [7.6 m] from the ground) can help separate species. A shrike's flight is usually low and undulating, and the tail is held nearly horizontal when perched. While moving between perches, shrikes commonly drop abruptly within a few feet from the ground and sweep upward with a sharp glide to a perch. This "galloping flight" produces a flickering white effect from the wing patches.

Total Range: Historically, the loggerhead shrike occurred throughout the United States from southern Canada to southern Mexico. Recently, continental downward trends have been recorded in this shrike's nesting range (Bystrak 1981; Robbins et al. 1986) and wintering range (Morrison 1981). This has been particularly pronounced at the northern parts of its range, east of the Mississippi River (Erdman 1970; Geissler and Noon 1981; Kridelbaugh 1981; Robbins et al. 1986; Tate 1986), where the subspecies *L. l. migrans* occurs (Miller 1931).

In the Canadian Maritimes, New England states, and Pennsylvania (all which may have been colonized by this shrike following the logging era), populations have disappeared completely (Milburn 1981). Quebec, New York, Virginia, and West Virginia only have sparse and declining populations. In the Great Lakes region, recent surveys have determined the following breeding populations in each state or province: Indiana with 98 pairs (Burton 1989); Minnesota with 32 pairs (Brooks and Temple 1986); Ohio with 10 to 20 pairs; Ontario with 50 to 100 pairs (Cadman et al. 1987); and Wisconsin with 5 pairs (Fruth 1988). Although western and southern populations are considered stable, Morrison (1981) found slight declines during winter months throughout both of these regions. A more severe decline was noted in the central southeastern United States where Michigan and other breeding populations of the loggerhead shrike generally winter (Burnside 1987).

State History and Distribution: Once considered an uncommon bird throughout the Lower Peninsula and rare in the eastern Upper Peninsula, this species has declined to near extirpation in the state. During the late 1950s, the breeding population began shifting north and west of its historical southeastern Lower Peninsula concentrations. This may have been a response to changing land-use patterns and increased human pressure. Following this northward expansion, its overall abundance began to exhibit a steady decline. By the late 1970s, a sharp drop in the population occurred and the loggerhead shrike has yet to recover.

During the 1980s, there have been only scattered nesting records from the following counties: Allegan, Alpena, Benzie, Grand Traverse, Huron, and Macomb. Although several summering individuals are recorded annually in the Lower Peninsula (as well as sightings in the Upper Peninsula), there was only one known nesting pair in the state from 1982 to 1985. Intensive searches from 1986 to 1988 found four nesting pairs in the latter two years, probably due to better censusing efforts versus an actual population increase. Many recent breeding pairs and summering individuals occur within several miles of the Great Lakes shoreline (Brewer et al. 1991), indicating an association with topographical features. This shrike may be occasionally found wintering in the southern third of Michigan (Payne 1983); it is usually replaced by the northern shrike, however, and winter identification must be verified carefully.

Habitat: Loggerhead shrike habitat is characterized as primarily open, agricultural areas interspersed with grassland habitat. It typically prefers pastures, old fields, and native prairie tracts with clumps of dense thickets, isolated trees, and hedgerows with thorny shrubs and trees (Brooks and Temple 1990). Barbed wire fence, hawthornes (*Crataegus* spp.), osage orange (*Maclura pomifera*), honey locust (*Gleditsia triacanthos*), and other thorny vegetation are frequently used for impaling prey. Historically, Michigan loggerhead shrikes were commonly associated with fields containing scattered hawthornes (Barrows 1912) and orchards, roadside fencerows, and abandoned farmlands (Cuthbert 1963). Shortgrass habitats are important for foraging breeding pairs, since prey probably is more accessible (Gawlik and Bildstein 1990).

The bulky nests are usually 8 to 15 feet (2.4 to 4.6 m) above the ground, interwoven within the dense foliage of a prominent shrub or small tree, such as eastern red cedar (*Juniperus virginianus*). In Michigan, from 1986 to 1988, loggerhead shrike nesting pairs were found utilizing the following habitats: nests in a spruce and red cedar hedgerow in an open field adjacent to intensive agricultural areas, nest in a Scotch pine (*Pinus sylvatica*) within an abandoned field of knapweed (*Centaurea* spp.) and brambles (*Rubus* spp.), nest in a spruce on a golf course, and nest in an apple tree (*Malus* spp.) in an abandoned yard and cornfield (Little 1987). As indicated in the literature, Little (1987) found Michigan shrikes in habitats with "low vegetation to forage in, multiple perches, and protective trees to nest in."

Ecology and Life History: Loggerhead shrikes arrive at their breeding grounds between mid-March and mid-April. Adult males arrive prior to the females, and nearly half of the males may return to the nesting area used in the previous year (Kridelbaugh 1983); the same nest may be used and rebuilt in successive years. Shrike pairs, particularly females, may change nest sites annually.

After a short courtship, both sexes begin nest construction in a thorny or dense shrub. An average clutch of four to six eggs is laid between mid-April and late June. The female is primarily responsible for incubation, which lasts 13 to 16 days. During this time, the male supplies food for the female and aggressively defends the nesting territory with wing-fluttering displays (Bent 1950; Smith 1973a). Both adults usually participate in feeding the young. Twenty days after hatching, an average of three young fledge; and two to four weeks later, the young are self-sufficient.

While young are still dependent on the adults for food, a second nesting is frequently initiated by the female, occasionally with another male (Haas and Sloane 1989). Double broods commonly occur in Michigan (Little 1986 and 1987). Whether the female renests or not, she usually abandons the male, which then assumes responsibilities for the fledglings (Kridelbaugh 1983). Paired adults eventually separate in the autumn and maintain individual winter territories until the following spring (Miller 1931).

The loggerhead shrike regularly preys on small mammals and birds, although larger insects serve as its primary food source. Availability of prey usually dictates the shrike's diet. In Michigan, the summer diet is commonly composed of insects (chiefly grasshoppers, crickets, and beetles), while mice and other small mammals are taken in the winter and early spring. Small birds, typically sparrows and warblers, are taken opportunistically. Larger species, such as northern cardinals (*Cardinalis cardinalis*)(Ingold and Ingold 1987) and mourning doves (*Zenaida macroura*)(Balda 1965), also are killed occasionally.

While hunting, the loggerhead shrike waits on an exposed tree limb, telephone wire, post, or other vantage point to locate prey. Once prey is sighted, the bird will either hover over or directly pounce on the prey. Larger prey are usually killed by striking at the back of the neck (Busbee 1976; Smith 1973b) with the highly developed, toothlike cutting edge of the upper bill (Cade 1967). Loggerhead shrikes prefer to hunt within a distance of 50 feet (15.2 m) from the perch (Morrison 1980), sometimes hopping on the ground to flush prey. When prey is captured, it is carried in the bill, occasionally transferred in midair to the shrike's feet. When feeding, a sharp projection is usually needed to impale and hold the prey due to the shrike's underdeveloped feet. Large prey are often impaled and stored. Prey less than approximately $1^{1}/_{2}$ inches (3.8 cm) long are swallowed whole (Craig 1978). In summer, males typically store food on thorns, frequently near the nest. The female uses this as a food source, or cache, which may increase fledging success by allowing the female to expend more time and energy with the young (Applegate 1977). The unique behavior of impaling food by shrikes has been studied in depth (Miller 1931; Wemmer 1969; Smith 1972; Smith 1973b; Busbee 1976).

Conservation/Management: Though still common in the western and southern part of its range, the loggerhead shrike has continued to decline east of the Mississippi River. New England shrike populations may have invaded the area after logging and clearing provided suitable habitat, but Midwest populations historically were present. The dependence on open country is one reason it is declining in the Northeast, since many areas are maturing from abandoned farms into forests (Milburn 1981). This does not explain its disappearance in areas with apparently suitable habitat (e.g., Michigan). Possible factors causing the loggerhead's decline include loss of specific breeding habitat components, reduction in an abundance of available prey (in spring and winter), removal of thorny shrubs and hedgerows (used for nesting, foraging, and food storage), high migratory mortality, and loss of available wintering habitat.

Since the shrike is at the upper level of the food chain, it may be significantly affected by environmental contaminants. Erdman (1970) suspected pesticide residues in eggshells were partially responsible for population declines. More recently, Anderson and Duzan (1978) revealed that eggshell thickness had been slightly reduced by DDE, but it apparently had little effect on that year's nesting success. The effects of DDE and other chemicals, such as endrin applied in orchards (Blus et al. 1983; Fleming et al. 1983), on shrike reproduction requires further study.

Although the reasons for the loggerhead shrike's decline are not known, it does require successional habitat, and, without proper management, its long-term survival in specifically designated areas can not be assured. More research is needed on its breeding and wintering grounds before active habitat management can be implemented for Michigan populations. Until then, strict protection of birds and current nest site areas, retention and increased planting of thorny shrubs, and public awareness are the main tools available for aiding this species' survival.

LITERATURE CITED

Anderson, W. L., and R. E. Duzan. 1978. DDE residues and eggshell thinning in loggerhead shrikes. *Wilson Bull.* 90:215–20.

Applegate, R. D. 1977. Possible ecological role of food caches of loggerhead shrike. *Auk* 94:391–92.

Balda, R. P. 1965. Loggerhead shrike kills mourning dove. *Condor* 67:359.

Barrows, W. B. 1912. *Michigan bird life*. Mich. Agric. Coll. Spec. Bull., E. Lansing.

Bent, A. C. 1950. Life histories of North American wagtails, shrikes, vireos, and their allies. *U.S. Natl. Mus. Bull.* 197.

Blus, L. J., C. J. Henny, T. E. Kaiser, and R. A. Grove. 1983. Effects on wildlife from use of endrin in Washington state orchards. *48th N. Am. Wildl. Conf.*,159–74.

Brewer, R., G. A. McPeek, and R. J. Adams, Jr. 1991. *The atlas of breeding birds of Michigan*. Mich. State Univ. Press, E. Lansing.

Brooks, B. L., and S. A. Temple. 1986. The breeding distribution of the loggerhead shrike in Minnesota: A preliminary report. *Loon* 58:151–54.

Brooks, B. L., and S. A. Temple. 1990. Habitat availability and suitability for loggerhead shrikes in the Upper Midwest. *Am. Midl. Nat.* 123:75–83.

Burnside, F. L. 1987. Long-distance movements by loggerhead shrikes. *J. Field Ornith.* 58:62–65.

Burton, K. M. 1989. 1989 loggerhead shrike progress update. *Indiana Dept. Nat. Resour.*, Unpubl. Rept.

Busbee, E. L. 1976. The ontogeny of cricket killing and mouse killing in loggerhead shrikes (*Lanius ludovicianus* L.). *Condor* 78:357–65.

Bystrak, D. 1981. The North American breeding bird survey. *Stud. Avian Biol.* 6:34–41.

Cade, T. J. 1967. Ecological and behavorial aspects of predation by a northern shrike. *Living Bird* 6:43–86.

Cadman, M. D., P. F. J. Eagles, and F. M. Helleiner. 1987. *Atlas of the breeding birds of Ontario*. Univ. Waterloo Press, Waterloo, Ont.

Craig, R. B. 1978. An analysis of the predatory behavior of the loggerhead shrike. *Auk* 95:221–34.

Cuthbert, N. L. 1963. *The birds of Isabella County, Michigan*. Edwards Brothers, Ann Arbor, MI.

Erdman, T. C. 1970. Current migrant shrike status in Wisconsin. *Passenger Pigeon* 32:144–50.

Fleming, W. J., D. R. Clark, and C. J. Henny. 1983. Organochlorine pesticides and PCB's: A continuing problem for the 1980s. *48th N. Am. Wildl. Conf.*, 186–99.

Fruth, K. J. 1988. The Wisconsin loggerhead shrike recovery plan. *Wisc. Dept. Nat. Resour.*, Unpubl. Rept.

Gawlik, D. E., and K. L. Bildstein. 1990. Reproductive success and nesting habitat of loggerhead shrikes in north-central South Carolina. *Wilson Bull.* 102:37–48.

Geissler, P. H., and B. R. Noon. 1981. Estimates of avian population trends from the North American breeding bird survey. *Stud. Avian Biol.* 6:42–51.

Haas, C. A., and S. A. Sloane. 1989. Low return rates of migratory loggerhead shrikes: Winter mortality or low site fidelity? *Wilson Bull.* 101:458–60.

Ingold, J. J., and D. A. Ingold. 1987. Loggerhead shrike kills and transports a northern cardinal. *J. Field Ornith.* 58:66–68.

Janssen, R. B. 1987. *Birds in Minnesota.* Univ. Minn. Press, Minneapolis.

Kridelbaugh, A. L. 1981. Population trend, breeding and wintering distribution of loggerhead shrikes (*Lanius ludovicianus*) in Missouri. *Trans. Missouri Acad. Sci.* 15:111–19.

Kridelbaugh, A. L. 1983. Nesting ecology of the loggerhead shrike in central Missouri. *Wilson Bull.* 95:303–8.

Little, J. M. 1986. The loggerhead shrike in Michigan: Its distribution and habitat needs. *Mich. Dept. Nat. Resour.*, Unpubl. Rept.

Little, J. M. 1987. Distribution and management potential of the loggerhead shrike in Michigan. *Mich. Dept. Nat. Resour.*, Unpubl. Rept.

Milburn, T. 1981. Status and distribution of the loggerhead shrike, *Lanius ludovicianus*, in the northeastern United States. *U.S. Fish Wildl. Serv.*, Unpubl. Rept.

Miller, A. H. 1931. Systematic revision and natural history of the American shrikes (Lanius). *Univ. Calif. Publ. Zool.* 38:11–242.

Morrison, M. L. 1980. Seasonal aspects of the predatory behavior of loggerhead shrikes. *Condor* 82:296–300.

Morrison, M. L. 1981. Population trends of the loggerhead shrike in the United States. *Am. Birds* 35:754–57.

Payne, R. P. 1983. A distributional checklist of the birds of Michigan. *Univ. Mich. Mus. Zool. Misc. Publ.* No. 164.

Robbins, C. S., D. Bystrak, and P. H. Geissler. 1986. The breeding bird survey: Its first fifteen years, 1965–1979. *U.S. Fish Wildl. Serv., Resour. Publ.* No. 157.

Smith, S. M. 1972. The ontogeny of impaling behavior in the loggerhead shrike, *Lanius ludovicianus* L. *Behaviour* 42:232–47.

Smith, S. M. 1973a. An aggressive display and related behavior in the loggerhead shrike. *Auk* 90:287–98.

Smith, S. M. 1973b. A study of prey-attack behavior in young loggerhead shrikes, *Lanius ludovicianus* L. *Behaviour* 44:113–41.

Tate, J. 1986. The blue list for 1986. *Am. Birds* 40:227–36.

Wemmer, C. 1969. Impaling behavior of the loggerhead shrike, *Lanius ludovicianus* Linnaeus. *Z. F. Tierpsychol.* 26:208–24.

Zimmerman, D. A. 1955. Notes on field identification and comparative behavior of shrikes in winter. *Wilson Bull.* 67:200–208.

Yellow-throated Warbler

· *Dendroica dominica* (Linnaeus)
· **State Threatened**

Map 24. Breeding evidence of yellow-throated warblers.
▲ = before 1983 (Barrows 1912); ● = 1983 to 1992 (Brewer et al. 1991).

Status Rationale: Extirpated from Michigan near the turn of the century, a population was discovered in the extreme southern Lower Peninsula in 1988 and since then singing males have been found in other locales. An estimated 20 to 25 nesting pairs occur in Michigan. Historical threats of habitat alteration have been temporarily alleviated with the maturation and protection of riverine floodplains.

Field Identification: This medium-sized warbler is appropriately named for its yellow throat and breast bib. The black-and-white head pattern, broad white eyebrow stripe, and solid slate-gray back are characteristic. Like many other species of *Dendroica*, it has black side streaks and two white wing bars. Breeding females, fall adults, and immatures are slightly duller in color. Atypical of wood warblers, the adults do not undergo much plumage transformation in autumn. Adults vary in length from $4^1/_2$ to $5^3/_4$ inches (11.4 to 14.6 cm). The western subspecies *D. d. albilora*, commonly known as the sycamore warbler, occurs in Michigan and is separated from the nominate subspecies (which has yellow lores) by its complete white eyebrow.

Because this species generally remains in the treetops during the breeding season, it is rarely seen. Detection and recognition of its song are crucial for determining its presence. The yellow-throated warbler has a loud, musical song that may be confused with several other passerines, particularly the indigo bunting (*Passerina cyanea*). The bunting, however, has a more varied song, typically given in paired notes. The primary song of the yellow-throated warbler is a series of slurred notes that becomes faster and drops in pitch and finally ends with an abrupt high note. A frequent associate, the Louisiana waterthrush (*Seiurus motacilla*), also has a vociferous, melodic song but differs by having three slurred introductory notes.

Total Range: The yellow-throated warbler is typically a species of the southeastern United States. Its primary range extends north along the Atlantic coast to Maryland, west to Oklahoma, and north to the southern Great Lakes region. This warbler has shown a significant increase in the eastern United States since the early 1970s (Robbins et al. 1986). Results from United States Fish and Wildlife Service Breeding Bird Survey show that the Cumberland Plateau in Kentucky and Tennessee supports the greatest abundance. This species is increasing in the Northeast, with recent confirmed nesting in West Virginia (Smith 1978), New York, and Pennsylvania (Andrle and Carroll 1988).

The western subspecies of the yellow-throated warbler breeds in the Mississippi Valley north to central Indiana and Ohio. Small, disjunct populations of this subspecies were historically present in extreme southern Michigan and adjoining states. Most populations in the southern Great Lakes region, however, were extirpated near the turn of the century. Long-term increases in Ohio (Peterjohn 1989) indicate a return to the southern Great Lakes region. The yellow-throated warbler winters along the Atlantic coast, north to South Carolina, and along the Gulf coast south to Central America. Michigan's birds probably winter along the Gulf coast (Griscom et al. 1957).

State History and Distribution: Michigan is at the northern limit of the yellow-throated warbler's range. Historically, known populations were restricted to suitable sycamore (*Platanus occidentalis*) floodplain habitat along the Raisin River Valley in Monroe County, the Huron River Valley in Washtenaw County, and the Kalamazoo River Valley in Kalamazoo County (Barrows 1912). Early observers noted its almost exclusive dependence on the upper, emergent branches of mature sycamores. After the turn of the century, this warbler virtually disappeared from Michigan, except for isolated observations in the 1970s.

Since the late 1980s, there has been a resurgence of the species. During this period, isolated singing males have been found in Branch, Cass, Hillsdale, Kalamazoo, Lenawee, Oakland, and St. Joseph counties, and several birds are observed annually in Berrien County (Brewer et al. 1991). In 1988 and 1989, systematic surveys along parts of Spring Brook and the South Branch of the Galien River documented 14 to 21 breeding territories (Evers 1994). This 5-mile (8-km) contiguous stretch of mature floodplain forest is currently the only known breeding population of the yellow-throated warbler in Michigan. An estimated total of 20 to 25 territories are known across the southernmost tier of counties.

Habitat: Michigan's yellow-throated warbler population is closely associated with mature sycamores. This tree is characteristic of bottomland and river floodplain forests and is limited to the southern Lower Peninsula (Barnes and Wagner 1981).

However, this habitat preference does not prevail throughout the warbler's range. In the South, it occurs in mature cypress swamps and southern pine forests (Mengel 1964). The recently discovered Berrien County population occurs along a meandering stream (averaging 20 to 30 feet [6.1 to 9.1 m] in width). This relatively undisturbed waterway is characterized by mature sycamores with mature silver maples (*Acer saccharinum*) and American basswood (*Tilia americana*) serving as the lower but continuous vegetative layer (Evers 1994).

Ecology and Life History: The yellow-throated warbler is one of the earliest to return to Michigan in the spring, arriving in the state from mid-April to mid-May. Breeding males sing loud, persistent songs, which may define territories, from the uppermost branches of mature trees, typically sycamores. Singing activity declines markedly by the end of June. Nests are generally placed in sycamores, far from the trunk and a substantial distance from the ground (nests are known to be over 100 feet [30.5 m] above the ground). An average of four eggs is laid, and incubation is by the female for approximately two weeks. Young probably fledge within another two weeks. Most individuals leave the breeding grounds by August, but there are few late autumn and winter records in Michigan.

The yellow-throated warbler is an opportunistic feeder that gleans or "flycatches" a wide range of insect species. This warbler's feeding habits are unusual; it has the ability to encircle tree branches at any position while searching for food.

Conservation/Management: The destruction of floodplain forests with mature sycamores resulted in the disappearance of this warbler in Michigan at the turn of this century. With the discovery of an established breeding population, protection of the yellow-throated warbler is possible within its historic range. Current threats still include development and alteration of habitat at suitable sites, as well as possible nest parasitism by brown-headed cowbirds (*Molothrus ater*). The cowbird negatively affects other forest-nesting species when large, contiguous tracts of forest are fragmented (Brittingham and Temple 1983).

Floodplain habitat in Berrien County needs to be protected to ensure the survival of the only known population of yellow-throated warblers. Active management should not be necessary as long as an adequate buffer zone is preserved. Other sites with single, singing males may be enhanced by limiting disturbance (e.g., selective cutting) within habitat containing large sycamore trees.

Southern Michigan bottomlands provide critical habitat for this warbler and other associated wildlife. In addition, riverine floodplain forests stabilize extensive natural draining systems. If planned conservation measures can be implemented in these areas, the yellow-throated warbler will remain a part of Michigan's forest community.

LITERATURE CITED

Andrle, R. F., and J. R. Carroll. 1988. *The atlas of breeding birds in New York state.* Cornell Univ. Press, Ithaca, NY.

Barnes, B. V., and W. H. Wagner, Jr. 1981. *Michigan trees.* Univ. Mich. Press, Ann Arbor.

Barrows, W. B. 1912. *Michigan bird life.* Mich. Agric. Coll. Spec. Bull., E. Lansing.

Brewer, R., G. A. McPeek, R. J. Adams, Jr. 1991. *The atlas of breeding birds of Michigan.* Mich. State Univ. Press, E. Lansing.

Brittingham, M. C., and S. A. Temple. 1983. Have cowbirds caused forest songbirds to decline? *BioScience* 33:31–35.

Evers, D. C. 1994. Expansion of the yellow-throated warbler in the southern Great Lakes region. Michigan Birds 1:3–9.

Griscom, L., et al. 1957. *The warblers of America.* Delvin-Adair Co., New York.

Mengel, R. M. 1964. The probable history of species formation in some northern wood warblers (Parulidae). *Living Bird* 3:9–43.

Peterjohn, B. G. 1989. *The birds of Ohio.* Ind. Univ. Press, Bloomington.

Robbins, C. S., D. Bystrak, and P. H. Geissler. 1986. The breeding bird survey: Its first fifteen years, 1965–1979. *U.S. Fish Wildl. Serv., Resour. Publ.* 157.

Smith, J. L. 1978. Northward expansion of the yellow-throated warbler. *Redstart* 45:56–58.

Prairie Warbler

- *Dendroica discolor* (Vieillot)
- **State Threatened**

Map 25. Breeding evidence of prairie warblers. Historically scattered through the Lower Peninsula (Barrows 1912); ● = 1983 to 1992 (Brewer et al. 1991).

Status Rationale: Since European settlement, Michigan's prairie warbler population has been uncommon and locally distributed. Recently, however, populations apparently have declined. If these declines are real, the loss of specific stages of optimal successional habitat may be responsible.

Field Identification: This medium-sized warbler has olive-green upperparts and yellow underparts with prominent black streaks confined to the sides. Chestnut-colored streaks on the back are apparent with close inspection, particularly on adult males. Two black streaks are present on the head, one through the eye, the other along the jaw. Sexual differences in plumage are minor; females have less pronounced black streaks. Immatures have faint rather than distinct head and side streaks. Males have a varied repertoire of songs with vocalizations that are thin and buzzy (e.g., *zee zee zee . . . zeet*) and ascend in scale, with a typical song consisting of 8 to 14 notes.

The characteristic tail bob of the prairie warbler distinguishes it from most other yellowish warblers. The palm (*Dendroica palmarum*) and Kirtland's (*D. kirtlandii*) warbler also pump their tail. In all plumages, the palm warbler lacks completely yellow underparts and the Kirtland's warbler has markedly contrasting yellow underparts, dark grayish upperparts, and a prominent eye ring. The plumage of immature prairie and Magnolia warblers are similar; however, the prairie warbler lacks white wing bars and a yellow rump.

Total Range: The prairie warbler is confined primarily to the southeastern United States. There are two subspecies: *D. d. discolor* ranges from eastern Oklahoma and northeastern Texas, east to the Atlantic coast and north to New England, southern Ontario, and Michigan. It is locally distributed throughout this range, being absent from much of the Midwest. Highest abundance occurs in the core of its range in the southern Piedmont and Virginia (Robbins et al. 1986). The other subspecies, *D. d. paludicola*, is a warbler of mangrove habitats along the ocean coastline.

The prairie warbler's presettlement distribution differed from its current range; populations were more restricted and disjunct within the formerly expansive deciduous forests (Nolan 1978). Recently, loss of scrub-growth habitats has been attributed to significant population declines, particularly in Arkansas, Georgia, Maryland, and North Carolina (Robbins et al. 1986). Declines are also evident for wintering migrants in the West Indies (Terborgh 1989). Populations of the more northern subspecies (*D. d. discolor*) are migratory and winter in Florida and the West Indies; small numbers occur in Mexico, Central, and South America (American Ornithologists' Union 1983). The more southern subspecies (*D. d. paludicola*) is mostly sedentary.

State History and Distribution: Michigan serves as the northernmost outpost of the prairie warbler, where it still occurs in disjunct populations. Populations never became established in the Upper Peninsula; the sole evidence of nesting is juveniles captured in July and August in Baraga County (Payne 1983). Because of its dependence on successional habitat, its range has changed predictably since its original discovery and subsequent population peak in the early 1900s. Recently, however, numbers have declined, particularly in former strongholds such as the north-central Lower Peninsula. Prairie warbler nesting pairs frequently were associated with Kirtland's warbler nesting populations in designated management areas of jack pine plains and oak barrens (Walkinshaw 1983). Surveys in the 1980s, however, indicate that prairie warbler populations now have disappeared from this region (Brewer et al. 1991).

Today, most populations and solitary singing males are restricted to habitats associated with the dune shoreline of Lake Michigan. The largest population is perhaps along the shoreline sand dune habitats in Mason and Benzie counties. Even in these habitats, however, the occurrence and abundance of the prairie warbler changes annually. For example, it disappeared as a summer resident of the sand dune shorelines in Berrien County as recently as the early 1970s (Payne 1983), although it still remains nearby in the Indiana Dunes State Park (Mlodinow 1984).

Habitat: The prairie warbler prefers scrub-shrub habitats, rather than prairies as its name implies. Large openings surrounding shrub islands are usually a major component of its breeding territory. Unfortunately, the age of the preferred breeding habitat (i.e., a relatively early stage of succession) coincides with rapid structural change in vegetation. Therefore, specific sites do not support long-term, self-sustaining populations; instead, a mosaic of suitable habitats is needed.

Optimal breeding habitats are usually associated with poor soils, such as jack pine plains (*Pinus banksiana*)(Walkinshaw 1983), farmed-over areas (Graber and Graber 1963), strip mines (Brewer 1958), and brush-grown sand dunes and burn areas (Walkinshaw 1959). Pine plantations (particularly Christmas tree farms), oak clear-cuts, and powerline rights-of-way also are used. All these habitat types (except strip mines) are used by territorial males in Michigan.

Ecology and Life History: The prairie warbler arrives in Michigan from early to mid-May (Walkinshaw 1959). Males migrate earlier than females and are three times more site faithful than females (Nolan 1978). Mean territory size ranges from 3^1/$_2$ to 5 acres (1.4 to 2.0 ha), depending on local population dynamics (Nolan 1978). The female is responsible for nest building; the male generally guards the nest site. In Michigan, active nests are known from late May to mid-July (Walkinshaw 1959) and may occur later for pairs that are renesting or attempting an occasional second brood. A sample of 18 nests, representing several populations in Michigan, averaged 27 inches (69 cm) in height and most were built in hazel (*Corylus* spp.) and pines (*Pinus* spp.); one aberrant nest site was in a bracken fern (*Pteridium aquilinum*) (Walkinshaw 1959).

The clutch size ranges from three to five eggs. Nolan (1978) found that nearly three-quarters of the clutches contained four eggs, and mean clutch size declined from more than four eggs in the beginning of the nesting cycle to three eggs near the end; the average incubation period is 12 days. Only the female incubates, but both sexes care for the young. The young leave the nest between 9 and 11 days after the first eggs of the clutch hatches, and are independent after 40 days. Unsuccessful pairs will renest, often dispersing to other areas (Jackson et al. 1989).

The southward migration in Michigan begins in August and lasts until mid-September (Walkinshaw 1959). Survivorship rates are calculated to be approximately 57% for the first year and average 65% for subsequent years (Nolan 1978). The average life span is 2^1/$_2$ years (Nolan 1978).

The prairie warbler gleans insects and spiders from vegetation. Young are fed mainly caterpillars (Nolan 1978).

Conservation/Management: Michigan is at the periphery of this warbler's range, and its populations are small and discontinuous. In addition, localities that are occupied are widely distributed, forcing populations to be self-sustaining or dependent on the intermittent influx of immigrants. For these reasons, the prairie warbler's population status is difficult to assess, and its disappearance from its stronghold in north-central Michigan and its rarity elsewhere is alarming.

This warbler is also highly susceptible to nest parasitism by the brown-headed cowbird (*Molothrus ater*), an open-country species that has significantly increased in the Midwest (Robbins et al. 1986). Cowbird nest parasitism on the prairie warbler is documented throughout the Midwest, including Michigan (Walkinshaw 1959). Natural predation also is an important factor affecting nest success (nearly 80% of all nesting attempts). Snakes, eastern chipmunks (*Tamia striatus*), and blue jays (*Cyanocitta cristata*) prey on nests with eggs and young (Nolan 1978).

Habitat management strategies are crucial for continuing self-sustaining populations. First, research is needed on the minimum size, vegetative structure, and physical limitations in the preferred habitat. Protection of current populations in natural and relatively stable habitats (e.g., sand dune shorelines) is of immediate concern. Control of cowbird numbers in the Kirtland's Warbler Management Area has nearly eliminated nest parasitism on the Kirtland's warbler; however, prairie warbler populations in this same area have disappeared.

Habitat changes, nest parasitism, and high predation rates are severely stressing Michigan's breeding populations. Prairie warbler populations require complete protection and intense management efforts to preserve the species for future generations.

LITERATURE CITED

American Ornithologists' Union. 1983. *Check-list of North American birds,* 6th ed. Am. Ornith. Union, Washington, D.C.

Barrows, W. B. 1912. *Michigan bird life.* Mich. Agric. Coll. Spec. Bull., E. Lansing.

Brewer, R. 1958. Breeding-bird populations of strip-mined land in Perry Co., Ill. *Ecology* 39:543–45.

Brewer, R., G. A. McPeek, and R. J. Adams, Jr. 1991. *Atlas of breeding birds in Michigan.* Mich. State Univ. Press, E. Lansing.

Graber, R., and J. Graber. 1963. A comparative study of bird populations in Illinois, 1906–1909 and 1956–1958. *Ill. Nat. Hist. Bull.* 28:383–528.

Jackson, W. M., S. Rohwer, and V. Nolan, Jr. 1989. Within-season breeding dispersal in prairie warblers and other passerines. *Condor* 91:233–41.

Mlodinow, S. 1984. *Chicago area birds.* Chicago Review Press, Chicago.

Nolan, V., Jr. 1978. The ecology and behavior of the prairie warbler *Dendroica discolor.* *Ornith. Monogr.* No. 26, Am. Ornith. Union.

Payne, R. B. 1983. A distributional checklist of the birds of Michigan. *Univ. Mich. Mus. Zool. Misc. Publ.* No. 164.

Robbins, C. S., D. Bystrak, and P. H. Geissler. 1986. The breeding bird survey: Its first fifteen years, 1965–1979. *U.S. Fish Wildl. Serv., Resour. Publ.* 157.

Terborgh, J. 1989. *Where have all the birds gone?* Princeton Univ. Press, Princeton, NJ.

Walkinshaw, L. H. 1959. The prairie warbler in Michigan. *Jack-Pine Warbler* 37:54–63.

Walkinshaw, L. H. 1983. *Kirtland's warbler: The natural history of an endangered species.* Cranbrook Inst. Science, Bloomfield Hills, MI.

Kirtland's Warbler

· *Dendroica kirtlandii* (Baird)
· **Federally and State Endangered**

Map 26. Breeding evidence of Kirtland's warblers. ▲ = before 1983 (Michigan DNR); ● = 1983 to 1992 (Michigan DNR).

Status Rationale: The Kirtland's warbler is the only bird currently known to nest exclusively within the state's boundaries. This warbler breeds in north-central Michigan, and fewer than 1,000 breeding individuals remain in the world. The breeding habitat is transitory and requires active management due to the species need for large stands of young jack pines.

Field Identification: This relatively large-sized wood warbler (adults are 5³/4 inches [14.6 cm] in length) has a yellow breast, black streaks typically confined to the sides, two white wing bars, and a heavily streaked blue-gray back. The adult female is less colorful than the male, having gray cheeks and paler streaked sides and breast. The loud song of this warbler is musical and composed of several low, staccato notes that end abruptly. The house wren (*Troglodytes aedon*) and northern waterthrush (*Seiurus noveboracensis*) have similar song variations.

The Kirtland's warbler's persistent tail-wagging habit is similar to that of the palm warbler (*Dendroica palmarum*) and is useful in distinguishing it from other dark-backed warblers. Palm warblers are rare nesting associates with the Kirtland's warblers and may be distinguished by the brown back. Other potentially confusing warblers with yellow breasts include (1) the Canada warbler (*Wilsonia canadensis*), which lacks the two white wing bars, the heavily streaked back, and has a necklace of short black stripes, and (2) the magnolia warbler (*Dendroica magnolia*), which differs with its large, white wing and tail patches and yellow rump. A species

199

occasionally associated with the Kirtland's warbler is the prairie warbler (*Dendroica discolor*), which is similarly patterned but has an olive-colored back and a distinctive black eye line over its yellow cheek.

Total Range: The breeding range of the Kirtland's warbler is currently confined to the north-central part of Michigan's Lower Peninsula. Although the Kirtland's warbler probably has had a restricted range for thousands of years, it is doubtful that it was ever in danger of natural extinction (Mayfield 1975). Recent evidence suggests it may have been initially limited to the southeastern United States during the last glacial advance and has since gradually moved northward, providing an explanation for its unique and concentrated migratory route (Mayfield 1988a).

Natural long-distance dispersal and colonization is expected for a species dependent upon transitory habitat. There is only one confirmed nesting of this species, however, outside of its present breeding grounds. In 1945, a pair was observed feeding a juvenile near Midhurst, Ontario (Speirs 1984). A breeding colony apparently existed near Renfew, Ontario, in 1916, but nesting was not confirmed (Harrington 1939; Speirs 1984). Since 1977, scattered territorial males have been observed in Ontario, Quebec, Wisconsin, and Michigan's Upper Peninsula (Probst 1985; Hoffman in Huron-Manistee National Forests 1989). Recent surveys in Ontario and Wisconsin have located singing males but no breeding pairs. In 1988, eight singing males were found in three northern Wisconsin counties.

The Kirtland's warbler is extremely elusive and seldom observed during migration (Stone 1986). Kirtland's warblers overwinter southeast of Florida in the 600-mile Bahama Archipelago. Recent United States Fish and Wildlife Service surveys have also found individuals on surrounding island chains.

State History and Distribution: In 1951, the first complete census of the warblers located 432 singing males (Mayfield 1953); in 1957, organized breeding habitat management efforts were initiated (Mayfield 1963; Radtke and Byelich 1963) and three, 4-square-mile sections were established by the Michigan Conservation Commission (now the Department of Natural Resources). In 1962, the National Forest Service became involved and dedicated a portion of the Huron National Forest for a Kirtland's Warbler Management Area. A second survey, in 1961, located a total of 502 singing males (Mayfield 1962), but, by 1971 (the third decennial survey), the warbler had declined dramatically with only 201 singing males counted (Mayfield 1972).

This alarming population decline and the advent of the federal (1973) and state (1974) endangered species acts prompted a tremendous increase in efforts by state and federal agencies. As a result, a Recovery Team was appointed to develop a recovery plan for the Kirtland's warbler. The primary objective of the plan is to establish and maintain a viable population of 1,000 pairs, which would require nearly 130,000 acres of habitat managed on a rotational basis (Byelich et al. 1985). Since warblers occupy stands between year 8 and 20 out of a 50-year timber rotation, only a small portion of the 130,000 acres is annually occupied.

Since the warbler's breeding grounds were found in 1903, 13 Michigan counties have supported nesting pairs. In 1988, singing males were found only in 6 Michigan counties. The bulk of the breeding population currently resides in Crawford and Oscoda counties. Other counties that recently supported singing males include Alcona, Iosco, Kalkaska, Montmorency, Ogemaw, and Roscommon. In 1982 and 1983, singing males were found near Gwinn, Marquette County, in the Upper Peninsula (Probst 1985). In 1992, 397 singing males were counted. Assuming a female is present for each singing male (and males have only one mate), the 1992 prenesting population was at least 794 individuals. There is some concern for this basic assumption of the breeding population estimate, since recent studies indicate a skewed sex ratio in favor of males (Probst and Hayes 1987).

By autumn, an estimated 1,000 adults and immatures depart for the Bahama Archipelago wintering grounds. The spring and autumn migration is confined to a concentrated corridor between the nesting and wintering sites. A proportionate number of observations occur near the end of each respective migratory route, indicating the Kirtland's warbler does most of its migratory flight without stopping (Mayfield 1988b). Autumn migrants probably move alone (Clench 1973; Sykes et al. 1989).

Habitat: The Kirtland's warbler is dependent upon large, open, relatively homogeneous stands of jack pine (*Pinus banksiana*) growing on the Grayling Sand substrate within the Au Sable River drainage. According to Walkinshaw (1983), 90% of the nests are associated with this soil type. This specific nesting habitat is further restricted because the nest is placed only among concealing ground vegetation, near jack pines at least 6 to 8 feet [1.8 to 2.4 m] tall (i.e., 8–20 years old) and in stands of 80 acres or more with numerous small grassy openings. Once jack pines reach a height greater than 18 feet (5.5 m), the lower branches begin to drop and the shade-intolerant ground cover changes in composition, thereby leading to unfavorable nesting conditions. Mayfield (1960) suggests that the presence of low, living branches (not the height of trees) is the most crucial habitat requirement. Kirtland's warblers are showing increasing use of jack pine plantations that have been specifically planted for them under a habitat management plan.

Jack pines generally need intense heat to open their cones and release the seeds. Burning of this warbler's habitat is generally required to clear areas for new growth and to prepare sites for regeneration of new jack pine stands, either through natural regeneration by seeds released from cones on residual trees or through the planting of seedlings. Although Buech (1980) found little significant difference between burned and unburned jack pine stands, fire probably contributes to soil and vegetative composition. Minimal stand size is also crucial. Anderson and Storer (1976) showed that jack pine stands larger than 200 acres significantly improve nesting success. Red pine (*Pinus resinosa*) stands are occasionally utilized (Orr 1975; Mayfield 1953). Buech (1980) suggested that nest sites are chosen in relation to the vegetative density and structure, not tree species. Therefore, this warbler will nest under various tree species (within jack pine stands), as long as its specific nesting niche is fulfilled.

Winter habitat requirements only recently have been studied and are probably not limiting. Individuals have been seen on the Bahama Archipelago in stunted scrub areas with dense thickets (Radabaugh 1974) and in pinelands with an open canopy and dense vegetative understory.

Ecology and Life History: Following the nearly 1,400 mile northward migration, the bulk of the male Kirtland's warbler population arrives on the breeding grounds around May 10, with females usually arriving after May 18. Males usually establish large (relative to other warbler species), 10- to 20-acre territories in loose aggregations through July (Walkinshaw 1983) and defend them both physically and vocally. Territory size decreases with increasing densities. Kirtland's warblers nest on the ground, usually near the base of young jack pines. The nest is typically sunken and well concealed by the surrounding vegetation. The sandy substrate used for nesting permits water to quickly percolate downward, preventing flooding of nests during sporadic summer rainstorms.

The first clutch, typically five eggs, is usually laid between May 21 and June 12. If unsuccessful, a second attempt averages four eggs (Walkinshaw 1983). The female incubates the eggs for 13 to 16 days, but both parents care for the young. Four to five weeks after hatching, young-of-the-year are independent. If young fledge before July, adults will often attempt a second nest (Radabaugh 1972; Walkinshaw and Faust 1974). Males are extremely territorial and, regardless of weather, sing persistently throughout the day. This behavior enables annual censuses of singing males in mid-June, which provide accurate counts of the species' total population. Immatures leave the nesting grounds from mid-August to early September and adults depart by late September (Sykes et al. 1989); the latest record on the breeding grounds is October 1 (Sykes and Munson 1989). Records are known on its wintering grounds beginning in August (Hundley 1967; Wallace 1968). During the spring and summer, Kirtland's warblers feed on insects on the ground, in pines, and in scrub oak. In winter, fruits and arthropods are eaten.

Conservation/Management: The current population of Kirtland's warbler is extremely small and vulnerable to a number of limiting factors. Although significant conservation methods have been employed, several serious threats remain. Problems that were originally identified as limiting the warbler's success are now alleviated or being addressed. These include intense brown-headed cowbird (*Molothrus ater*) brood parasitism, human intrusion, and loss of nesting habitat. Cowbirds became a problem when they expanded northward in the late 1800s following deforestation in the Great Lakes region (Brittingham and Temple 1983). Cowbirds lay one or more eggs in a host's nest, and their young typically overpower the smaller nestlings. Since the Kirtland's warbler lacks an effective defense mechanism against the cowbird, it is an extremely vulnerable host (Mayfield 1961; Walkinshaw 1972).

In 1972, the United States Fish and Wildlife Service initiated a cowbird trapping program. This highly efficient and successful annual program continues today. Parasitism is currently reduced to negligible levels, and warbler fledging success is nearly identical to historic rates. The control of the cowbird (Kelly and DeCapita

1982) has temporarily prevented the Kirtland's warbler's predicted extinction (Mayfield 1975).

Proper habitat management within the breeding range is an important factor for the warbler's survival. A rotation of managed habitat began in 1980. Nearly 130,000 acres (58% state forest and 42% national forest) are currently managed in 24 general areas (17 on state forests and 7 on national forests) that are subdivided into 1,000- to 2,000-acre management units. Within these units, five blocks are generally cut at 10-year intervals on a 50-year rotation cycle. This system is designed to provide large tracts, totaling 36,000 to 40,000 acres of jack pine, that will annually offer suitable habitat for the warblers. About 2 million jack pines are annually planted, providing about 1,500 acres of suitable breeding habitat.

The Kirtland's warbler population has stabilized, and recently has shown dramatic increases. However, other limiting factors possibly impeding the warbler's recovery are being addressed. Anderson and Storer (1976) evaluated several variables that may affect the Kirtland's warbler's success and found a significant relationship between trees and snags remaining after clearing and decreased warbler success. Perches aid cowbirds and avian predators with vantage points.

The establishment of populations outside the current major breeding grounds is part of the Kirtland's Warbler Recovery Plan (Byelich et al. 1985). Actual transfer of adults and cross-fostering are two commonly used reestablishment techniques. Recent studies by Ohio State University show low success (with surrogate warblers) in the establishment of nesting pairs through the capture of individuals in autumn, holding over winter, and then releasing in the spring (Bocetti in Huron-Manistee National Forests 1989). Brewer and Morris (1984) tested the feasibility of cross-fostering with a similar warbler species as the donor and a sparrow species as the foster parents. They had limited success and found the chipping sparrow (*Spizella passerina*) to be the best foster parent based on similar habitat needs. Emergency efforts (e.g., captive breeding) designated by the Recovery Plan would be implemented if the population declines to fewer than 100 pairs.

Visitors are currently allowed to view the birds only in selected managed units during guided tours originating from the Grayling Holiday Inn or National Forest Service District Ranger Office in Mio. Otherwise, the areas are off-limits during the May 1 through August 15 breeding season. Sykes et al. (1989) suggest complete closure of areas with more than ten singing males until mid-September to protect postbreeding individuals (the major areas are currently off-limits through September 10).

The current Recovery Plan maintains the primary objective as the reestablishment of a self-sustaining Kirtland's warbler population with a minimum level of 1,000 pairs. To reach this goal, five objectives have been determined and are currently being reviewed and addressed. With the many accomplishments and massive amount of effort that has been expended and with prospects of a large amount of new habitat available, long-term survival of the Kirtland's warbler is promising.

LITERATURE CITED

Anderson, W. L., and R. W. Storer. 1976. Factors influencing Kirtland's warbler nesting success. *Jack-Pine Warbler* 54:105–15.

Brewer, R., and K. R. Morris. 1984. Cross fostering as a management tool for the Kirtland's warbler. *J. Wildl. Mgmt.* 48:1041–45.

Brittingham, M. C., and S. A. Temple. 1983. Have cowbirds caused forest songbirds to decline? *BioScience* 33:31–35.

Buech, R. R. 1980. Vegetation of a Kirtland's warbler breeding area and 10 nest sites. *Jack-Pine Warbler* 58:58–72.

Byelich, J., M. E. DeCapita, G. W. Irvine, N. I. Johnson, W. R. Jones, H. Mayfield, R. E. Radtke, and W. J. Mahalak. 1985. Kirtland's warbler recovery plan. *U.S. Fish Wildl. Serv.*, Twin Cities, MN.

Clench, M. H. 1973. The fall migration route of Kirtland's warbler. *Wilson Bull.* 85:417–28.

Harrington, P. 1939. Kirtland's warbler in Ontario. *Jack-Pine Warbler* 17:95–97.

Hundley, M. H. 1967. Recent winter records of the Kirtland's warbler. *Auk* 84:425–26.

Huron-Manistee National Forests. 1989. At the crossroads—extinction or survival. *Proc. Kirtland's warbler symposium*, Lansing, MI.

Kelly, S. T., and M. E. DeCapita. 1982. Cowbird control and its effect on Kirtland's warbler reproductive success. *Wilson Bull.* 94:363–65.

Mayfield, H. F. 1953. A census of the Kirtland's warbler. *Auk* 70:17–20.

Mayfield, H. F. 1960. *The Kirtland's warbler.* Cranbrook Inst. Science, Bloomfield Hills, MI.

Mayfield, H. F. 1961. Cowbird parasitism and the population of the Kirtland's warbler. *Evolution* 15:174–79.

Mayfield, H. F. 1962. 1961 decennial census of the Kirtland's warbler. *Auk* 79:173–82.

Mayfield, H. F. 1963. Establishment of preserves for the Kirtland's warbler in the state and national forests of Michigan. *Wilson Bull.* 75:216–20.

Mayfield, H. F. 1972. Third decennial census of Kirtland's warbler. *Auk* 89:263–68.

Mayfield, H. F. 1975. The numbers of Kirtland's warblers. *Jack-Pine Warbler* 53:39–47.

Mayfield, H. F. 1988a. Where were Kirtland's warblers during the last ice age? *Wilson Bull.* 100:659–60.

Mayfield, H. F. 1988b. Do Kirtland's warblers migrate in one hop? *Auk* 105:204–5.

Orr, C. D. 1975. 1974 breeding success of the Kirtland's warbler. *Jack-Pine Warbler* 53:59–66.

Probst, J. R. 1985. Summer records and management implications of Kirtland's warbler in Michigan's Upper Peninsula. *Jack-Pine Warbler* 63:9–16.

Probst, J. R., and J. P. Hayes. 1987. Pairing success of Kirtland's warblers in marginal vs. suitable habitat. *Auk* 104:234–41.

Radabaugh, B. E. 1972. Double-broodness in the Kirtland's warbler. *Bird Banding* 43:55.

Radabaugh, B. E. 1974. Kirtland's warbler and its Bahama wintering grounds. *Wilson Bull.* 86:374–83.

Radtke, R., and J. Byelich. 1963. Kirtland's warbler management. *Wilson Bull.* 75:208–15.

Speirs, D. H. 1984. The first breeding record of Kirtland's warbler in Ontario. *Ontario Birds* 2:80–84.

Stone, A. E. 1986. Migration and wintering records of Kirtland's warbler: An annotated bibliography. *U.S. Fish Wildl. Serv.*, Unpubl. Rept., Athens, GA.

Sykes, P. W., Jr., and D. J. Munson. 1989. Late record of Kirtland's warbler on the breeding grounds. *Jack-Pine Warbler* 67:101.

Sykes, P. W., Jr., C. B. Kepler, D. A. Jett, and M. E. DeCapita. 1989. Kirtland's warblers on the nesting ground during the post-breeding period. *Wilson Bull.* 101:545–58.

Walkinshaw, L. H. 1972. Kirtland's warbler—endangered. *Am. Birds* 26:3–9.

Walkinshaw, L. H. 1983. *Kirtland's warbler: The natural history of an endangered species.* Cranbrook Inst. Science, Bloomfield Hills, MI.

Walkinshaw, L. H., and W. R. Faust. 1974. Some aspects of Kirtland's warbler breeding biology. *Jack-Pine Warbler* 52:64–75.

Wallace, G. J. 1968. Another August record of Kirtland's warbler on its wintering grounds. *Jack-Pine Warbler* 46:7.

Trumpeter Swan

· *Cygnus buccinator* **Richardson**

· **State Extirpated**

The trumpeter swan formerly ranged throughout much of North America; however, European settlement and market hunting of migratory flocks severely altered its distribution and abundance. Eastern populations were decimated prior to 1800, and, by the end of the nineteenth century, it was limited to a handful of areas (Banko 1960), surviving only in small numbers in Alaska and western Canada. There was also a remnant, nonmigratory population in the Yellowstone National Park area.

Unregulated shooting every year and throughout its range was the major reason for the trumpeter swan's decline. Habitat degradation from wetland drainage in the Great Plains also played an important role in the demise of breeding populations.

A census in 1932 found only 69 adults and 12 cygnets remaining in the contiguous United States, with an unknown (at that time) isolated northern population. In 1935, important habitat was protected specifically to aid in the trumpeter's survival—the Red Rock Lakes National Wildlife Refuge in Montana. This and subsequent management and conservation projects enabled the trumpeter swan to reverse its decline toward extinction. Due to these measures and an incomplete, but impressive, Alaskan count of over 2,800 birds in 1968 (Hansen et al. 1971), it was not placed on the federal endangered species list.

Historically, this swan nested through much of Alaska, western and central Canada, the northern Rocky Mountains, and the Great Plains. Eastward extensions of its breeding range included Missouri, Wisconsin, and Ohio and probably southern Ontario, Quebec, and Michigan (Rogers and Hammer 1981). Few records have been confirmed in Michigan although late summer reports of swans reportedly were common around the Lake St. Clair marshes, an area with suitable breeding habitat. Several authorities now include Michigan as part of the trumpeter swan's historical breeding range (Bellrose 1976; Rogers and Hammer 1981; Lumsden 1984).

During migration, the trumpeter swan was formerly observed in each of the four major flyways of North America. Wintering areas were divided into five regions; (1) Chesapeake Bay area south to North Carolina, (2) Mississippi River to the Texas coast, (3) San Francisco Bay area, (4) the coast of British Columbia to Washington, and (5) in the thermal region in and around Yellowstone National Park, Wyoming (Banko 1960). Only the latter two regions still contain wintering birds.

Today, the trumpeter swan resides in increasing and expanding populations of over 12,000 individuals in Alaska, with an additional 2,000 birds in Canada and the contiguous United States. Although Saskatchewan populations remain low, most Alaskan and western Canadian populations are expanding (McKelvey et al. 1988). Several releases by state and provincial conservation agencies are planned or are in progress (Matteson et al. 1988).

Although no trumpeter subspecies are recognized and it exhibits little genetic variability (Barrett and Vyse 1982), the trumpeter swan's current breeding population is divided into three distinct populations. A migratory group nests in coastal and interior Alaska and British Columbia, as well as the Toobally Lakes area in Yukon Territory, the Grande Prairie region in central British Columbia and Alberta, and very small isolated populations in southern Alberta and Saskatchewan. The Pacific coast population is migratory and winters on or near the Pacific coast from southern Alaska to northern Oregon. The Tristate population breeds in Montana, Idaho, and Wyoming and winters in the vicinity of Red Rock lakes and Yellowstone National park. In winter, this population also consists of migrants from Canada.

Restoration efforts have expanded this range. Several releases have been made in the following National Wildlife Refuges: Malheur in Oregon, Turnbull in Washington, Ruby Lake in Nevada, National Elk Range in Wyoming, Lacreek in South Dakota, Mingo in Missouri, as well as outside National Wildlife Refuges in Minnesota, Nebraska, and Ontario. The South Dakota restoration effort has been especially successful (300 individuals in 1990), dispersing into Nebraska and Missouri. Nesting in South Dakota began in 1963 and was the first recorded breeding record east of the Rocky Mountains since the late 1800s (Monnie 1966). Trumpeter swans occurring east of the Rocky Mountains are part of the interior population; they numbered nearly 600 individuals in 1990, including approximately 20 free-flying swans in Michigan.

Full protection is now provided for the trumpeter swan but habitat deterioration, lead posioning (Munroe 1925), and limited wintering areas are threats to expanding populations. The latter concern is crucial since interior populations are limited by the amount of open water in the winter. Current protected areas with limited wintering habitat (e.g., Yellowstone area) contain the majority of the population under crowded conditions, thus providing an opportunity for disease outbreaks and starvation during severe winters.

The trumpeter swan is the world's largest waterfowl, weighing 20 to 30 pounds (9 to 14 kg) with a wingspan of approximately 8 feet (24 m) in males (smaller in females), and an average length of $5^1/2$ feet (1.7 m). Adults are entirely white with a black bill that has a narrow, pinkish-red edged mandible, visible at close range. Immatures are dusky brown with a pink, black-based bill. The trumpeter swan's distinctive and descriptive voice is a loud, deep and low-pitched trumpeting call, usually followed by three higher pitched calls.

The similar, but smaller, tundra swan that annually migrates through Michigan in spring and autumn typically has a conspicous yellow spot in front of its eye. The tundra swan's call is a muffled and mellow high-pitched note, similar to a snow goose (Chen caserulescens).

Trumpeter swans pair for life and usually do not nest until the fourth year.

The nests (sometimes 5 feet [1.5 m] in diameter) are usually placed within emergent vegetation or on top of muskrat houses in large, shallow lakes. Beaver ponds also are utilized and may be linked to this swan's success in some areas (King and Conant 1981). An average of four to six eggs are laid between mid-April and mid-May, followed by five weeks of incubation. Young are able to fly in 3^1/$_2$ to 4 months.

Michigan is attempting to establish two self-sustaining trumpeter swan populations. Early attempts, between 1986 and 1988, used a cross-fostering technique, introducing trumpeter swan eggs into mute swan (*Cygnus olor*) nests at the Allegan State Game Area, Allegan County. The mute swan is an exotic species with a feral population of over 1,700 individuals, mainly occurring in counties adjacent to northern Lake Michigan (Payne 1983). Adult mute swans act as foster parents until young trumpeter swans are independent. Mixed results from the cross-fostering program necessitated changing to the release of two-year-olds. Between 1989 and 1990, 17 individuals were released in Kalamazoo and Barry counties. Sixteen Alaskan trumpeters were released on the Seney National Wildlife Refuge in 1991 and 23 in 1992. Two pairs of trumpeters successfully nested in Michigan in 1992. One pair, believed to be from the original 1987 release, produced two cygnets on a private marsh in southwestern Michigan. A pair of three-year-old birds produced four cygnets on the Seney National Wildlife Refuge.

Michigan's current objective is to release 40 individuals per year for at least three years until reaching a goal of 200 individuals and 30 nesting pairs in Michigan. It is hoped that an established trumpeter swan population will out-compete and replace the nonnative mute swan, a species that causes serious disruptions in wetland ecosystems.

LITERATURE CITED

Banko, W. E. 1960. The trumpeter swan: Its history, habits, and population in the United States. *U.S. Fish Wildl. Serv., N. Am. Fauna No. 63.*

Barrett, V. A., and E. R. Vyse. 1982. Comparative genetics of three trumpeter swan populations. *Auk* 99:103–8.

Bellrose, F. C. 1976. *Ducks, geese and swans of North America.* Wildl. Mgmt. Inst., Stackpole Books, Harrisburg, PA.

Hansen, H. A., P. E. Shepard, J. G. King, and W. A. Troyer. 1971. The trumpeter swan in Alaska. *Wildl. Mongr. No. 26.*

King, J. G., and B. Conant. 1981. The 1980 census of trumpeter swans on Alaskan nesting habitats. *Am. Birds* 35:789–93.

Lumsden, H. G. 1984. The pre-settlement breeding distribution of trumpeter, *Cygnus buccinator*, and tundra swans, (*C. columbianus*), in eastern Canada. *Can. Field-Nat.* 98:415–24.

McKelvey, R. W., K. J. McCormick, and L. J. Shandruk. 1988. The status of Trumpeter Swans, *Cygnus buccinator*, in western Canada, 1985. *Can. Field-Nat.* 102:495–99.

Matteson, S. W., T. A. Andryk, and J. Wetzel. 1988. Wisconsin Trumpeter Swan Recovery Plan. *Passenger Pigeon* 50:119–30.

Monnie, J. B. 1966. Reintroduction of the trumpeter swan to its former prairie breeding range. *J. Wildl. Mgmt.* 30:691–96.

Munroe, J. A. 1925. Lead poisoning in trumpeter swans. *Can. Field-Nat.* 39:160–62.

Payne, R. P. 1983. A distributional checklist of the birds of Michigan. *Univ. Mich. Mus. Zool. Misc. Publ.* No. 164.

Rogers, P. M., and D. A. Hammer. 1981. Ancestral breeding and wintering ranges of the trumpeter swan (*Cygnus buccinator*) in the eastern United States. *Trumpeter Swan Soc.*, Unpubl. Rept.

Greater Prairie Chicken

- *Tympanuchus cupido* (Linnaeus)
- **State Extirpated**

The greater prairie chicken has experienced widespread habitat loss throughout its range. Early population expansions in the Midwest, following widespread clearing, have receded due to natural succession and intensive modern farming. Although Michigan's southwestern counties originally contained prairie chickens, these natural prairie habitat pockets are now gone.

Adult prairie chickens average 17 inches (43 cm) long with a wingspan of 28 inches (71 cm). Males differ markedly from the females during the courtship. While on the booming or display grounds, the males inflate large, orange neck sacs, and hold the elongated head feathers (pinnae).

Historically, this bird was distributed throughout the Great Plains, Midwest, southeastern Texas, and the East coast. The limits of its presettlement distribution are hard to define, particularly along the northern borders (Johnsgard and Wood 1968). The conversion of native prairie to modern farming techniques and grazing dramatically reduced the prairie chicken's range. Now it occurs in sizable populations only in Kansas, Nebraska, Oklahoma, and South Dakota, and scattered isolated pockets are still present in the Great Plains, Midwest, and southeastern Texas.

Before Michigan's forests were cleared, greater prairie chickens probably only occurred in the two southernmost tiers of counties. Prior to European settlement, oak savannas and scattered upland prairies were historically extensive in lower Michigan (Chapman 1984). The early destruction of Michigan's prairies was compensated by the widespread lumbering practices of the late 1800s. The following fires cleared remaining brush and subsequently provided increased and expanded suitable habitat for prairie chickens.

By 1930, prairie chickens ranged across the state, including most of the Upper Peninsula. After the early 1930s, however, it began to decline. The decline was primarily attributed to reforestation of the ravaged land in the north, and more intensive land-use practices in the southern portion of the state. By the late 1950s, the prairie chicken was extirpated in the Upper Peninsula and fewer than 1,000 birds survived in the northern half of the Lower Peninsula.

The decline continued through the mid-1970s until only one colony survived in northeastern Osceola County. In 1970, the state designated a management area to sustain a viable prairie chicken population. However, this remnant flock of 25 to 50 birds also experienced a downward trend, even though habitat apparently was available. In 1981, an estimated 20 birds still survived, but during the following year the flock disappeared from their spring booming grounds.

Optimal prairie chicken habitat can be described as a homogeneous expanse of vegetation approximately 20 inches high (51 cm)(during the spring) within a two-square-mile (3.2-square-km) area, with a minimum block size of 160 acres (65 ha) and at least one-half mile (.8 km) wide (Kirsch 1974). For the existing marginal habitat in Michigan, a minimum area of 4 square miles (10.4 square km) would probably provide an area large enough for a viable, self-sustaining prairie chicken population. The booming grounds are usually selected on a rise of land (Ammann 1957) that provides easy observation of the surrounding area. On the booming grounds, it prefers grass cover less than 6 inches (15 cm) tall, with little woody cover. The most limiting habitat requirement of the prairie chicken is the presence and quality of open areas for nesting and brood-rearing habitat (Kirsch 1974; Hamerstrom et al. 1957).

By February, flocks of males begin to gather on the booming grounds. In April, during peak male display, females begin to visit the breeding grounds. The "booming" usually occurs at dawn and dusk, and may be heard for several miles. The female lays an average clutch of 12 to 14 eggs in a ground nest, which is usually constructed in the general vicinity of the booming grounds. The female incubates the eggs for 23 to 26 days, and broods typically remain with females for six to eight weeks. Spring and summer movements are minimal, but mobility increases in autumn and winter. Dispersal of the first-year individuals is particularly high, during which mortality rates may exceed 50% by late autumn (Bowman and Robel 1977). The prairie chicken's year-round diet is about 75% plant food and 25% animal food (e.g., small insects).

The life cycle of this grouse revolves around the availability of open habitat that adequately provides booming grounds, nesting and brooding cover, and winter feeding areas. Due to the destruction of the prairie chicken's original prairie habitat and subsequent favorable land clearing practices, its disappearance was not immediate.

The apparent failure to manage suitable habitat may have been related, instead, to the lack of proper habitat manipulation techniques. The prairie chicken management area in Osceola County, which harbored the last Michigan prairie chickens, was maintained with tree and brush removal, limited burns, mowing, and approved specialized herbicides to maximize habitat suitability. More than 1,000 acres were managed and protected from adverse human disturbances, with several adjacent tracts providing marginal open habitat. The sensitive booming grounds were protected and buffered from the public. Although these methods typically support viable prairie chicken populations, large expanses of suitable habitat were limited, the total "preserve"population only numbered approximately 50 birds, and the northern climate, may have all contributed to the demise of this bird.

Reintroduction attempts would need to be made to recover the prairie chicken in Michigan. Either large blocks of land (e.g., 4,000 to 6,000 acres) should be set aside for continual clearing, or optimally, in a pattern of scattered units of 2,000 to 2,500 acres within a 10- to 15-square-mile area (Sanderson and Edwards 1966). Grasses such as redtop (*Agrostis alba*) provide excellent nest and brooding habitat, and can be maintained by controlled burning (Kirsch 1974). Adjacent farming communities can help by setting aside grassy refuges in idle lands or along ditches and fencerows (Yeatter 1963). No fewer than 100 prairie chickens per year over two to three years are required to sustain a population. This is primarily due to the birds ten-year cyclic fluctuations (Hamerstrom and Hamerstrom 1955) combined with high juvenile mortality rates and random dispersal from release sites (if reintroduced).

LITERATURE CITED

Ammann, G. A. 1957. The prairie grouse of Michigan. *Mich. Dept. Conserv., Tech. Bull.*

Bowman, T. J., and R. J. Robel. 1977. Brood break-up, dispersal, mobility, and mortality of juvenile prairie chickens. *J. Wildl. Mgmt.* 41:27–34.

Chapman, K. A. 1984. *An ecological investigation of native grassland in southern lower Michigan.* M.S. thesis, W. Mich. Univ., Kalamazoo.

Hamerstrom, F. N., Jr., and F. Hamerstrom. 1955. Population density and behavior in Wisconsin prairie chickens (*Tympanuchus cupido pinnatus*). *Trans. Inter. Ornith. Congr.* 11:459–66.

Hamerstrom, F. N., Jr., O. E. Mattson, and F. Hamerstrom. 1957. A guide to prairie chicken management. *Wisc. Conserv. Dept. Tech. Bull.* 15.

Johnsgard, P. A., and R. E. Wood. 1968. Distributional changes and interaction between prairie chickens and sharp-tailed grouse in the Midwest. *Wilson Bull.* 80:173–88.

Kirsch, L. M. 1974. Habitat management considerations for prairie chickens. *Wildl. Soc. Bull.* 2:124–29.

Sanderson, G. C., and W. R. Edwards. 1966. Efforts to prevent the extinction of the prairie chicken in Illinois. *Trans. Ill. State Acad. Sci.* 59:326–33.

Yeatter, R. E. 1963. Population responses of prairie chickens to land-use changes in Illinois. *J. Wildl. Mgmt.* 27:739–57.

Whooping Crane

- *Grus americana* (Linnaeus)
- Federally Endangered
- State Extirpated

Since historic times, the whooping crane is known to have bred only in three disjunct populations (United States Fish and Wildlife Service 1985). The most accepted estimate for the size of these three populations is between 1,300 and 1,400 individuals (Allen 1952). The principal range of the major breeding population extended from northern Illinois, northwestward through parts of Iowa, Minnesota, North Dakota, and into the southern portions of Canada's three prairie provinces. The other migratory population, which was discovered in 1954 (Allen 1956), was isolated in Wood Buffalo National Park, Northwest Territories, Canada. The only other breeding population was believed to be nonmigratory, surviving up to 1948 in southwestern Louisiana. Former wintering areas for these populations included the southern Atlantic seaboard, the Texas-Louisiana coast, and the central intermountain region in Mexico.

From an extreme low of 16 individuals in 1943, wild whooping crane populations have achieved over a ninefold increase to 151 individuals in 1989. In addition, more than 50 individuals are in captivity at the Patuxent Wildlife Research Center in Laurel, Maryland, the International Crane Foundation in Baraboo, Wisconsin, and the San Antonio Zoological Garden in Texas.

Today, there are two wild populations of whooping cranes. The largest group, and the only one to survive widespread human depredation, breeds in Wood Buffalo National Park and migrates through the Great Plains to Aransas National Wildlife Refuge, Texas (a 2,500-mile, [4,000 km] journey).

To make the species less vulnerable to natural or human-related catastrophes, an introduction into the Grays Lake National Wildlife Refuge, Idaho, is being attempted (Drewien and Bizeau in Temple 1978). Resident sandhill cranes (*Grus canadensis*) are used as foster parents for the transplanted whooping crane eggs. Subsequently, the young whooping cranes follow their sandhill crane foster parents to their major wintering grounds 800 miles (over 1,200 km) south in Bosque del Apache National Wildlife Refuge, New Mexico. Although approximately 300 eggs have been placed in nests since 1975, fledged young have not yet bred successfully. Their eventual establishment depends on whether or not the adults will naturally reproduce. In 1989, the Federal Whooping Crane Recovery Team decided to end cross-fostering techniques at this site.

In the Great Lakes region, the whooping crane apparently was observed regularly (Barrows 1912). However, only Pleistocene fossils have been found in Michigan (Brodkorb in Olson 1972). Individuals may have been taken in Washtenaw and Livingston counties in the late 1800s, but this has not been substantiated by photos or specimens (Barrows 1912). In 1989, the sighting of an adult during migration in Isle Royale, Keweenaw County, marks the first known observation in Michigan.

Although this crane's historical Michigan existence is questionable, there are studies currently seeking to determine the feasibility of establishing a self-sustaining eastern U.S. flock in the wetlands of Seney National Wildlife Refuge, Schoolcraft County. Controversy regarding the success of establishing migratory populations, however, has determined the next release site will be in Florida. Conditions are suitable to release subadult whooping cranes in the central Upper Peninsula, should another population need to be established.

Although whooping crane populations are protected, their low numbers continue to warrant concern, especially since the slow growth of the population was due to a decline in the mortality rate rather than an increase in recruitment (Miller et al. 1974). In the past, hunting and collecting coupled with habitat modification and general human disturbance contributed to the decline of this species' naturally limited populations (U. S. Fish and Wildlife Service 1985). Today, the two separate flocks are likely inhibited by natural limitations (e.g., predation and weather). Electric wires also cause some mortality; 13 of 18 individuals have died from hitting wires. Apparently, whooping cranes are relatively unaffected by organochlorine pesticide residues (Anderson and Kreitzer 1971; Lamont and Reichel 1970).

This crane is the tallest North American bird, averaging a height of 5 feet (1.5 m), with males generally being larger than females. The average weight is 15 pounds (6.8 kg), length is $4^1/2$ feet (1.4 m), and wingspan is $6^1/2$ to $7^1/2$ feet (2.0 to 2.3 m). The adult plumage is basically white except for the characteristic black wing tips and bare red skin in its head region. Albino sandhill cranes lack the black wing tips. Immatures are washed with a rust color (particularly near the head), on a mostly white body plumage, in contrast to immature sandhills, which have a gray body irregularly mottled with a brown coloration.

Most of the Wood Buffalo National Park breeding population return to begin nest construction in late April. Whooping cranes are monogamous, maintain life-long pair bonds, and display considerable site faithfulness to their breeding territories. Nesting territories vary greatly in size, from a few hundred acres to many square miles (Kuyt in Lewis and Masatomi 1981). Nests are typically constructed in marshy areas of muskeg habitat. Normally two eggs are laid in late April and early May and hatch one month later. Both parents share with the incubation and brood-rearing duties. Invariably, only one whooper chick survives (Novakowski 1966). Researchers take advantage of this fact and remove one egg from many of the wild nests to augment captive populations and release programs (U.S. Fish and Wildlife Service 1985).

By mid-September, family groups begin to migrate to the Texas coast. Today, migrants typically pass through the Great Plains, however individuals have been seen flying east to Minnesota (Bellefeville 1986) and in Michigan in 1989. Arriving between late October and mid-November, family groups procure a feeding and roosting territory and defend it until their departure in late March to mid-April (Blankinship in Lewis 1976).

Additional aspects of the whooping crane's winter ecology have been published in Bishop and Blankinship in Lewis 1981. Captive individuals of this long-lived species are known to reach at least 35 years of age (McNulty 1966). These cranes are generally omnivorous, feeding on invertebrates, small vertebrates, and such vegetable matter as acorns and grain.

LITERATURE CITED

Allen, R. P. 1952. The whooping crane. *Natl. Audubon Soc., Res. Rept. 3.* 2.

Allen, R. P. 1956. The whooping crane's northern breeding grounds. *Natl. Audubon Soc., Suppl. Res. Rept. 3.*

Anderson, D. W., and J. F. Kreitzer. 1971. Thickness of 1967–69 whooping crane eggshells compared to that of pre-1910 specimens. *Auk* 88:433–34.

Barrows, W. B. 1912. *Michigan bird life.* Mich. Agric. Coll. Spec. Bull., E. Lansing.

Bellefeville, D. 1986. Whooping cranes in Mahnomen County. *Loon* 58:45.

Lamont, T., and W. Reichel. 1970. Organochlorine pesticide residues in whooping cranes and everglade kites. *Auk* 87:158–59.

Lewis, J. C. (ed.). 1976. *Proceedings International Crane Workshop.* Oklahoma State Univ. Press, Stillwater.

Lewis, J. C. (ed.). 1981. *Proceedings 1981 Crane Workshop.* Natl. Audubon Soc., Tavernier, FL.

Lewis, J. C., and H. Masatomi. 1981. *Crane research around the world.* Int. Crane Found., Baraboo, WI.

McNulty, F. 1966. *The whooping crane: The bird that defies extinction.* E.P. Dutton Co., New York.

Miller, R. S., D. B. Botkin, and R. Mendelssohn. 1974. The whooping crane (*Grus americana*) population of North America. *Biol. Conserv.* 6:106–11.

Novakowski, N. S. 1966. Whooping crane population dynamics on the nesting grounds. Wood Buffalo National Park, Northwest Territories, Canada. *Can. Wildl. Serv., Res. Rept. Ser.* No. 1.

Olson, S. L. 1972. A whooping crane from the Pleistocene of north Florida. *Condor* 74:341.

Temple, S. A. (ed.). 1978. *Endangerd Birds—Management techniques for preserving threatened species.* Univ. Wisc., Madison.

U.S. Fish and Wildlife Service. 1985. Whooping crane recovery plan. *U.S. Fish and Wildl. Serv.*, Albuquerque, NM.

Eskimo Curlew

- *Numenius borealis* (Forster)
- **Federally Endangered**
- **State Extirpated**

The once abundant Eskimo curlew occurred in tremendous flocks in parts of the United States during its migration to South America. It originally nested within the Arctic Circle in northern Canada and possibly Alaska. By late summer, most of the population moved down to a feeding and staging area at the southern tip of Labrador. From this point, it would make a 7,000-mile (over 11,000-km) trip to the pampas (grasslands) of Argentina for the winter. Autumn storms commonly forced individuals and flocks to the New England coastline (MacKay 1892). In April, it reached the Texas coast and continued to move through the Great Plains reaching its breeding grounds in May (Aldrich 1978).

Although Michigan was not part of the Eskimo curlew's typical migratory route, small numbers may have made autumn flights through the Great Lakes region via James Bay (Cooke 1910; Hagar and Anderson 1977). This curlew was considered to be common in Detroit food markets in the mid-1800s (Barrows 1912), but there are only two confirmed state records, including individuals from Kalamazoo and Wayne counties (Payne 1983). Two specimens were reportedly taken in the St. Clair Flats, St. Clair County in 1883 (Barrows 1912). The eskimo curlew formerly existed in dense flocks so impressive that John James Audubon was reminded of the abundant (but now extinct) passenger pigeon (*Ectopistes migratorius*)(Banks 1977).

Sport hunting for this bird was popular but apparently had little effect on the entire population until the late 1800s. The first indication of a population decline was during spring migrations through the prairie states between 1875 and 1880 (Banks 1977). A decade later, it was apparent, during autumn migration, that the population had collapsed, and by 1891 this species had disappeared. Rare, but regular reports of eskimo curlews were made until 1915. For the next decade there were no sightings and the species was considered extinct until a few birds were found on the Argentina wintering grounds (Swenk 1926). The Eskimo curlew continued to be extremely scarce into the mid-1900s. The 1945 appearance of two individuals in Texas were the first records for 40 years and apparently marked a turning point for the species. Between 1945 and 1985, 25 of the 41 years had confirmed known sightings (Gollop et al. 1986). Most were autumn sightings in Ontario and New England, such as an August observation of two individuals at North Point on James Bay, Ontario, in 1976 (Hagar and Anderson 1977).

Spring observations have been relatively regular on Galveston Island, Texas (Blankinship and King 1984). A nearby Texas site, Atkinson Island, was the location of a flock of 23 individuals observed on 7 May 1981 (Blankinship and King

1984), undoubtedly the largest concentration of Eskimo curlews since its virtual disappearance in 1891. Recently, several breeding ground observations have been made, including the Mackenzie River Delta, a former nesting area (Aldrich 1978). In 1983, breeding was confirmed for the first time in nearly a century. An adult with one young was found in Alaska. Current probable breeding areas include part of the Northwest Territories, Yukon, Alaska, and possibly across the Bering Strait into Siberia (Gollop et al. 1986). The curlew's decline probably was not solely the result of sport hunting (and later market hunting, beginning in the 1880s). Banks (1977) considered habitat loss, storms, nesting failure, and changing climatic factors to have also infuenced this species, to an extent that it nearly became extinct. These factors may still be repressing the existing remnant population, due to its slow response to complete protection.

The eskimo curlew is a medium-sized shorebird, 12 to 13 inches long (30 to 33 cm) with a wingspan of 26 to 30 inches (66 to 76 cm). It is buffy brown with a slender, slightly down-curved bill, buffy eyebrow, and uniformly dark primaries with cinnamon underwing linings. The similar whimbrel (*Numenius phaeopus*) is a larger shorebird with a longer, more deeply curved bill, conspicuous striping on the crown, barred primaries, and no cinnamon underwing linings. A closely related and similar species, the little curlew (*Numenius minutus*), has been confirmed in North America but occurs primarily in Siberia. Farrand (1977) discusses the subtle differentiating characteristics of these two species. On its treeless Arctic tundra home, eskimo curlews usually lay four eggs in late May to early June (Aldrich 1978). The length of the incubation and fledging period is not known, but the breeding season is apparently over by late July.

In 1987, the official beginning of a federal recovery team occurred to aid this species to recovery. Michigan observers should realize this species may occur in Michigan during migration. All potential sightings should be photographed and reported with thorough written details.

LITERATURE CITED

Aldrich, J. 1978. Eskimo curlew. *U. S. Fish Wildl. Serv.*, Unpubl. Rept.
Banks, R. C. 1977. The decline and fall of the eskimo curlew, or why did the curlew go extaille? *Am. Birds* 31:127–34.
Barrows, W. B. 1912. *Michigan bird life.* Mich. Agric. Coll. Spec. Bull., E. Lansing.
Blankinship, D. R., and K. A. King. 1984. A probable sighting of 23 eskimo curlews in Texas. *Am. Birds* 38:1066–67.
Cooke, W. W. 1910. Distribution and migration of North American shorebirds. *Bur. Biol. Surv. Bull.* No. 35.
Farrand, J. 1977. What to look for: Eskimo and little curlews compared. *Am. Birds* 31:135–36.
Gollop, J. B., T. W. Barry, and E. H. Iverson. 1986. Eskimo curlew: A vanishing species? *Saskatchewan Nat. History Soc., Spec. Publ.* No. 17.
Hagar, J. A., and K. S. Anderson. 1977. Sight record of eskimo curlew. *Am. Birds* 31:135–36.
MacKay, G. H. 1892. Habits of the eskimo curlew (*Numenius borealis*) in New England. *Auk* 9:16–21.
Payne, R. P. 1983. A distributional checklist of the birds of Michigan. *Univ. Mich. Mus. Zool. Misc. Publ.* No. 164.
Swenk, M. H. 1926. The eskimo curlew in Nebraska. *Wilson Bull.* 38:117–18.

Lark Sparrow

· *Chondestes grammacus* (Say)
· **State Extirpated**

The lark sparrow is a western species that extended its range through the Midwest and mid-Atlantic states by following the cutting and clearing of forests. Historically, populations occurred in the upland prairie and oak savanna of southernmost Michigan. Nonetheless, the lark sparrow has retreated and is now extirpated from Michigan.

The adult lark sparrow is unmistakable, with its striking head pattern and chestnut ear patches, a black central breast spot, and a rounded tail with conspicuous white corners. It averages 6 inches (15 cm) in length with a wingspan of 11 inches (28 cm). The variable, melodic song of the lark sparrow is characterized by clear buzzes and trills broken by pauses.

The lark sparrow breeds from extreme south-central Canada to northern Mexico, and from the Pacific coast east to the Appalachians. Its range expanded east with the clearing of land during European settlement, but it now appears to have withdrawn to its historical distribution due to reforestation in the mid-Atlantic states. United States Breeding Bird Surveys also have shown significant declines in the Great Plains (Robbins et al. 1986). Texas's Central Plateau contains the greatest breeding density in the United States. Wintering grounds range from the southern United States south to Central America.

Before European settlement, the lark sparrow most likely occurred in the natural prairies and extensive oak savannas of southern Michigan (Campbell 1940; Campbell 1968; Chapman 1984; Mayfield 1988). Following the logging era, this sparrow spread throughout southern Michigan, and, by the late 1800s, it was locally common (especially in southeastern Michigan) and nested in 11 southern lower Michigan counties north to St. Clair and Kent counties (Barrows 1912; Zimmerman and Van Tyne 1959; Payne 1983). Occasional migrants were reported in the northern Lower Peninsula and Upper Peninsula (Barrows 1912; Payne 1983). During this invasion period, the sparrow's population fluctuated annually and was widely scattered. During intensive surveys between 1983 and 1988, no summer observations were made (Brewer et al. 1991).

In southeastern Michigan, where it was most widespread, Sutton (1960) found it nesting only in open, sandy areas with scattered young oaks. This habitat is similar to former Michigan preferred habitats in the extensive oak openings of southern lower Michigan (Chapman 1984). The lark sparrow's spring migration period for northern populations is from mid-April to mid-May. It nests in open pasture lands, roadsides, and cultivated fields near orchards and woodlots. A clutch of four to five eggs is laid between late May and mid-June; above ground nests are less vulnerable to predators (Newman 1970). The female incubates approximately 12 days, and young are fledged within 9 to 10 days. Double brooding may occur in southern areas of its range. By September, most lark sparrows have left the nesting areas and, in late autumn and winter, they congregate in large flocks. The lark sparrow feeds on the ground in small flocks, even during the nesting season. Grass and weed seeds comprise the bulk of its diet; insects, particularly grasshoppers, are also taken.

In Michigan, the creation of extensive farmland habitat temporarily replaced the loss of upland prairies and oak savannas. Beginning in the mid-1900s, the intensive Midwest land-use patterns and overall decline in farmland acreage eventually forced this species to abandon these areas. Marginal habitats, such as abandoned fields, may be unable to maintain populations of this primarily prairie species because predation rates are higher in successional habitats (Zimmerman 1984). Large tracts of monocultures, modern farm practices, and forest succession eventually eliminated most Midwest populations.

This species could conceivably recolonize its former Michigan haunts. It is still observed on migration, and neighboring populations in Ohio and Indiana are not too distant. However, long-term population stability will depend upon proper management of large openings to sustain this species in its native Michigan range.

LITERATURE CITED

Barrows, W. B. 1912. *Michigan bird life.* Mich. Agric. Coll. Spec. Bull., E. Lansing.

Brewer, R., G. A. McPeek, R. J. Adams, Jr. 1991. *The atlas of breeding birds in Michigan.* Mich. State Univ. Press, E. Lansing.

Campbell, W. W. 1940. Birds of Lucas County. *Toledo Mus. Sci., Bull.* 1.

Campbell, W. W. 1968. *Birds of the Toledo area.* Toledo Blade Co., Toledo, OH.

Chapman, K. A. 1984. *An ecological investigation of native grassland in southern lower Michigan.* M.S. thesis, W. Mich. Univ., Kalamazoo.

Mayfield, H. F. 1988. Changes in bird life at the western end of Lake Erie (pt. 1). *Am. Birds* 42:393–98.

Newman, G. A. 1970. Cowbird parasitism and nesting success of lark sparrows in southern Oklahoma. *Wilson Bull.* 82:304–9.

Payne, R. P. 1983. A distributional checklist of the birds of Michigan. *Univ. Mich. Mus. Zool. Misc. Publ.* No. 164.

Robbins, C. S., D. Bystrak, and P. H. Geissler. 1986. The breeding bird survey: Its first fifteen years, 1965–1979. *U.S. Fish Wildl. Serv., Resour. Publ.* 157.

Sutton, G. M. 1960. The nesting fringillids of the Edwin S. George Reserve, southeastern Michigan (Part 5). *Jack-Pine Warbler* 38:2–15.

Zimmerman, D. A., and J. Van Tyne. 1959. A distributional check-list of the birds of Michigan. *Univ. Mich. Mus. Zool. Occ. Pap.* 608.

Zimmerman, J. L. 1984. Nest predation and its relationship to habitat and nest density in dickcissels. *Condor* 86:68–72.

Passenger Pigeon

- *Ectopistes migratorius* (Linnaeus)
- Extinct

Once presumed to be infinitely abundant, the passenger pigeon became extinct in the wild at the turn of the century (Schorger 1955). An estimated 3 to 5 billion passenger pigeons comprised 25% to 40% of the total U.S. bird population. Formerly occurring throughout much of the United States and southern Canada west to the Great Plains, its decline began in the New England states in 1851. Following three decades of widespread market hunting and habitat destruction, this species' nesting range was limited to the Great Lakes region in the late 1870s; only a few colonies and several million birds survived. The most effective method of harvesting was by trapping with the use of food baits and tethered live pigeons (Brewster 1889).

Due to the passenger pigeons' highly gregarious breeding habitats (Craig 1911; Schorger 1955; Brisbin 1968), it required a large population for successful nesting. By the 1870s, the apparently enormous population may have been below this minimum threshold, starting a rapid population decline. Beside market hunting and the need for large numbers for self-sustaining populations, habitat destruction contributed to the decline. The demise of the remaining viable nesting colonies in Wisconsin and Michigan may be related to the Great Chicago Fire in 1871.

To rebuild and supply food for the city, many of these pigeons and their remaining beech-oak-maple forest nesting and feeding habitat was destroyed. Only a few scattered and small colonies survived into the 1890s. The last passenger pigeon, a captive in the Cincinnati Zoo, died in 1914. Historically, the passenger pigeon was an abundant bird in Michigan, nesting in large colonies in the Lower Peninsula. Near the end of its existence, the passenger pigeon maintained large colonies in northern lower Michigan until the late 1870s. This area was the last stronghold of the species. In 1878, a large colony survived in Emmet County, inhabiting a continuous area estimated to be 28 to 40 miles (45 to 64 km) long and 3 to 10 miles (5 to 16 km) wide (Mershon 1907). The last viable nesting colony in Michigan (and throughout its range) was in 1881 in Grand Traverse County, and the last confirmed nesting of the passenger pigeon was near the headwaters of the Au Sable River in 1896 (Barrows 1912). The final known occurrence of this extinct species in Michigan was a specimen obtained in Wayne County in 1898.

Literature Cited

Barrows, W. B. 1912. *The birds of Michigan.* Mich. Agric. Coll. Spec. Bull., E. Lansing.

Brewster, W. 1889. The present status of the wild pigeon (*Ectopistes migratorius*) as a bird of the United States, with some notes on its habits. *Auk* 6:285–91.

Brisbin, I. L. 1968. The passenger pigeon. A study on the ecology of extinction. *Modern Game Breeding* 4:13–20.

Craig, W. 1911. The expressions of emotion in the pigeons: The passenger pigeon (*Ectopistes migratorius*, Linn.). *Auk* 28:408–27.

Mershon, W. B. 1907. *The passenger pigeon.* The Outing Publishing Co., New York.

Schorger, A. W. 1955. *The passenger pigeon: Its natural history and extinction.* Univ. Wisc. Press, Madison.

Reptiles and Amphibians

Endangered and Threatened

Kirtland's snake (*Clonophis kirtlandii*)
Eastern fox snake (*Elaphe vulpina gloydi*)
Copperbelly water snake (*Nerodia erythrogaster neglecta*)
Marbled salamander (*Ambystoma opacum*)
Smallmouth salamander (*Ambystoma texanum*)

An Introduction to Reptiles and Amphibians

James H. Harding
Michigan State University

Public attitudes concerning amphibians (frogs, toads, salamanders) and reptiles (turtles, lizards, snakes) have all too often been based on ignorance, superstition, or prejudice. Until recently, these animals generally have been ignored, or even persecuted, by wildlife biologists and resource managers. Amphibians and reptiles are now known to play an important role in temperate, as well as tropical, ecosystems. One study (Burton and Likens 1975) found that, in a small 89-acre (36-ha) New Hampshire woodland, salamanders were more numerous than either birds or small mammals, and that salamander biomass was 2.6 times the biomass of birds during the peak of the breeding season. The most abundant species found was the red-backed salamander (*Plethodon cinereus*), a common amphibian in Michigan woodlands as well.

Some "herps" (from herpetology), particularly the larger frogs, certain snakes, and basking turtles, are obvious even to the casual observer. Many other species are shy and secretive and live in habitats that make them difficult to observe. Thus, it is easy to underestimate their numbers in a cursory survey. Conversely, it is also easy to become complacent and assume that their populations are healthy, based on earlier records, when in fact they may be in serious trouble. A number of Michigan amphibians and reptiles are now listed as endangered, threatened, or a species of special concern, and they are protected by state law. Many other species also have declined, and it is certain that there will be additions to these lists in the near future.

Our assessment of herp population trends in Michigan is hampered by a lack of early, baseline surveys. Few comprehensive studies on amphibian and reptile distribution have been published for the state. Certain herps (such as many frogs and salamanders) have highly variable populations over time, due to the effect of local climate on reproductive success. Others, such as some turtle species, are long-lived animals with slower population turnover and replacement rates. Thus, obtaining a proper perspective on herp ecology and population structure is difficult without long-term research programs. Noteworthy long-term herpetological studies have taken place at the University of Michigan's E. S. George Reserve over the last three decades. This research has produced important ecological data, particularly on turtles and amphibians (Collins and Wilber 1979; Tinkle et al. 1981; Congdon et al. 1983; Congdon et al. 1987).

The only complete summary of Michigan herpetology (Ruthven et al., 1928)

225

was published more than 60 years ago. Based on published sources and museum records, Pentecost and Vogt (1976) summarized the distribution of amphibians and reptiles in the Lake Michigan drainage basin. More recent publications, intended for popular use, include booklets on turtles (Harding and Holman 1990), snakes (Holman et al. 1989), and amphibians (Harding and Holman 1992). There are some noteworthy publications for neighboring states that are useful in Michigan due to the broad overlap in native species. These include Minton (1972) for Indiana, Vogt (1981) for Wisconsin, and Pfingsten and Downs (1989) for Ohio salamanders.

The Historical Record

The presettlement record for Michigan reptiles and amphibians is unfortunately quite limited. Because Michigan was greatly affected by extremes in climate and glaciation during the late Pleistocene period, we can be sure that the herpetological "slate" was wiped clean prior to the last glacial retreat, between 12,000 and 14,000 years before the present. Only one reptile, the painted turtle (*Chrysemys picta*), is presently known from Michigan in the last Pleistocene. Two amphibian records (American toad and northern leopard frog) for the period from 12,000 to 13,000 years ago in southern Michigan appear to be valid.

As the climate moderated after the end of the Pleistocene (10,000 years before the present), reptiles were able to recolonize the state. Reliable records for herps occur in deposits dating from about 5,000 B.P. at several sites in the southern Lower Peninsula (Holman 1988 and personal communication). These are mostly turtles; snakes were probably present as well, but their fragile remains are much less likely to be preserved. There are numerous records of herps from archaeological sites, especially turtles, which were used as food and for ceremonial and trade items (Adler 1968). The postcolonial record is poor. What was known of amphibian and reptile distribution up to the early decades of the twentieth century is included in Ruthven et al. (1928).

Because of the effects that human activities have had on natural communities since the late 1800s, we are in great need of data on the past and present distribution of amphibians and reptiles in Michigan. The scarcity of baseline surveys often necessitates a certain reliance on anecdotal reports, popular articles, and the memories of naturalists and sportspersons about "how things used to be." Such information can be useful, provided that it is based on accurate species identification. Collection data from university and museum collections can be cautiously utilized to help build a picture of former herp distributions. It cannot be relied upon in determining the present situation, considering the age of much of the museum material and the rapidity with which human activities can impact herp populations. Some parts of the state are better known than others, with greater gaps in our knowledge occuring farther from the population centers of the southern Lower Peninsula.

Much of the recent herpetological research in Michigan has been local and focused on one or a few species. These studies are important, but there is a critical need for more comprehensive faunal surveys. One recent herpetological survey in the state was undertaken by a team from the University of Michigan Museum of Zoology on behalf of the Michigan Department of Natural Resources Endangered Species Program in 1977. The survey targeted species listed as rare, threatened, or endangered at that time; it was accomplished by a direct mail questionnaire, a search of museum records, a public appeal for information, and field visits of known or reported localities. Sites in the Upper Peninsula were not visited. Their final report to the DNR (Tinkle et al. 1979) shed some light on the distributions and populations of the species concerned, but also highlighted the need for additional field surveys over a longer period of time. In 1988, the Michigan Natural Features Inventory initiated a survey of frog and toad breeding populations, using the identification of calling males as the primary means of identification. This survey is incomplete at this time.

The Human Factor

Human activities in Michigan have had significant, sometimes dramatic, effects on amphibian and reptile populations. Although native Americans certainly utilized these creatures, human impacts were probably localized and minimal until the large-scale settlement by Europeans began in the early nineteenth century. The clearing of the original forests for timber and agriculture had mixed results for herps. Those species preferring wooded habitats (e.g., spotted salamander, black rat snake, eastern box turtle) presumably declined; other species needing more open habitats (e.g., northern leopard frog, chorus frog, red-bellied snake, smooth green snake) would have benefited. Although supporting data is scarce, we might predict that human activities into the first decades of the twentieth century had probably caused a shift in relative species abundance, but had not yet seriously threatened any native amphibian or reptile. After the mid-1940s, a number of changes occurred that seem to have tipped the balance against several species. These include the following.

1. An accelerated conversion of land from a "natural" state or from agricultural use to residential and industrial use.
2. Accelerated drainage of wetlands (the DNR estimates that about 75% of Michigan's wetlands had been destroyed by the middle of this century).
3. The widespread use of chemical pesticides, including such persistent chlorinated hydrocarbons as DDT, and the introduction of new industrial chemicals into the environment.
4. The use of increasingly intensive agricultural methods.
5. Growing demand for recreational opportunities and better mobility has brought greater human use of public lands and waters (including the use of off-road vehicles).
6. The demand for live reptiles and amphibians collected for the pet trade, for food, and for use as laboratory specimens.

The loss of habitat is undoubtedly the major factor in the decline of most herp species. Particularly important is the continued destruction of wetland habitats in the state, since many reptiles and nearly all amphibians depend on wetlands for all or part of their life cycle. Recent legislation, such as the Wetland Protection Act (PA 203 of 1979), has slowed this loss. However, drainage of small ponds and wetlands less than 5 acres is not regulated by the act, and these are of critical concern in the preservation of such declining species as the spotted turtle, marbled salamander, and Blanchard's cricket frog. Many amphibians depend on the smaller, seasonal wetlands for breeding purposes. In addition, significant degradation of herp habitats by improper use of off-road vehicles is a growing problem in many parts of Michigan.

Less understood is the effect of pesticides and other forms of chemical pollution on amphibians and reptiles. Certainly, direct application of certain chemicals is harmful or fatal to various species, particularly amphibians with their more permeable skin. Reptiles, however, are not immune to the effects of toxic chemicals. Minton (1972) documented the death of smooth green snakes by pesticides. Such insectivorous species may also accumulate pesticides and their metabolites through consumption of contaminated prey. (The green snake has suspiciously disappeared from much of its former range in the southern Lower Peninsula.) Hall (1980) reviewed published data on the effects of contaminants on reptiles. Additional research is needed to document the long-term effect of chemical residues on herps.

A related problem is the impact of acid precipitation on herps. Because of their dependence on aquatic environments, amphibians are undoubtedly more susceptible to acidification than are reptiles; however, there have been few direct studies of the latter. Pierce (1985) summarized research on amphibians and noted that there is considerable variation in acid tolerance in these animals. Two Michigan species that reportedly have comparatively low acid tolerance, the spotted salamander and the leopard frog, have experienced declines in parts of their Michigan ranges in recent years. The leopard frog was formerly the most common frog in the state, but it declined dramatically in the late 1960s and 1970s and practically disappeared from some areas. A similar decline was noted in other states as well (e.g., Wisconsin [Hine et al. 1981]), and much speculation centered on various environmental pollutants as a causative factor. No conclusive studies have been published, however, and the data on acid tolerance merely raise one more hypothesis to be considered. Leopard frog numbers have recovered in some parts of the state in the last few years but not in others, and the mystery remains.

The increasing use of Michigan's natural areas, public and private, for recreation has brought more people into contact with native reptiles and amphibians, often with unfortunate results. Direct persecution of snakes has undoubtedly caused local declines in several of the larger species, notably the northern water snake, the endangered copperbelly water snake, eastern hognose snake, and blue racer. Basking turtles have been frequent targets for thoughtless people using firearms (the shooting of reptiles is now illegal under Michigan law). Large numbers of herps are killed on roads each year, with a portion of this carnage being intentional or avoidable.

The direct exploitation of herps by private collectors and commercial animal dealers can be a significant cause of population reduction, especially when aimed at rarer, more vulnerable species—often those in the greatest demand. In Michigan, the spotted turtle, wood turtle, and eastern box turtle have been frequent targets along with some of the larger snakes, such as the threatened eastern fox snake. These species are now protected under Michigan law, but enforcement is difficult and poaching is a continuing threat. A few amphibians, particularly the larger, more colorful salamanders (e.g., spotted salamander and tiger salamander) are in demand by the pet trade, and salamanders and frogs are also collected for fish bait. The taking of the larger turtles and frogs for human food was, until 1988, poorly regulated, and preferred species (especially the snapping turtle and bull frog) suffered significant overharvesting in many parts of the state.

Current Legislation and Regulation

Five Michigan amphibians and reptiles are currently listed under Michigan's Endangered Species Act (Public Act 203 of 1974 as amended). Under this act any species listed as endangered or threatened is protected from capture or exploitation, but no direct provision is made for the protection of critical habitats for these species.

Some species such as the eastern box turtle are protected under state regulations resulting from Public Act 373 of 1988. This legislation amended Michigan's Sport Fishing Law (PA 165 of 1928) and extended state ownership and regulatory authority to reptiles and amphibians. Prior to this act, the state had demonstrated little regulatory interest in these animals. The act and its new regulations also established special license requirements for persons taking herps on a commercial basis and instituted additional regulations on the taking of herps, including daily capture and total possession limits, limitations on mode of capture, and season and size limits on certain species. Full protection was granted to a number of rare and vulnerable species; these were essentially those listed as species of special concern by the DNR but not afforded protection under the Endangered Species Act. Because the regulations on herps can—like those covering game fish, birds, and mammals— be changed from year to year, interested persons should contact the Michigan Department of Natural Resources Fisheries Division for details.

Summary of Listed Species

The two reptiles currently listed as endangered are the copperbelly water snake and Kirtland's snake. The copperbelly water snake, a subspecies of the plainbelly water snake, is found in southern Indiana, Illinois, and westen Kentucky, and in disjunct colonies farther north in Ohio, Indiana, and the south-central part of Michigan's Lower Peninsula. The species may have entered Michigan from Indiana via the St. Joseph River valley. Copperbellies prefer wooded river edges and backwater sloughs and swamps, though they often traverse adjacent uplands. They are threatened by habitat loss and direct persecution.

Kirtland's snake is a small, secretive, burrowing snake that inhabits moist meadows and open woodlands, often near water. They formerly were quite common in grassy, vacant lots near ponds or ditches in certain urban areas within their limited range (western Pennsylvania to central Illinois, north to Michigan, and south to the Indiana-Kentucky border). In Michigan, Kirtland's snake has been found in a limited number of locations in the southern counties of the Lower Peninsula. The species may be a relict of a warmer period several thousand years ago when grasslands spread through parts of the lower Great Lakes area. The reasons for this snake's rarity are poorly understood. It is possible that a combination of natural and human-caused factors are limiting Kirtland's snake populations.

One snake and two frogs are currently listed as threatened. The eastern fox snake is a characteristic species of the marshes and associated habitats along the Great Lakes shoreline in eastern Michigan from Saginaw Bay to western Lake Erie. This snake has lost large portions of its habitat to drainage and development schemes. Its large size, orangish head, and blotchy coloration has led to much direct persecution by humans who erroneously think it venomous. Additionally, there has been some exploitation of the fox snake for the pet trade.

The marbled salamander is known from only three Michigan locales in Berrien, Van Buren, and Allegan counties. This salamander is more common farther south and east of Michigan, and the Michigan records may represent relict populations resulting from post-glacial dispersal during warmer times. The small-mouth salamander is a common woodland species of the eastern United States that just enters Michigan in the southeastern counties of the Lower Peninsula. The loss of moist, wooded habitat is a threat to this salamander's existence in the state.

The spotted turtle, currently considered a species of special concern, inhabits shallow marshy ponds, bogs, seepage areas in damp meadows, and small, sedge-dominated wetlands. Formerly found in scattered colonies throughout the southern and western Lower Peninsula, this turtle has lost much of its habitat to agricultural drainage and tiling projects and to residential and industrial development. Remaining populations of spotted turtles, centered mostly in southwestern Michigan, have been exploited by pet trade collectors in recent years. They are now protected under new state regulations, and several Michigan herpetologists have recommended threatened status for this turtle.

Three other Michigan reptiles, two turtles and one snake, are also species of special concern due to serious reductions in their numbers as well as vulnerability to persecution and exploitation. The wood turtle is found in and near rivers and streams in the northern Lower Peninsula and the Upper Peninsula. This habitat is less degraded than that of the related spotted turtle in southern Michigan, but local populations of wood turtles have been seriously impacted by incidental collecting (e.g., by canoeists and other recreationists). A perception that these turtles are intelligent and particularly desirable as pets has led to increased collecting for the commercial pet trade in recent years. The wood turtle, like many turtle species, is a slow maturing (14 years +) animal with high nest and hatchling mortality and low annual recruitment, balanced by a long adult reproductive life. Thus, the safest management posture is to assume that no "harvestable surplus" of adult turtles exists, since the removal of mature turtles from a population (beyond natural background mortality levels) will predictably result in a declining population.

The eastern box turtle and the black rat snake both inhabit woodlands in the southern Lower Peninsula and have declined as their habitat has been converted to agricultural and urban uses. These reptiles also incur direct mortality from human activities. The black rat snake is often killed by fearful or ignorant people, while box turtles suffer terrible losses when attempting to cross roads. These reptiles are often collected as pets, a practice now illegal under state regulations.

Of the two "special concern" frogs presently listed, the boreal chorus frog is a peripheral species that reaches its greatest abundance to the west of the state. In Michigan, they are only known from Isle Royale. Blanchard's cricket frog was, until recently, a familiar frog of river bottoms, marsh mud flats, and bog lake shorelines in the southern and western Lower Peninsula. Over the last few years, naturalists and herpetologists have become aware that this tiny frog has apparently disappeared from many places where it was once common.

Recommendations

There are several steps that need to be taken to ensure that the diverse herpetofauna native to Michigan can continue to exist into the next century. These include the following.

I. Conduct status surveys of all native species of amphibians and reptiles.
   a. Determine the historical and recent range by researching professional publications and museum records as well as careful consideration of unpublished field notes and verbal reports by competent observers.
   b. Establish the present range by surveying knowledgeable naturalists, biologists, sportspersons, and others who spend much time in the field, and contract direct field surveys of former and potentially new locales for each species.
II. Identify the habitat requirements for each species, and identify the existing critical habitats for all listed species and those with special habitat needs.
III. Continue and expand the regulatory and statutory protection for native herps.
   a. Eliminate the commercial exploitation and sale of all native amphibians and reptiles.
   b. Determine the suitability of "personal use" collecting on a species-by-species basis.
   c. Grant full and unambiguous protection to all species that are declining, vulnerable to exploitation, or protected in any portion of their range, whether in or out of the state. (This action assists law enforcement in other states or countries by preventing the "laundering" of illegally collected animals.)
   d. Inventory native herps now held by hobbyists and private breeders and establish annual report procedures for the breeding and disposition of protected species.

IV. Increase public education concerning the values of amphibians and reptiles, both ecologically and economically. Museums, nature center, private conservation groups and sporting groups, and state agencies can all assist in this effort.

V. Include amphibians and reptiles in all land-use and regulatory decisions concerning wildlife. State biologists, foresters, and conservation law enforcement officers should be trained to recognize amphibians and reptiles and to know and enforce laws and regulations pertaining to these animals.

LITERATURE CITED

Adler, K. 1968. Turtles from archeological sites in the Great Lakes Region. *Mich. Archaeol.* 14:147–63.

Burton, T., and G. Likens. 1975. Salamander populations and biomass in the Hubbard Brook Experimental forest, New Hampshire. *Copeia* 1975: 541–46.

Collins, J. P., and H. M. Wilber. 1979. Breeding habits and habitats of the amphibians of the Edwin S. George Reserve, Michigan, with notes on the local distribution of fishes. *Univ. Mich. Mus. Zool. Occ. Pap.* No. 686.

Congdon, J. D., G.L. Breitenback, R.C. van Loben Sels, and D.W. Tinkle. 1987. Reproduction and nesting ecology of snapping turtles (*Chelydra serpentina*) in southeastern Michigan. *Herpetologica* 43:439–54.

Congdon, J. D., D. W. Tinkle, G.L. Breitenbach, and R. C. van Loben Sels. 1983. Nesting ecology and hatching success in the turtle *Emydoidea blandingi. Herpetologica* 39:417–29.

Hall, R. J. 1980. Effects of environmental contaminants on reptiles: A review. *USDI/USFWS Sp. Sci. Rep.-Wildl.* No. 228. Washington D.C.

Harding, J. H., and J. A. Holman. 1990. Michigan Turtles and Lizards. *Mich. State Univ. Coop. Ext. Serv. Publ.* E-2234.

Harding, J. H., and J. A. Holman. 1992. Michigan Frogs, Toads, and Salamanders. *Mich. State Univ. Coop. Ext. Serv. Publ.* E-2350.

Hine, R., B. L. Les, and B. F. Hellmich. 1981. Leopard frog populations and mortality in Wisconsin, 1974–1976. *Wisc. Dept. Nat. Resour., Tech. Bull.* No. 122.

Holman, J. A. 1988. The status of Michigan's Pleistocene herpetofauna. *Mich. Acad.* 20:125–32.

Holman, J. A., and J. H. Harding, M. M. Hensley, and G. R. Dudderar. 1989. Michigan Snakes. *Mich. State Univ. Coop. Ext. Serv. Publ.* E-2000.

Minton, S.A., Jr. 1972. Amphibians and Reptiles of Indiana. *Ind. Acad. Sci. Monogr.* No. 3.

Pentecost, E. D., and R. C. Vogt. 1976. Environmental status of the Lake Michigan region. Vol 16: Amphibians and reptiles of the Lake Michigan drainage basin. *Argonne Nat. Lab.,* Argonne, IL.

Pfingsten, R. A., and F. L. Downs, eds. 1989. Salamanders of Ohio. *Ohio Biol. Survey Bull.* n.s., 7(2).

Pierce, B. A. 1985. Acid tolerance in amphibians. *BioScience* 35:239–43.

Ruthven, A.G., C. Thompson, and H.T. Gaige. 1928. *The herpetology of Michigan.* Univ. Mich., Univ. Mich. Handbook Ser. No. 3.

Tinkle, D. W., J. D. Congdon, and P. C. Rosen. 1981. Nesting frequency and success: Implications for the demography of painted turtles. *Ecology* 62: 1426–32.

Tinkle, D. W., P. E. Feaver, R.W. Van Devender, and L. J. Vitt. 1979. A survey of the status, distribution, and abundance of threatened and endangered species of amphibians and reptiles: *Mich. Dept. Nat. Resour.,* Unpubl. Rept.

Vogt, R.C. 1981. *Natural History of Amphibians and Reptiles in Wisconsin.* Milwaukee Publ. Mus., Milwaukee.

Kirtland's Snake

· *Clonophis kirtlandii* (Kennicott)
· **State Endangered**

Map 27. Confirmed individual Kirtland's snakes. ▲ = before 1981 (Michigan DNR); ● = 1981 to 1989 (Michigan DNR).

Status Rationale: The Kirtland's snake is apparently declining throughout its limited range and is being reviewed for federal protection. The status of Michigan's population is not well known due to this snake's secretive behavior and sporadic distribution. It is known in only two southwestern Michigan counties; the success of a 1986 reestablishment attempt in a third county has not been evaluated.

Field Identification: The Kirtland's snake is one of the smaller species of snake found in Michigan. This nonpoisonous snake is easily recognized by its bright red belly, which is conspicuously edged with two parallel rows of black spots. Adults average 14 to 18 inches (36 to 46 cm) in length (maximum length, 24^1/$_2$ inches [62 cm]) and are reddish to dark brown with four rows of dark, faded blotches on the back. The head is mostly black with a light, cream-colored throat. Back scales are keeled (midline ridge that makes the scales rough to the touch), and the anal plate is divided.

Three other Michigan snakes have red or orange bellies, but none possesses the two rows of black belly spots. The northern ringneck snake (*Diadophis punctatus*) has a more yellowish orange belly with a ring encircling its neck; the red-bellied snake (*Storeria occipitomaculata*) has three white spots immediately behind the head; and adult copperbelly water snakes (*Nerodia erythrogaster neglecta*) are larger and have a solid black back with no dark blotches. Young copperbelly water snakes have patterned backs but lack the diagnostic belly. When

233

disturbed, the Kirtland's snake often flattens its body. Similar behavior may be observed with the eastern hog-nosed snake (*Heterodon platyrhinos*), which may have a reddish back pattern but distinctly lacks a red or orange belly.

Total Range: The Kirtland's snake occurs in disjunct populations throughout its restricted range in the north-central Midwest. It is probably a relict prairie species (Conant 1978), as indicated by its close association with the Prairie Peninsula in the Upper Midwest (Transeau 1935). Its distribution extends from Ohio to western Illinois, and from southern Michigan to extreme north-central Kentucky. This species has not been reported from its range peripheries in northeastern Missouri (Jones 1967), southeastern Wisconsin, and western Pennsylvania since the mid-1900s. Two individuals documented from the Delaware Valley along the Atlantic Coastal Plain (Conant 1943) are probably not representative of a natural population (Wilsmann and Sellers 1988). Between 1980 and 1987, the Kirtland's snake had been found only in 48 different sites in five states, including two counties in Michigan (Wilsmann and Sellers 1988).

State History and Distribution: In Michigan, this species has been reported from only 14 sites in the following southern counties: Berrien, Cass, Kalamazoo, Lenawee, Ottawa, Van Buren, and Washtenaw. State herpetofauna surveys by the University of Michigan in the late 1970s did not locate the Kirtland's snake (Tinkle et al. 1979). In 1981, two individuals were found near Benton Harbor, Berrien County, the first documented observations since a 1976 Ottawa County record. In 1985 and 1987, individuals were again located in Berrien County. Also significant was the discovery of a population of the Kirtland's snake in a fen in Allegan County in 1986. The two sites harbor the only known naturally occurring populations in Michigan. In September, 1986, 20 adult Kirtland's snakes were captured in Indianapolis, Indiana, and released in an enclosure at the Wolf Lake Fish Hatchery, Van Buren County (Sellers 1987).

Habitat: Prior to European settlement, the Kirtland's snake probably inhabited prairie fens, wet meadows, and the grassy openings of forested wetlands (e.g., tamarack swamps) adjacent to ponds, streams, and other bodies of water. These habitat types are generally associated with loose, organic-rich soil, well suited for the fossorial habits of the Kirtland's snake as well as for agriculture. Conant (1943) showed it had successfully adapted and colonized suitable open areas near urban centers where it became locally abundant. Minton (1972) also found open grassy areas in parks, cemeteries, and vacant lots in city and residential areas harboring significant populations. Today, urban and suburban populations probably contain the best known examples of self-sustaining populations; however, they are also the most vulnerable to extirpation (Wilsmann and Sellers 1988).

One of the difficulties with the assessment of this species' distribution, abundance, and habitat preference is its nocturnal and fossorial behaviors. In Michigan, limited observations since the 1970s indicate that this species utilizes fens, floodplains and other open forested wetlands, and marshes (Wilsmann and Sellers 1988). Recent Michigan individuals located in Berrien County were found under wet, matted leaves (which provide an abundance of earthworm prey) adjacent to a stream. Kirtland's snakes frequently use open, grassy areas with numerous chimney crayfish (*Cambarus diogenes*) burrows. Use of these tunnels may be extensive year-round and probably serve as hibernacula sites. Because of the preference for moist areas, these snakes are frequently found under logs, boards, or other objects within suitable wetland habitats.

Ecology and Life History: The Kirtland's snake spends the winter in an underground hibernaculum and typically does not awaken until April. In April and May, the Kirtland's snake is most visible (Conant 1938) due to increased posthibernation food requirements, lack of dense herbaceous vegetation, and mating activities. Unlike other reptiles, this species rarely basks in the sun. Even during periods of activity, it generally remains under cover of debris or leaf litter. The Kirtland's snake is live bearing (ovoviviparous), and the young are usually born in late summer or early autumn. An average of seven young are born yearly (Tucker 1976), each approximately 6 inches (15 cm) in length (Conant 1943).

This snake is highly secretive and most easily found foraging during mild spring or autumn nights following periods of rainfall. Its main food source is earthworms; primary alternatives may include slugs and leeches (Tucker 1977).

Conservation/Management: Due to habitat degradation throughout much of its restricted range, the Kirtland's snake has declined dramatically, even though it appears to have adapted to disturbed habitats, such as grassy parks within urban centers. Urbanization, wetland drainage, alteration of water systems, and development of natural areas (e.g., fens) continue to reduce habitat availability. Since it occurs in widely disjunct populations, it is vulnerable to local extinction. This includes weather perturbations, chemical contamination (e.g., agricultural pesticide runoff), mowing and vehicular traffic (Dalrymple and Reichenbach 1984), and overcollection for commercial and scientific purposes. The ongoing and unabated threat of removing individuals and populations for captive purposes is potentially significant to the survival of this species in many regions, including Michigan.

Due to its precarious status, intensive surveys are underway nationwide to provide an accurate evaluation of its status. Since loss of habitat is of primary concern, preservation of open, grassy areas adjacent to wetlands and bodies of water are needed to protect current populations. Fens, wet meadows, and open tamarack swamps are particularly important habitats in southwestern Michigan. Active management of these areas may be intermittently required: using methods such as controlled burns, to which this fossorial snake is well adapted. Construction of leaf piles and selective placement of waterproof cover (e.g., rubber mats) offer moist areas and an abundance of prey and are optimal methods for locating this species.

Information is lacking on the population ecology of the Kirtland's snake. Individuals within known populations need to be studied to gather basic biological data unavailable in captive situations. Strict protection of known populations and continued searches for other populations will provide an opportunity for the long-term survival of the Kirtland's snake in Michigan.

LITERATURE CITED

Conant, R. 1938. On the seasonal occurrences of reptiles in Lucas County, Ohio. *Herpetologica* 1:137–44.

Conant, R. 1943. Studies on North American water snakes-I: *Natrix kirtlandii* (Kennicott). *Am. Midl. Nat.* 29:313–41.

Conant, R. 1978. Distributional patterns of North American snakes: Some examples of the effects of Pleistocene glaciation and subsequent climatic changes. *Bull. Maryland Herp. Soc.* 14:241–59.

Dalrymple, G. H., and N. G. Reichenbach. 1984. Management of an endangered species of snake in Ohio, USA. *Biol. Conserv.* 30:195–200.

Jones, J. M. 1967. A western extension of the known range of Kirtland's watersnake. *Herpetologica* 23:66–67.

Minton, S. A. 1972. Amphibians and reptiles of Indiana. *Indiana Acad. Sci., Monogr.* No. 3.

Sellers, M. A. 1987. Final report of the 1987 survey of Michigan's two endangered reptiles. *Mich. Dept. Nat. Resour.*, Unpubl. Rept.

Tinkle, D. W., P. E. Feaver, R. W. Van Devender, and L. J. Vitt. 1979. A survey of the status, distribution and abundance of threatened and endangered species of reptiles and amphibians. *Mich. Dept. Nat. Resour.*, Unpubl. Rept.

Transeau, E. N. 1935. The prairie peninsula. *Ecology* 16:423–37.

Tucker, J. K. 1976. Observations on the birth of a brood of Kirtland's water snake, *Clonophis kirtlandii* (Kennicott) (Reptilia, Serpentes, Colubridae). *J. Herp.* 10:53–54.

Tucker, J. K. 1977. Notes on the food habits of Kirtland's water snake, *Clonophis kirtlandii*. *Bull. Maryland Herp. Soc.* 13:193–95.

Wilsmann, L. D., and M. A. Sellers. 1988. *Clonophis kirtlandii* rangewide survey. *U.S. Fish Wildl. Serv.*, Unpubl. Rept.

Eastern Fox Snake

· *Elaphe vulpina gloydi* Conant
· **State Threatened**

Map 28. Confirmed individual eastern fox snakes.
▲ = before 1977 (Michigan DNR); ● = 1977 to 1989 (Michigan DNR).

Status Rationale: The eastern fox snake is limited to open wetland areas along Lakes Erie, St. Clair, and Huron. Continued habitat loss and harassment by humans currently threatens the four known populations in Michigan. Sites still inhabited contain relatively large populations, thereby allowing time to research this species' requirements and to prevent its disappearance from Michigan.

Field Identification: The eastern fox snake is a boldly marked, nonpoisonous species with dark blotches patterned on a yellowish to light brown background. Alternating with the larger back markings are smaller ones along the sides. The head is reddish and the underside is yellowish with dark markings. The scales are keeled and the anal plate is divided. Adults measure an average of 3 to 5 feet (0.9 to 1.5 m). Immatures are paler in color than adults.

Several snake species with similarly patterned backs may be confused with the eastern fox snake. The most easily confused species is the black rat snake (*Elaphe o. obsoleta*). Although adults are distinctive, the young are as strongly spotted as the eastern fox snake. Ideally, the number of ventral scutes should be counted— 216 or fewer in the eastern fox snake, and 221 or more in the black rat snake (Conant 1975). The northern water snake (*Nerodia s. sipedon*) has distinct cross-bands instead of a blotchlike pattern. The eastern hog-nosed snake (*Heterodon platyrhinos*) has an upturned snout and, like the reddish-colored eastern milk snake (*Lampropeltis t. triangulum*), occurs in sandy habitats. The rattles of the

237

massasauga (*Sistrurus catenatus*) are visible at any age. At first sight, the eastern fox snake could be initially mistaken for a rattlesnake when vibrating its tail in dry leaves.

Total Range: The fox snake has a disjunct population. The more common western subspecies occurs in the western Great Lakes basin (including the Upper Peninsula) west to Nebraska; the eastern subspecies has a more limited range. It is restricted to the shores of southern Lake Huron and western Lake Erie in Ohio, Ontario, and Michigan. The eastern subspecies has not been documented outside this relatively small geographic area.

State History and Distribution: Historically, the eastern fox snake was known to occur in six southern Michigan counties along Lakes Erie, St. Clair, and Huron. Individuals also have been recorded along the Raisin, Detroit, and Clinton rivers, and large populations are known from the lower Shiawassee River basin. Four isolated populations were documented in a 1986 survey of the eastern fox snake in southern Michigan (Weatherby 1986). Two sites were in Monroe County along western Lake Erie, including the general regions surrounding the Erie and Point Mouillee State Game areas. A third occupied site, in the vicinity of the St. Clair Flats Wildlife Area, contains the largest acreage of suitable habitat, although it is increasingly vulnerable to alteration and degradation. The fourth site is located inland, along the Shiawassee River, Saginaw County, and selected adjacent tributaries. Populations in this area occur within the Shiawassee National Wildlife Refuge and State Game Area.

Habitat: This snake is an inhabitant of emergent wetlands along the Great Lakes shorelines and associated large rivers and impoundments. Habitats with herbaceous vegetation, such as cattails (*Typha* spp.), are preferred. Individuals frequent artificially created dikes and other elevated sites and are most easily observed at these vantage points (Conant 1938; Weatherby 1986). Although primarily an open wetland species, individuals occasionally may be found in drier habitats (Catling and Freedman 1980; Weatherby 1986).

Ecology and Life History: Eastern fox snakes become active as spring temperatures warm. Individual movements are most profound in May and June, during its mating behavior peak. This species is active throughout much of the daylight hours; during intense heat, nighttime excursions are more frequent (Kraus and Schuett 1982). Life history information for the eastern fox snake species is limited and mostly conjecture, based on related species.

It probably breeds annually, beginning at two years of age (approximately 3 feet [0.9 m] in length). Egg laying is initiated in June and may extend to August for some populations. An average of 15 eggs are laid per clutch (range of 7 to 29) generally hatching in autumn (Wright and Wright 1957). Availability of nesting sites is crucial, since the unguarded eggs require rotting wood piles, stumps, and hollow logs for protection and moisture. Even then, less than 50% of the eggs hatch (Mattlin 1948; Weatherby 1986). Young-of-the-year also have a high mortality rate and generally remain under cover. Limited home range studies have shown individual movements to vary up to several hundred feet (Rivard 1976; Freedman and Catling 1979). A recent study indicated limited territoriality by the eastern fox snake. At three Michigan locations, researchers found adult male individuals to be spaced at least 165 feet (50 m) apart on dikes and other narrow strips of land confined by water (Weatherby 1986). As characteristic of this genus, the eastern fox snake constricts prey in its strong coils. Small mammals and birds are the most common prey.

Conservation/Management: The eastern fox snake has a limited and declining range. It is currently threatened in Michigan due to human-related threats at the four known remaining sites and continued habitat loss of Great Lakes marshes. Although habitat of the four populations is partially protected on state and federal land, public access and use is relatively unrestricted. Much of the emergent wetland habitat along the Great Lakes has been altered. Several remaining tracts are large but are being fragmented by recreation and urban development, channelization, draining, and other factors degrading the quality of wetlands. Fluctuating water levels of the Great Lakes also increase the value of undisturbed wetlands and apply pressure to concentrated eastern fox snake populations as suitable habitat becomes unavailable. Other limiting factors include traffic-related fatalities and polluted waterways.

Public education at known sites is needed to discourage illegal harassment. Enforcement of the Endangered Species Act to protect this species from commercial collection also is required. Management of emergent wetlands should include limiting disturbance on dike areas (e.g., restricting mowing between mid-June and mid-October) and microhabitat enhancement, such as providing hollow logs for protected incubation sites (Weatherby 1986). The "cleaning up" of woody debris was suggested as one limitation to reproductive success by Tinkle et al. (1979). The recent application of radio-telemetry with larger snakes associated with wetland habitats (Reinert and Kodrich 1982; Weatherhead and Charland 1985) will play a significant role in documenting the ecological requirements of the eastern fox snake and saving it in Michigan.

LITERATURE CITED

Catling, P. M., and B. Freedman. 1980. Variation in distribution and abundance of four sympatric species of snakes at Amherstburg, Ontario. *Can. Field-Nat.* 94:19–27.

Conant, R. 1938. The reptiles of Ohio. *Am. Midl. Nat.* 20:1–284.

Conant, R. 1975. *A field guide to reptiles and amphibians of eastern and central North America.* Houghton Mifflin Co., Boston.

Freedman, B., and P. M. Catling. 1979. Movements of sympatric species of snakes at Amherstburg, Ontario. *Can. Field-Nat.* 93:399–404.

Kraus, F., and G. W. Schuett. 1982. A herpetofaunal survey of the coastal zone of northwest Ohio. *Kirtlandia* 1982:21–54.

Mattlin, R. 1948. Observations on the eggs and young of the eastern fox snake. *Herpetologica* 4:115–16.

Reinert, H. K., and W. R. Kodrich. 1982. Movements and habitat utilization by the massasauga, *Sistrurus catenatus catenatus*. *J. Herp.* 16:162–171.

Rivard, D. H. 1976. *The biology and conservation of eastern fox snakes (Elaphe vulpina gloydi)*. M.S. thesis, Carlton Univ., Ottawa.

Tinkle, D. W., P. E. Feaver, R. W. Van Devender, and L. J. Vitt. 1979. A survey of the status, distribution and abundance of threatened and endangered species of reptiles and amphibians. *Mich. Dept. Nat. Resour.*, Unpubl. Rept.

Weatherby, C. A. 1986. Michigan Nature Conservancy *Elaphe vulpina gloydi* and *Clonophis kirtlandii* 1986 contracted survey. *Mich. Nat. Conserv.*, Unpubl. Rept.

Weatherhead, P. J., and M. B. Charland. 1985. Habitat selection in an Ontario population of the snake, *Elaphe o. obsoleta*. *J. Herp.* 19:12–19.

Wright, A. H., and A. A. Wright. 1957. *Handbook of snakes of the United States and Canada*. Comstock Publ. Assoc., Ithaca, NY.

Copperbelly Water Snake

- *Nerodia erythrogaster neglecta* (Conant)
- State Endangered

Map 29. Confirmed individual copperbelly water snakes. ▲ = before 1982 (Michigan DNR); ● = 1982 to 1989 (Michigan DNR).

Status Rationale: The water snake is a common and widespread southern species, but the northern copperbelly subspecies is declining rapidly and has a disjunct and limited distribution. Michigan's three remaining populations are small and vulnerable to human disturbance.

Field Identification: The large, stout-bodied, nonpoisonous copperbelly water snake is easily recognized by its uniform, dark-colored back and bright orange-red belly and chin. The average adult's total body length is 30 to 48 inches (76 to 122 cm); attaining a maximum of 56 inches (142 cm). Juveniles are lighter colored with dark, saddlelike blotches alternating along the backs and sides and fusing into bands toward the head. This pattern disappears with age and is usually absent when individuals reach sexual maturity (Conant 1949). The belly is pale orange for young less than 20 inches (51 cm). The scales are keeled and the anal plate is usually divided.

Michigan's three other orange or red-bellied snakes can be easily distinguished from adult and juvenile copperbelly water snakes (also known as redbellies). The northern ringneck snake (*Diadophis punctatus*) has a yellowish orange ring encircling its neck. The small (normally 8 to 10 inches [20 to 25 cm]), brownish, red-bellied snake (*Storeria occipitomaculata*) has three white spots immediately behind the head, and the state-endangered and secretive Kirtland's snake (*Clonophis kirtlandii*) can be differentiated by its two parallel rows of distinct black spots on the underside. Furthermore, these three snakes typically are not associated with the copperbelly water snake's forested wetland habitats. A species of snake that frequently coexists in habitats with the copperbelly water snake is the common, northern water snake (*Nerodia sipedon sipedon*). This large snake has a distinctly patterned back, and its yellowish-to-reddish belly is interrupted by dark crescent markings in young and adult individuals (unlike the copperbelly water snake's solid-colored belly). Young copperbelly water snakes are the most difficult to discern from other snake species.

Total Range and Taxonomic Status: This species occurs throughout much of the southern United States, from the Atlantic coast (north to Delaware) and along the Gulf coast (except southern Florida), north to Kansas, Missouri, southern Illinois, southern Michigan, and western Tennessee. The copperbelly subspecies occurs in southeastern Illinois, western Kentucky, southwestern and northeastern Indiana, and south-central Michigan. Disjunct populations may exist in northwestern Illinois, north-central Tennessee, southeastern Indiana, and central and northwestern Ohio.

Formerly placed in the genus *Natrix*, subspecies *N. e. erythrogaster* (Conant 1934; Grant and Necker 1943), it is now classified as the northern race of the four recognized subspecies of *Nerodia erythrogaster* (Conant 1949). The extent of the copperbelly water snake's historic range never may be known, because it was not identified and accepted as a species separate from the northern water snake until the 1930s.

State History and Distribution: The copperbelly water snake was first discovered in Eaton County in 1903 (Clark 1903) but was not confirmed again until 1933 (Clay 1934). Confusion regarding its taxonomic status was probably responsible for several undocumented sightings. Nonetheless, confirmed observations of this species were rare and scattered until systematic surveys were initiated in the 1980s.

This snake currently lives in small colonies confined to the southernmost county tier of the Lower Peninsula. A statewide herpetofauna survey in the late 1970s failed to find any individuals, and, consequently, the subspecies was considered to be extirpated in the state (Tinkle et al. 1979). Intensive state-supported studies from 1982 through 1988 have provided crucial information on the status and ecological needs of the copperbelly water snake in Michigan. Since 1982, individuals have been found at seven general locations, (one in Cass and St. Joseph counties and six in Hillsdale County), each of which were known to have harbored this species previously (Sellers 1986 and 1987). Among these seven sites, however, only two have confirmed reproducing populations. Individuals have been found at other areas disjunct from known historic sites but are probably representative of this species highly mobile behavior. Connecting waterways are frequently used as migratory corridors.

Habitat: This species prefers wooded floodplains and scrub-shrub wetlands adjacent to lakes, ponds, and slow-moving rivers. Areas that are sparsely vegetated and that contain clear, deep waters are avoided (Minton 1972). During the spring, it is commonly found in quiet, shallow, warm ponds surrounded by lowland woods and shrubs (Winn and Gillingham 1987). Emergent woody vegetation and debris piles near bodies of water serve as valuable basking sites. The copperbelly water snake is most easily found before and during the spring breeding season.

In summer, when seasonal ponds are receding, the copperbelly water snake usually migrates, via wooded avenues, to more permanent bodies of water (Sellers 1984). These migration corridors are an important component for a self-sustaining population. If severe drought conditions occur, individuals may wander from dried waterbeds to search for more favorable sites (e.g., artificial structures), or they will become inactive, taking refuge underground or within large debris piles until suitable conditions return, such as rain and/or cooler temperatures (60–80°F [Sellers 1984]). Areas used for winter hibernacula are usually upland scrub-shrub woodland slopes near summer habitats. Throughout the year, copperbelly water snakes require large areas and are known to range over at least 40 acres (Sellers 1986).

Ecology and Life History: Copperbelly water snakes typically awaken from their winter hibernacula in April. Mating usually occurs in May in floodplain habitat and around the wooded edges of swamps and lakes. This species is one of the ten Michigan snakes that bear live young instead of laying eggs. In September or October, the one to two dozen young are born, averaging 9 inches (23 cm) in length. Adult females immediately shed their skin after birth. By early November, most of the population has returned to hibernacula until the following spring. Gravid females are usually the last individuals to seek protected winter sites.

This snake frequently forages within the waters of emergent wetlands and swamps during the day, searching for aquatic vertebrates, such as tadpoles, frogs, and salamanders, as well as invertebrates, such as insect larvae and crayfish. A recent study in southern Michigan found that large tadpoles are a common prey (Winn and Gillingham 1987). Foraging techniques include waiting or actively seeking prey.

Conservation/Management: The primary threats to this species are habitat degradation and other human-related threats. Coupled with these limitations is an already restricted and discontinuous distribution. Its affinity for forested and scrub-shrub wetlands adjacent to ponds, lakes, and streams has made it particularly vulnerable to widespread wetland draining and residential development. The copperbelly water snake does not require mature floodplains but can survive in self-sustaining populations in lowland second-growth regions (Sellers 1984). In addition to aquatic and riparian habitat destruction, other limiting factors include illegal collecting for captive purposes, indiscriminate killing of snakes by misinformed people, and roadways that cause high mortality and act as major barriers to seasonal movements.

To help keep this species from disappearing in Michigan, its wetland habitat needs to be protected. Michigan's largest population in Hillsdale County is protected; however, more needs to be done to ensure its long-term viability. Active management of lowland forests and other riparian systems can be beneficial for copperbelly water snakes. Selective cutting of wooded areas every 20–30 years can improve habitat conditions. Logging or clearing should occur between November and March to avoid disrupting individuals. By preserving seasonal pools and shrub swamps, constructing or permitting debris piles, retaining a wide vegetative zone along waterways, restricting grazing along riparian and shoreline areas from April to June, and minimizing mowing, populations of the copperbelly water snake can survive in minimally disturbed areas (Sellers 1984).

LITERATURE CITED

Clark, H. L. 1903. The water snakes of southern Michigan. *Am. Nat.* 37:1–23.
Clay, W. M. 1934. Rediscovery of the red-bellied watersnake, *Natrix sipedon erythrogaster* (Forster), in Michigan. *Copeia* 1934:95.
Conant, R. 1934. The red-bellied water snake, *Natrix sipedon erythrogaster* (Foster), in Ohio. *Ohio J. Sci.* 34:21–30.
Conant, R. 1949. Two new races of *Natrix erythrogaster. Copeia* 1949:1–15.
Grant, C., and W. L. Necker. 1943. *Natrix erythrogaster erythrogaster* in the northeastern part of its range. *Herpetologica* 2:83–86.
Minton, S. A. 1972. Amphibians and reptiles of Indiana. *Indiana Acad. Sci., Monogr.* No. 3.
Sellers, M. A. 1984. Field notes on confirmed sightings of the northern copperbelly (*Nerodia erythrogaster neglecta*) in Michigan 1978–83, with related discussion. *Mich. Nat. Feat. Invent.*, Unpubl. Rept.
Sellers, M. A. 1986. Final report summary of the 1986 status survey of historic and potential state sites for the northern copperbelly water snake (*Nerodia erythrogaster neglecta*, Conant) and Kirtland's snake (*Clonophis kirtlandii*, Cope), Michigan endangered species. *Mich. Nat. Feat. Invent.*, Unpubl. Rept.
Sellers, M. A. 1987. Final report of the 1987 survey of Michigan's two endangered reptiles. *Mich. Dept. Nat. Resour.*, Unpubl. Rept.
Tinkle, D. W., P. E. Feaver, R. W. Van Devender, and L. J. Vitt. 1979. A survey of the status, distribution, and abundance of threatened and endangered species of reptiles and amphibians. *Mich. Dept. Nat. Resour.*, Unpubl. Rept.
Winn, G. D., and J. C. Gillingham. 1987. Habitat selection and movements of the copperbelly water snake, *Nerodia erythrogaster neglecta*, in southern Michigan. *Mich. Dept. Nat. Resour.*, Unpubl. Rept.

Marbled Salamander

· *Ambystoma opacum*
  (Gravenhorst)

· **State Threatened**

Map 30. Confirmed individual marbled salamanders.
▲ = before 1988 (Michigan DNR); ● = 1988 to 1989
(Michigan DNR).

Status Rationale: The marbled salamander is a secretive species that only occurs in extreme southwestern Michigan. Although commonly found throughout the southeastern United States, its status in Michigan is not well known; it may be threatened with local extinction due to intensive habitat modification.

Field Identification: The strikingly marked marbled salamander has silvery crossbands along the entire length of its back. These markings are variable in size and shape and tend to be darker in females. The crossbands typically interconnect, particularly in the front half of the body. The belly is solid black. This salamander averages 3^1/$_2$ to 4^1/$_4$ inches (8.9 to 10.8 cm) in length, occasionally reaching 5 inches (12.7 cm). The back pattern of subadults is much less defined.

There are two other Michigan salamanders with distinctly light-colored back markings, but both species have yellow blotches instead of silvery crossbands. The spotted salamander (*Ambystoma maculatum*) has rounded spots that are nonconnecting. The eastern tiger salamander (*Ambystoma tigrinum*) has profuse yellow mottling throughout its body, including the belly.

Total Range: The marbled salamander is primarily a species of the southeastern United States ranging north to northern Indiana, Ohio, and southern New England.

245

Disjunct populations occur along southern Lake Michigan, including extreme southwestern Michigan.

State History and Distribution: This secretive species is known only from a few scattered records in three counties of southwestern Michigan. Early records are from Berrien and Allegan counties. More recent records include three individuals collected near Wolf Lake, Van Buren County, in 1966 and individuals known from a site in Allegan County. Michigan's populations are isolated from the more southern populations, possibly relicts from long-term climatic changes or more recent human-related habitat degradation.

Habitat: The marbled salamander occurs in a variety of habitats, depending on the season. In the spring and summer, adults typically inhabit upland areas, such as sandy deciduous woodlands (Minton 1972). In these dry habitats, individuals usually seek shelter under logs or in underground tunnels of other animals (Bishop 1941). In autumn, marbled salamanders migrate to lowland, forested habitats for breeding. Ephemeral ponds are typical egg-laying sites. This is an optimal time to locate individuals, as they may congregate in groups awaiting suitable egg-laying conditions.

Ecology and Life History: Unlike other Michigan salamanders that breed in the spring, the marbled salamander lays its eggs in autumn. More northern populations (such as in Michigan), probably breed earlier (Anderson and Williamson 1973). Microhabitat temperatures at approximately 60°F appear to be preferred for egg laying (Anderson and Williamson 1973). An average of 100 eggs are laid in a nest, constructed by the female in a protected depression (Smith 1961) or at the edge of a shallow body of water (Bishop 1941). The female remains with the eggs until rains flood the nest site. Eggs then hatch within two weeks (Oliver 1955). If eggs are not inundated by water, they will remain dormant until the following spring.

Nest site selection has been intensively studied for the marbled salamander (Graham 1971; Petranka and Petranka 1981). Evidence indicates that females actively select nest sites that favor egg and larvae survivorship. These microhabitats are at the mean water levels of dried pool depressions, which are typically inundated by autumn rains. Eggs deposited at the edges of pools are more subject to freezing. The aquatic larvae that do hatch in autumn overwinter with little growth and metamorphose into the terrestrial adults in late spring (Noble and Brady 1933) or in June and July (Smith 1961). This nocturnal salamander typically inhabits underground tunnels and logs. Its primary prey are earthworms, insects, shell-less mollusks, and other small, soft organisms (e.g., crickets).

Conservation/Management: The marbled salamander currently has a small, disjunct Michigan distribution limited to widely scattered sites. Severe habitat degradation and fragmentation has affected this species' survival. Forests in southwestern Michigan have been reduced in size, and wetland destruction from draining, channelization, and filling has greatly limited suitable habitat for the marbled salamander. The continued fragmentation of habitats used by this species stress existing populations and subject them to local extinction through natural catastrophic events (e.g., droughts). Road density and placement can also extirpate or severely depress migratory salamander populations. Recently, water quality has become important for survival, because amphibians may be particularly susceptible to various forms of pollution. Survival to metamorphosis can be relatively low (Graham 1971; Stenhouse 1987) and can be confounded by other limiting factors. Acidification is known to affect the survival of other members of the genus *Ambystoma* (Punzo 1983).

Identification and conservation of sites with current marbled salamander populations must be made. Specially constructed and placed culverts, fencing, and signs also may serve in reducing high adult mortality along migration routes. Salamanders are early-warning indicators that aquatic environments may be contaminated and should be viewed as crucial components of habitats by land managers involved with environmental impact statements. Protection of connected forested uplands and wetlands with intermittent pools would assure suitable habitats for the marbled salamander in southwestern Michigan.

LITERATURE CITED

Anderson, J. D., and G. K. Williamson. 1973. The breeding season of *Ambystoma opacum* in the northern and southern parts of its range. *J. Herp.* 7:320–21.
Bishop, S. C. 1941. *The salamanders of New York.* NY State Mus. Bull. No. 324.
Graham, R. E. 1971. *Environmental effects on deme structure, dynamics, and breeding strategy of Ambystoma opacum (Amphibia: Ambystomatidae), with a hypothesis of the probable origin of the marbled salamander life-style.* Ph.D. diss., Rutgers Univ., New Brunswick, NJ.
Minton, S. A. 1972. Amphibians and reptiles of Indiana. *Ind. Acad. Sci. Monogr.* No. 3.
Noble, G. K., and M. K. Brady. 1933. Observations of the life history of the marbled salamander, *Ambystoma opacum* (Gravenhorst). *Zoologica* 11:89–132.
Oliver, J. A. 1955. *The natural history of North American amphibians and reptiles.* D. VanNostrand Co., Princeton, NJ.
Petranka, J. W., and J. G. Petranka. 1981. On the evolution of nest site selection in the marbled salamander, *Ambystoma opacum. Copeia* 1981:387–91.
Punzo, F. 1983. Effects of environmental pH and temperature on embryonic survival capacity and metabolic rates in smallmouth salamander, *Ambystoma texanum. Bull. Environ. Contam. Toxicol.* 31:467–773.
Smith, P. W. 1961. The amphibians and reptiles of Illinois. *Ill. Nat. Hist. Surv., Bull.* No. 28.
Stenhouse, S. L. 1987. Embryo mortality and recruitment of juveniles of *Ambystoma maculatum* and *Ambystoma opacum* in North Carolina. *Herpetologica* 43:496–501.

Smallmouth Salamander

· *Ambystoma texanum* (Matthews)
· **State Threatened**

Map 31. Confirmed individual smallmouth salamanders.
▲ = before 1988 (Michigan DNR); ● = 1988 to 1989
(Michigan DNR).

Status Rationale: The northern range limits of this little-known species are in southeastern Michigan. Encompassing four counties, each of the eight known populations is threatened by habitat disturbance and genetic dilution by hybridization.

Field Identification: The smallmouth salamander is recognized by its slate-colored back with lichenlike patches of silvery-gray extending to the sides. This pattern can be variable in both intensity and distribution. Smaller individuals are more brightly marked; older ones may become nearly black. As indicated by its common name, this salamander has a small, blunt nose. Adults average a length of 4^1/$_2$ to 5^1/$_2$ inches, occasionally reaching 8 inches.

Several other *Ambystoma* species may be confused with the smallmouth salamander, particularly species in the Jefferson complex (*A. jeffersonianum*), which recently underwent taxonomic revision. Within this taxomically similar complex, four species were recognized by Uzzell (1964), including the silvery (*A. platineum*), Tremblay's (*A. tremblayi*), and blue-spotted (*A. laterale*) salamanders in Michigan. These three species differ by their relatively longer toes and snout and individuals generally lack the heavily speckled back markings. Recent studies have found increasing hybridization between the smallmouth and blue-spotted (*A. laterale-texanum*) and silvery salamanders (*A. platineum-texanum*), necessitating laboratory examination to identify pure individuals from hybrids (Morris and Brandon 1984; Kraus 1985; Ducey et al. 1988). Electrophoretic identification has been successfully used (Bogart et al. 1985).

Total Range: This species primarily occurs in the south-central United States. It ranges from the Texas and Alabama Gulf coasts north to southern Iowa, northern Ohio, and extreme southeastern Michigan. Historic records exist for Pelee Island, Ontario, in Lake Erie (Uzzell 1962). The smallmouth salamander does not occur east of the Appalachian Mountains and is absent from the Missouri and Arkansas Central Highlands.

State History and Distribution: The smallmouth salamander is historically known from four southeastern Michigan counties: Hillsdale, Monroe, Washtenaw, and Wayne. Until a 1988 survey, the extent of Michigan's population was not well known. In 1988, active populations were located in each of these four counties (Ducey et al. 1988). Another population was found in a Hillsdale County protected preserve in 1989. Eight populations of the smallmouth salamander are currently known, although only the sole Washtenaw County site lacks hybrid *Ambystoma* salamanders (Ducey et al. 1988). The hybrid *A. laterale-texanum* competes with pure strains of the smallmouth salamander and eventually may genetically dilute *Ambystoma texanum* populations in Michigan.

Habitat: As with many members in the genus *Ambystoma*, the smallmouth salamander's habitat preferences vary seasonally. During the spring breeding season, adults congregate in forested floodplains at ephemeral and oxbow ponds. Open wetlands with clay-based soils appear to be particularly favorable spring breeding areas. After breeding, adults move to adjacent upland and lowland wooded areas, followed by first-year individuals in midsummer (Smith 1961). Individuals generally remain underground throughout the rest of the year, resurfacing before the ice melts from pond edges in late winter or early spring.

In many parts of its range, individuals seasonally migrate from upland to lowland habitats (Williams 1973). Recent evidence in Michigan indicates smallmouth salamander populations are more sedentary, possibly spending the entire year within wet, clay-based soils adjacent to the breeding areas (Ducey et al. 1988). After breeding, adults frequently reside in self-constructed burrows and possibly use abandoned crayfish tunnels.

Ecology and Life History: Spring rains generally disrupt smallmouth salamanders from the hibernaculum. This salamander has a variable reproductive behavior throughout its range; some populations breed in streams and others in ponds (Petranka 1982a). Adults congregate at ephemeral ponds immediately following periods of heavy rainfall and mate. Courtship behavior of this salamander has been extensively studied and includes mutual nudging by both sexes prior to spermataphore deposition (Wyman 1971; Arnold 1972; Garton 1972; Petranka 1982b). Between 200 and 600 eggs are attached singly or in scattered groups of 5 to 10 eggs to vegetation or detritus in shallow water. Although water temperatures may not be crucial to the timing of egg deposition (Petranka 1984), the two to six week hatching process is dependent upon water temperature. The ephemeral breeding habitat types are preferred because they lack fish populations, which can be major predators of eggs and larvae (Petranka 1983).

Growth of the larvae is dependent upon water levels, availability of dissolved oxygen, and density of individuals. In Michigan, underdeveloped larvae die if ponds evaporate in May or early June (Ducey et al. 1988). The aquatic larvae generally develop by midsummer and metamorphize into terrestrial adults. First-year individuals probably remain in proximity to natal ponds. Unlike other *Ambystoma* salamanders, little agonistic behavior (or territoriality) of adults has been observed in this species (Martin et al. 1986). Prey for larvae include zooplankton and such aquatic invertebrates as mosquito larvae. Adults are nocturnal and fossorial, feeding primarily upon earthworms and slugs.

Conservation/Management: Habitat degradation and displacement through hybridization are the primary factors pressuring Michigan's smallmouth salamander populations. Floodplain forests and associated ephemeral ponds are currently limited and are continually being fragmented and altered in southeastern Michigan. Suitable wetland breeding habitats have been, and continue to be, drained and altered. The lowering of water levels at preferred breeding sites may cause early pond evaporation and subsequent high larval mortality.

Competition by hybrid *Ambystoma laterale-texanum* individuals reduces the viability of smallmouth salamander populations. Hybridization is increasing since the widespread blue-spotted salamander (*Ambystoma laterale*) is adaptable to changing habitats, invading areas once solely occupied by smallmouth salamanders. Other major threats include pollution, fertilizer and pesticide runoff, and the continuing isolation and extirpation of smallmouth salamander populations. This salamander's hatching success and larvae metabolic rates are affected when pH levels of pond water fall below 6.0 or exceed 10.0 (Punzo 1983).

A key solution for maintaining this species in Michigan is habitat protection. The preservation of optimal breeding sites and their related drainage, buffered by upland habitats, would provide areas for self-sustaining populations. Furthermore, evidence now indicates that Michigan populations are more sedentary and not as migratory as observed in other regions. Further surveys augmenting the successful efforts by Ducey et al. (1988) are needed in other counties, such as Hillsdale and Lenawee. Nighttime searches should be made during early spring rains at potential breeding habitats. Research examining the extent of displacement by hybrids on smallmouth salamander populations also must continue. Providing that the known sites can be adequately protected and monitored, Michigan's smallmouth salamanders will continue to be a viable wildlife component of our state's southeastern wetlands for years to come.

LITERATURE CITED

Arnold, S. J. 1972. *The evolution of courtship behavior in salamanders.* Ph.D. diss., Univ. Mich., Ann Arbor.

Bogart, J. P., M. J. Oldham, and S. J. Darbyshire. 1985. Electrophoretic identification of *Ambystoma laterale* and *Ambystoma texanum* as well as their diploid and triploid interspecific hybrids (Amphibia: Caudata) on Pelee Island, Ontario. *Can. J. Zool.* 63:340–47.

Ducey, P. K., F. Kraus, G. Schneider, and R. Kessie. 1988. An investigation into the status of the small-mouthed salamander, *Ambystoma texanum*, in Michigan. *Mich. Dept. Nat. Resour.*, Unpubl. Rept.

Garton, J. S. 1972. Courtship of the small-mouthed salamander, *Ambystoma texanum*, in southern Illinois. *Herpetologica* 28:41–45.

Kraus, F. 1985. Unisexual salamander lineages in northwestern Ohio and southeastern Michigan: A study of the consequences of hybridization. *Copeia* 1985:309–24.

Martin, D. L., R. G. Jaeger, and C. P. LaBat. 1986. Territoriality in an *Ambystoma* salamander? Support for the null hypothesis. *Copeia* 1986:725–31.

Morris, M. A., and R. A. Brandon. 1984. Gynogenesis and hybridization between *Ambystoma platineum* and *Ambystoma texanum* in Illinois. *Copeia* 1984:324–37.

Petranka, J. W. 1982a. Geographic variation in the mode of reproduction and larval characteristics of the small-mouthed salamander (*Ambystoma texanum*) in the east-central United States. *Herpetologica* 38:475–85.

Petranka, J. W. 1982b. Courtship behavior of the small-mouthed salamander (*Ambystoma texanum*) in central Kentucky. *Herpetologica* 38:333–36.

Petranka, J. W. 1983. Fish predation: A factor affecting the spatial distribution of a stream-breeding salamander. *Copeia* 1983:624–28.

Petranka, J. W. 1984. Breeding migrations, breeding season, clutch size, and oviposition of stream-breeding *Ambystoma texanum*. *J. Herpet.* 18:106–12.

Punzo, F. 1983. Effects of environmental pH and temperature on embryonic survival capacity and metabolic rates in the smallmouth salamander, *Ambystoma texanum*. *Bull. Environ. Contam. Toxicol.* 31:467–73.

Smith, P. W. 1961. The amphibians and reptiles of Illinois. *Ill. Nat. Hist. Surv., Bull.* No. 28.

Uzzell, T. M. 1962. The small-mouthed salamander, new to the fauna of Canada. *Can. Field-Nat.* 76:182.

Uzzell, T. M. 1964. Relations of the diploid and triploid species of the *Ambystoma jeffersonianum* complex (Amphibia, Caudata). *Copeia* 1964:257–300.

Williams, P. K. 1973. *Seasonal movements and population dynamics of four sympatric mole salamanders, genus Ambystoma.* Ph.D. diss., Mich. State Univ., E. Lansing.

Wyman, R. L. 1971. The courtship behavior of the small-mouthed salamander, *Ambystoma texanum*. *Herpetologica* 27:491–98.

Endangered and Threatened

Lake sturgeon (*Acipenser fulvescens*)
Mooneye (*Hiodon tergisus*)
Bigeye chub (*Hybopsis amblops*)
Ironcolor shiner (*Notropis chalybaeus*)
Silver shiner (*Notropis photogenis*)
Weed shiner (*Notropis texanus*)
Pugnose minnow (*Opsopoeodus emiliae*)
Southern redbelly dace (*Phoxinus erythrogaster*)
Creek chubsucker (*Erimyzon oblongus*)
River redhorse (*Moxostoma carinatum*)
Northern madtom (*Noturus stigmosus*)
Sauger (*Stizostedion canadense*)
Eastern sand darter (*Ammocrypta pellucida*)
Channel darter (*Percina copelandi*)
River darter (*Percina shumardi*)
Redside dace (*Clinostomus elongatus*)

Extirpated

Paddlefish (*Polyodon spathula*)
Arctic grayling (*Thymallus arcticus*)

Extinct

Longjaw cisco (*Coregonus alpenae*)
Blackfin cisco (*Coregonus nigripinnis*)
Deepwater cisco (*Coregonus johannae*)
Blue pike (*Stizostedion vitreum glaucum*)

An Introduction to Fish

Gerald Smith
University of Michigan

The Laurentian Great Lakes and their tributaries contain about one-fifth of the world's fresh water. As might be expected, a rich freshwater fauna thrives in this enormous and diverse aquatic habitat. Glaciers retreated from the Michigan landscape only 10,000 to 14,000 years ago, leaving thousands of lakes as well as thousands of miles of rivers and streams. Since that time, 151 species of fish (136 native) have colonized Michigan waters, mostly from the south and east (Bailey and Smith 1981). More recently, 27 species have gained access directly or indirectly through cultural intervention. During this same cultural period, and largely because of it, 3 species of Great Lakes fish were driven to total extinction and another 2 species and 1 subspecies were extirpated from Michigan waters by habitat degradation and overexploitation. Nineteen species hover at the edge of local or total extinction. Many alien species are illegally or officially being considered for introduction.

Currently, Michigan's fish species face challenges from (1) erosion, silt, mill refuse, and other pollution originating from deforestation; (2) agricultural silt and pesticides as well as destruction of habitat by draining; (3) overexploitation; (4) urbanization, with industrial and municipal pollution; (5) predation and competition from introduced alien species; and (6) climatic change. Erosion, mill refuse, drainage, and overexploitation were historically the worst threats; urbanization, pollution, siltation, pesticides, the introduction of aliens, and climatic change are the current and future threats.

The degradation of habitat by deforestation and siltation has been responsible for significant changes in the ranges of most of the river and stream fish discussed in the subsequent accounts. Siltation of gravel spawning sites and elimination of food organisms probably have been the major reasons causing the local extinction of the bigeye chub, the creek chubsucker, the ironcolor shiner, the weed shiner, the southern redbelly dace, the redside dace, the river darter, and possibly the channel darter. Industrial and municipal pollution also harmed these species, as well as the river redhorse, northern madtom, and silver shiner.

The effect of siltation is subtle but widespread, especially in the southern part of the state. Species that require clean water are stressed and a few have been eliminated. The effects of industrial pollution may be dramatic and frightening, however. The 1978 study of the Raisin River fish (Smith et al. 1981) encountered about 5 miles of lifeless, metallic blue-green water in the Saline River below the town of Milan in Monroe County. Normally, even filthy, sewage-laden water has a few hardy fish. Even the badly silted Saline River above Milan had 10–15 species. But below the city, the water was poisonous.

History

In the late 1800s, pollution by sawmill refuse caused extensive damage to fish habitats and spawning grounds in the Great Lakes and their tributary waters ([Michigan] State Board of Fish Commissioners 1887, 1890, and 1905). Extirpation of the arctic grayling from Michigan's inland waters was one of the most dramatic effects of pollution associated with deforestation and lumber mills. Major changes in cisco populations, such as those of the blackfin cisco, also date from this era (Smith 1964). Undoubtedly, more species than the grayling and ciscoes were adversely affected by lumber-associated degradation of the watershed. The former abundance of Great Lakes fisheries is legendary. Charlevoix, in his voyage to North America 1721, speaks of the Michillimackinas.

> The indians live entirely by fishing, and there is perhaps no place in the world where they are in greater plenty. The most common sort of fish in the three lakes which discharge themselves into these straits are the herring, the carp [suckers], the goldfish [redhorses], the pike, the sturgeon, the attikumaig or whitefish, and especially the trout. There are three sorts of these taken, among which is one of monstrous size, and in so great quantitives that the Indian with his spear will strike to the number of fifty sometimes in the space of three hours, but the most famous of all is the whitefish, and nothing of the fish kind can exceed it (Whitaker 1893).

Uncontrolled fishing with gill nets and pound nets also played a major role in the depletion of trout and whitefish in the Great Lakes in the late nineteenth and early twentieth centuries.

More destructive than the gill nets were the pound nets—huge underwater traps that held schools of fish live until they could be pulled onto fishing vessels. According to Whitaker, this fishing technique began around 1850.

> It has been one of the most destructive of fish devices and is responsible for the great decay of the fisheries which has been observable during the last twenty years. From this date on the pound-net fishing increased beyond all conception. It is not infrequently the case that pound-nets are set in gangs reaching out from the shore a distance of 3 or more miles, and the destruction of fish by this method of fishing is immense (Whitaker 1893).

Whitaker (1893, 179) estimated that during the 1880s, 5,000 tons, or 4,000,000 fish per year were being killed and marketed. Reports from the State Board of Fish Commissioners from this time describe their biennial appearances in Lansing to recommend legislation to regulate the fishery. Some commercial fishermen also appeared before the legislature to plead for limits on mesh sizes and seasons, lest the wanton killing of fish destroy the fishery and leave the next generation without a means of livelihood. The legislation was effectively blocked, but by whom? Certainly, fishermen did not have that kind of influence. The political power tells us much about environmental problems in our system.

As described by the Commissioners in 1897:

There was a time when the fisherman, the man directly interested in the catching of fish and in the money to be made from the business, was the man most deeply interested in the pursuit. With the decay of the fisheries came changed conditions. Times began to be hard for the fisherman. He had no money to invest in new twine. Then came the man of capital who had money to loan, who was making money in the wholesale-way out of the fishing industry. This man never fished for himself. He stayed ashore. He never endured the hardships of the calling. His interest in the fisheries began and ended with the profits that might be made by buying and marketing fish. His hands were not roughened by exposure. He did not participate in the personal hazards of the calling of the fisherman. As a rule, he had no personal knowledge of the details of the handling of nets, and such information as he possessed was wholly theoretical, yet, before committees, whose business led them to investigate these abuses, his voice has ever been loudest. . . .

By 1895, the Commission reported, the length of gillnets, pound nets, and other nets being fished in Michigan reached over 3,000 miles (Twelfth Biennial Rept. 1897, 14). That year, numerous letters from commercial fishermen documented the decreases in stocks of whitefish and lake trout and urged immediate legal protection of immature fish and closure to fishing during the spawning season (Twelfth Biennial Rept. 1897, 62). The Commission concluded:

If there is any one thing in the state that deserves protection, it is the commercial fisheries. It is not only a matter of the greatest concern to our present population, but it is of vital interest to those who are to come after us . . . . (Twelfth Biennial Rept. 1897, 14–17)

The 1897 legislature's response to the problem was to cut the budget of the Commission (Sixteenth Biennial Rept. 1905, 6). The budget was restored after the Commission changed its philosophy in 1899:

We need not be concerned whether one hundred or one million parent fish are captured at any given time or place, if such capture or killing is coincident with provisions for fully restoring or replacing them . . . . if we can get possession of the matured spawn from only one out of several hundred fish caught on spawning grounds, more young fish may be returned to such grounds than would hatch there were fishing entirely suspended. This marvelous gain in the production of young is the result merely of more perfect fertilization of the ova and of its withdrawal . . . from the presence of its natural enemies (Thirteenth Biennial Rept. 1899, 16–17).

Thus the commitment was made to attempt to restore Michigan fisheries by hatchery production and introduction of alien species rather than by prudent management of habitat and harvesting.

The twentieth century saw the culmination of the effects of deforestation and overexploitation. Arctic grayling declined and became extinct in the Lower Peninsula in the early 1900s and in the Upper Peninsula around 1932. Eight species of *Coregonus* were found in the Great Lakes as recently as the early 1950s, but, of these, 27 populations formed by the 8 species in five lakes have declined and 15 have become extinct.

The population explosion of the alewife, an alien species that came to totally dominate Lakes Huron and Michigan in the 1950s and 1960s also contributed to the decline and extinction of lake herring and ciscoes through competition. The deepwater and blackfin cisco finally became extinct in the Great Lakes during this period; kiyis and shortjaw ciscoes remain only in Lake Superior; the shortnose cisco was allowed to go commercially extinct through management neglect. The lake herring was reduced severely, but is increasing again as alewives decline.

The explosive increase in the sea lamprey in the 30 years following its first appearance in Lake Erie in 1921 was one of the most destructive events in the history of the Great Lakes. Sea lampreys decimated the lake trout population in the lower lakes but seem not to have caused the extinction of any species. Of the many species of large fish that sea lampreys attack, only commercially exploited populations were significantly reduced.

It appears that the actual last occurrences marking the final extinctions lagged behind the apparent causes of extinction by several decades, perhaps because of interactions among the different causes of decline. The pattern suggests that the current changes in local ranges and species abundances should be monitored carefully to determine primary and secondary causes of potential extinctions. Lists of declining fish in this region and throughout North America are provided by Robins et al. (1990) for extinct species and Williams et al. (1990) for endangered and threatened species.

The decline of many environmentally sensitive minnows and other stream fishes in southern Michigan occurred during mid-century and is documented in the subsequent species accounts. These declines are correlated with changes in agricultural practices that increased soil runoff. Most of the species accounts include sensitivity to siltation; many of the stressed fish species occur in agricultural areas located in the southern part of the Lower Peninsula. Carp foraging activities also stir up and suspend silt. When this effect is added to the widespread agricultural input of sediment, the frequent resuspension and redistribution of silt covers vegetation and gravel habitat for aquatic insects, eliminating both nutrients and oxygen availability for eggs and larvae.

Of the 136 species of fish native to Michigan waters, 84 are judged to have sufficiently large breeding populations. On the other hand, 18 species are at least locally depleted, 6 are threatened, 13 are endangered, 2 are locally extinct, and 3 are extinct. Twenty-one species have been deliberately introduced into the state and 6 have colonized the state's waters through constructed waterways. Some of these, for example the salmonids, have contributed significantly to the fishery of the state; some, like the alewife and carp, have become multimillion dollar problems.

The Michigan Department of Natural Resources carries out a vigilant check on actual and potential introductions. Endangered and threatened species are also monitored. Most active monitoring efforts are aimed primarily at game species, however, and environmentally sensitive species are frequently eradicated locally as part of projects to improve the habitat for game species.

In addition, reestablishment projects have been directed primarily to enhance populations of game species. Despite the necessary attention to individual endangered species, nongame fish are not likely to make their greatest ecological contribution as isolated species. Nor is their survival likely to be assured solely through species-by-species protection. Although ecological monitoring necessarily focuses on individual abundances, it is the fish community's relationship to quality habitat that has the greatest importance to environmental management. Fish provide a direct and comprehensive means of evaluating water quality for multiple use. Nongame fish are sensitive indicators of water quality and the health of aquatic ecosystems. The long-term message from the distribution and local abundance of fish is that, although municipal pollution may be reduced, loss of wetlands habitat and siltation are increasing threats to ecosystem health. Increased study of nongame fish promises valuable information for local as well as global resource and environmental management.

LITERATURE CITED

Bailey, R. M., and G. R. Smith. 1981. Origin and geography of the fish fauna of the Laurentian Great Lakes basin. *Can. J. Fish. Aquat. Sci.* 38:1539–61.
[Michigan] State Board of Fish Commissioners. 1887. *Sixth Biennial Rept.*, Lansing.
[Michigan] State Board of Fish Commissioners. 1890. *Ninth Biennial Rept.*, Lansing.
[Michigan] State Board of Fish Commissioners. 1897. *Twelfth Biennial Rept.*, Lansing.
[Michigan] State Board of Fish Commissioners. 1899. *Thirteenth Biennial Rept.*, Lansing.
[Michigan] State Board of Fish Commissioners. 1905. *Sixteenth Biennial Rept.*, Lansing.
Robins, C. R., R. M. Bailey, C. E. Bond, J. R. Brooker, E. A. Lachner, R. N. Lea, and W. B. Scott. 1990. A list of common and scientific names of fishes from the United States and Canada. *Am. Fish. Soc., Spec. Publ.* No. 20.
Smith, G. R., J. N. Taylor, and T. W. Grimshaw. 1981. Ecological survey of fishes in the Raisin River drainage, Michigan. *Mich. Acad. Sci.* 13:275–305.
Smith, S. H. 1964. Status of the deepwater cisco population of Lake Michigan. *Trans. Am. Fish. Soc.* 93:155–63.
Whitaker, H. 1893. Early history of the fisheries on the Great Lakes. Pp. 172–81 *in* [Michigan] State Board of Fish Commissioners, *Tenth Biennial Rept.*, Lansing.
Williams, J. E., J. E. Johnson, D. A. Hendrickson, S. Contreras-Balderas, J. D. Williams, M. Navarro-Mendoza, D. E. McAllister, and J. E. Deacon. 1990. Fishes of North America, endangered, threatened, or of special concern: 1989. *Fisheries* 14:2–20.

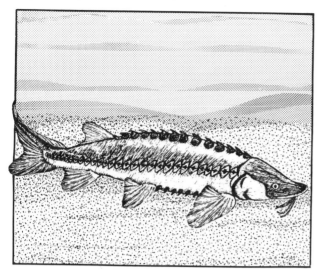

Lake Sturgeon

· *Acipenser fulvescens* Rafinesque
· **State Threatened**

Map 32. Lake sturgeon populations. Historically widespread in Great Lakes and large inland river and lake systems; recently disjunct in Great Lakes and some large river and lake systems (Michigan DNR); ● = known inland viable populations, 1992 (Michigan DNR).

Status Rationale: This once common and widespread species experienced a dramatic population decline by the turn of the century and now has a restricted Michigan distribution. Because of its low reproductive rate, full recovery of the lake sturgeon will require long-term protection and management.

In-hand Identification: The ancient lake sturgeon can not be confused with any other native fish in Michigan. It has a characteristic body form with a conical snout ahead of a ventral protrusible mouth, four snout barbels, five rows of bony plates on the sides and back, and a sharklike (heterocercal) tail. This bottom-dwelling fish may reach a length of 8 feet (2.4 m) and weigh over 300 pounds (136 kg); although for adults 15 to 35 years old, the average length ranges from 45 to 60 inches (114 to 152 cm) with a weight of 20 to 55 pounds (9.1 to 25.0 kg) (Priegel and Wirth 1975).

Total Range: Lake sturgeons range from the North, inhabiting rivers draining into Hudson Bay in Saskatchewan and Manitoba, east to the St. Lawrence estuary, south throughout the Great Lakes and the upper and middle Mississippi River drainages. Disjunct populations occur in the Tennessee, Ohio, and lower Mississippi River drainages. The lake sturgeon is now rare throughout much of its range, particularly in the southern regions (Deacon et al. 1979). Canadian populations are also becoming vulnerable (Houston 1987). Within the United States, Michigan and Wisconsin are the last strongholds for this species.

State History and Distribution: Throughout the Great Lakes and its major tributaries, the lake sturgeon historically was an abundant fish. Intensive exploitation during the late 1800s severely reduced this species by the early 1900s. Although still vulnerable in much of Michigan (some authorities consider it to only occupy 5% of its original range), recent studies indicate that populations have at least stabilized and are possibly increasing.

Adult lake sturgeon currently occur throughout the Great Lakes; however, the number of known coastal spawning areas are limited (Organ et al. 1979; Goodyear et al. 1982; Herdendorf et al. 1986). In Lake Michigan, sturgeon spawn along the southeastern shoreline and possibly in Grand Traverse Bay. In Lake Huron, spawning is restricted to the Saginaw Bay region, from Sand Point in Huron County to Fish Point in Tuscola County. Lake St. Clair and several areas along the Detroit River now serve as major spawning grounds for the lake sturgeon in the Great Lakes. The history of the lake sturgeon in Lake Erie is probably indicative of its decline throughout the Great Lakes. Between 1889 and 1895, an 80% reduction of the sturgeon population occurred due to heavy fisheries pressure (Scott and Crossman 1973). The most recently occupied spawning area within Michigan's Lake Erie shoreline (near Stony Point, Monroe County) was active until the 1970s (Goodyear et al. 1982). There are no known coastal spawning sites in Lake Superior.

Significant lake sturgeon populations occur in inland waters as well. River and inland lake systems currently known to contain spawning populations include the Menominee River, Menominee County (Thuemler 1985); Sturgeon River, Baraga and Houghton counties (Auer 1987 and 1988); Indian River and Indian Lake, Alger and Schoolcraft counties (Bassett 1982); and the Cheboygan River system, including Burt, Mullet, and Black lakes (Hay-Chmielewski 1987). A release was made in Manistique Lake, Mackinac and Luce counties, in 1983, but its success is not yet determined. Occasionally, individuals will occur (and spawn) in major tributaries of the Great Lakes, although most rivers have been artificially dammed for decades. Rivers with former populations of spawning sturgeon include the AuSable from Lake Huron; the Kalamazoo, St. Joseph, and Galien from Lake Michigan; and the St. Marys and Tahquamenon from Lake Superior (Goodyear et al. 1982).

Habitat: The preferred habitats for the lake sturgeon are the Great Lakes shorelines and large shallow lakes and rivers. Individuals followed by biotelemetry in the Cheboygan River system (Black Lake) preferred water depths from 27 to 41 feet (8.2 to 12.5 m) in summer and 21 to 26 feet (6.4 to 7.9 m) in winter (Hay-Chmielewski 1987). The presence of these relatively shallow depths indicates the importance of foraging habitat, such as muck bottoms, that provide abundant prey generally absent in deeper waters. Shallow areas also are crucial as spawning and nursery habitats. Lake spawning may occur in rocky areas around islands (Baker 1980) or wave-swept shorelines with gravel to rubble substrates. River spawning occurs in clear water over large rubble bottoms; in the Detroit River, several spawning sites ranged in water depth from 10 to 30 feet (3.0 to 9.1 m) (Organ et al. 1979). Tributary streams of large lakes and rivers also may be utilized for spawning.

Ecology and Life History: In Michigan, lake sturgeon generally migrate from larger bodies of water to smaller or shallower waters to spawn in May and June. Populations originating from Lake Superior may spawn into July (Goodyear et al. 1982). Individuals frequently return to the same sites to spawn (Folz and Meyers 1985). Optimal water temperature during spawning is 53°F (12°C) (Priegel and Wirth 1977; Baker 1980), although temperatures may range up to 64°F (18°C) (Scott and Crossman 1973). Several hundred thousand adhesive eggs are scattered on the substrate and abandoned. During the spawning period, lake sturgeon do not feed (Scott and Crossman 1973). While on their river spawning grounds, adults will frequently jump out of the water (Trautman 1981).

Within five to eight days, young hatch and remain on the gravel beds absorbing the yolk sac for the next eight to ten days before moving from the immediate hatching area (Auer 1988). Growth is relatively rapid during the first ten years, but then slows considerably. After hatching, natural mortality of lake sturgeon is low due to their overall moderate growth rate, large size, and armored-scale protection.

At approximately 2 years of age a lake sturgeon averages a length of 12 inches (30.5 cm). Sexual maturity of males is attained at 15 years of age, at a length of approximately 45 inches (114 cm). Females do not mature until 25 years of age, when they are more than 50 inches (127 cm) long. Female sturgeons only spawn once every 4 to 6 years, while males may spawn every 2 years (Baker 1980). This late maturity and infrequent reproduction results in a very low potential for rapidly increasing populations of the lake sturgeon.

Lake sturgeon live longer than any other freshwater fish in the Great Lakes. Individuals over 150 years of age have been documented (MacKay in Baker 1980). This species can be aged by the number of annular rings from a cross section of the first ray of the pectoral fin (Currier 1951). With older individuals, this technique may vary approximately 5 years from the actual age.

Sand or muck bottoms with an abundant supply of bottom-dwelling organisms for food are crucial for self-sustaining lake sturgeon populations. This bottom feeder primarily depends upon insect larvae and crustaceans (e.g., crayfish), small-sized items that need to occur in abundance to support lake sturgeon populations (Harkness and Dymond 1961; Hay-Chmielewski 1987). Trautman (1981) suggested the destruction of large mussel beds contributed to the decline of this species.

Conservation/Management: The decline of this once common fish to the currently low population level can be attributed to a combination of factors, including dam construction that blocks spring spawning runs, habitat destruction of feeding and spawning grounds, overexploitation, and pollution of rivers and lakes. Additionally, the human-assisted invasion of the sea lamprey (*Petromyzon marinus*) formerly affected lake sturgeon populations. Sea lamprey populations are now strictly controlled, although recently this exotic species has been locally increasing and is once again a potential threat. The lake sturgeon, an extremely long-lived species, may accumulate significant levels of fat-soluble toxins (Baker 1980). Mortality is also known from commercial gill nets (Auer 1988).

The lake sturgeon is characterized by a low reproductive rate, which makes it vulnerable to high fisheries demands (Probst and Cooper 1954). Therefore, management of this species is necessary to sustain harvestable populations as well as to aid in the recovery of threatened ones. Efforts to raise sturgeons in fish hatcheries for stocking purposes have been successful (Anderson 1984), although the outcome of stocking remains to be determined. Releases have been made in Michigan waters with more planned in the future.

Research with radio telemetry has been used for a better understanding of individual movements, habitat use, foraging preferences, and other ecological needs (Bassett 1982; Auer 1987; Hay-Chmielewski 1987). Legal harvest of the lake sturgeon (e.g., spear fishing) is currently allowed in Michigan under a system that only affects the surplus of local populations. However, commercial fishing is banned and sport fisheries are restricted to a 50-inch (127-cm) minimum size limit. This size is preferable to maintain breeding populations (Priegel 1973). No more than 5% of the individuals should be taken in order to maintain self-sustaining populations in rivers (Priegel 1973) and inland lakes (Priegel and Wirth 1975).

With improvement and protection of spawning areas, stocking lakes and rivers that have suitable spawning and foraging habitats, controlling illegal harvest (particularly on the spawning grounds), and carefully monitoring the annual harvest, the lake sturgeon can be maintained at self-sustaining levels in Michigan.

LITERATURE CITED

Anderson, E. R. 1984. Artificial propagation of lake sturgeon *Acipenser fulvescens* (Rafinesque), under hatchery conditions in Michigan. *Mich. Dept. Nat. Resour.*, Unpubl. Rept.
Auer, N. A. 1987. Evaluation of a lake sturgeon population. *Mich. Dept. Nat. Resour.*, Unpubl. Rept.
Auer, N. A. 1988. Survey of the Stugeon River Michigan lake sturgeon population. *Mich. Dept. Nat. Resour.*, Unpubl. Rept.
Baker, J. P. 1980. *The distribution, ecology, and management of the lake sturgeon (Acipenser fulvescens Rafinesque) in Michigan*. M.S. thesis, Univ. Mich., Ann Arbor.
Bassett, C. 1982. Management plan for lake sturgeon (*Acipenser fulvescens*) in the Indian River and Indian Lake, Alger and Schoolcraft counties, Michigan. *Mich. Dept. Nat. Resour.*, Unpubl. Rept.
Currier, J. P. 1951. The use of pectoral fin rays for determining age of sturgeon and other species of fish. *Can. Fish. Cult.* 11:10–18.
Deacon, J. F., G. Kubetich, J. D. Williams, and S. Contreras. 1979. Fishes of North America, endangered, threatened, or of special concern. *Fisheries* 4:29–44.
Folz, D. J., and L. Meyers. 1985. Management of the lake sturgeon, *Acipenser fulvescens*, population in the Lake Winnebago system, Wisconsin. Pp. 135–46 *in* F. P. Binkowski and S. I. Doroshov (eds.), *North American Sturgeons*. Junk Publ., Netherlands.
Goodyear, C. D., T. A. Edsall, D. M. Ormsby Dempsey, G. D. Moss, and P. E. Polanski. 1982. Atlas of the spawning and nursery areas of Great Lakes fishes (vol. 1–9). *U.S. Fish Wildl. Serv.*, FWS/OBS-82/52.
Harkness, W. J., and J. R. Dymond. 1961. The lake sturgeon: The history of its fishery and problems of conservation. *Ont. Dept. Lands For., Fish Wildl. Branch*, Toronto.

Hay-Chmielewski, E. M. 1987. *Habitat preferences and movement patterns of the lake sturgeon (Acipenser fulvescens) in Black Lake, Michigan.* M.S. thesis, Univ. Mich., Ann Arbor.

Herdendorf, C. E., C. N. Raphael, and E. Jaworski. 1986. The ecology of Lake St. Clair wetlands: A community profile. *U.S. Fish Wildl. Serv., Biol. Rept.* 85(7.7).

Houston, J. J. 1987. Status of the lake sturgeon, *Acipenser fulvescens,* in Canada. *Can. Field-Nat.* 101:171–85.

Organ, W. L., G. L. Towns, M. O. Walter, R. B. Pelletier, and D. A. Riege. 1979. Past and presently known spawning grounds of fishes in the Michigan coastal waters of the Great Lakes. *Mich. Dept. Nat. Resour., Tech. Rept.* No. 79-1.

Priegel, G. R. 1973. Lake sturgeon management on the Menominee River. *Wisc. Dept. Nat. Resour., Tech. Bull.* No. 67.

Priegel, G. R., and T. L. Wirth. 1975. Lake sturgeon harvest, growth, and recruitment in Lake Winnebago, Wisconsin. *Wisc. Dept. Nat. Resour., Tech. Bull.* No. 83.

Priegel, G. R., and T. L. Wirth. 1977. The lake sturgeon: Its life history, ecology and management. *Wisc. Dept. Nat. Resour., Publ.* No. 4-3600(77).

Probst, R. T., and E. L. Cooper. 1954. Age, growth, and production of the lake sturgeon (*Acipenser fulvescens*) in the Lake Winnebago, Wisconsin. *Trans. Am. Fish. Soc.* 84:207–27.

Scott, W. B., and E. H. Crossman. 1973. *Freshwater fishes of Canada.* Fish. Res. Board Can., Bull. 184.

Thuemler, T. F. 1985. The lake sturgeon, *Acipenser fulvescens,* in the Menominee River, Wisconsin-Michigan. *Env. Biol. Fishes* 14:73–78.

Trautman, M. B. 1981. *The fishes of Ohio.* Ohio State Univ. Press, Columbus.

Mooneye

· *Hiodon tergisus* Lesueur
· **State Threatened**

Map 33. Confirmed individuals. Historically from Lakes Michigan, Huron, St. Clair, and Erie (Michigan DNR). No records known since 1978.

Status Rationale: The historic distribution and abundance of the mooneye in Michigan is not well known. However, this species has declined in the Great Lakes, and water quality is probably the primary limiting factor.

In-hand Identification: The mooneye is a thin-bodied, silvery fish with adults ranging from 11 to 15 inches (28.0 to 38.1 cm) in length. The snout is short and rounded, eyes are yellowish, and the anal fin has a long base. It is a member of the family Hiodontidae, which is characterized by prominent teeth on the jaws, a small head, medium-sized terminal mouth, and large eyes. The back, dorsal, and caudal fins are light brown. The lateral line is complete.

The mooneye has frequently been mistaken for other species by commercial and sport fisheries (Van Oosten 1961) but can be unquestionably identified by the rows of large teeth. Members of the herring family (Clupeidae), such as the alewife (*Alosa pseudoharengus*) and gizzard shad (*Dorosoma cepedianum*), are superficially similar to the mooneye but have minute teeth. They may be separated from the mooneye by the saw-toothed, edged midline of the narrow belly and the lack of a complete lateral line. Species in the family Salmonidae are distinguished by their adipose fin. Some salmonids, like the kiyi (*Coregonus kiyi*), that have comparatively large eyes can be distinguished by the lack of large teeth.

265

Total Range: This species has a distribution in north-central North America. The mooneye ranges from the southern and western tributaries of Hudson Bay, south through much of the Mississippi River valley. Populations also occur in the Ohio River drainage, lower Great Lakes basin, and St. Lawrence River–Lake Champlain drainage. The mooneye appears to be absent from the Lake Superior watershed; early 1900 records (Radforth 1944) are suspect and unconfirmed (Van Oosten 1961).

State History and Distribution: The historic range and abundance of the mooneye in the Great Lakes is difficult to assess because commercial fishery records are unreliable and often erroneous due to the confusion of the mooneye with other species (Van Oosten 1961). Before 1900, the mooneye was known to be abundant in Lake Erie and the Maumee River in Ohio (Trautman 1981). At this time, it migrated in considerable numbers into large, clear rivers during the spawning season. As the turbidity increased in the Maumee River, the mooneye declined and was virtually extirpated by the mid-1900s (Trautman 1981). No Michigan records are known for the mooneye in the Maumee River system in the southeastern part of the state. Confirmed records exist for Lakes Michigan, Huron, St. Clair, and Erie. Historically, mooneyes were frequently caught in Lake St. Clair (Van Oosten 1961); a 1977 record in Anchor Bay indicates populations may still be present. Currently, mooneyes are probably most prevalent in Lake Erie and rare in Lakes Michigan and Huron.

Habitat: The mooneye inhabits the Great Lakes shorelines and, at least in Ohio, their larger tributaries (Trautman 1957). Clear, shallow waters with minimal currents are preferred. The mooneye is generally absent at depths below 35 feet (10.7 m) (Van Oosten 1961). Individuals confirmed in Michigan's Great Lakes occurred within 1 mile of the shoreline. In Michigan, this species has not been documented in rivers or streams flowing into the Great Lakes, although these habitats are used for spawning in other regions.

Ecology and Life History: This species is relatively long lived. Males begin breeding in the third year; females usually take one to two years longer to sexually mature (Glenn and Williams 1976; Wallace and Buchanan 1989). Females average greater lengths and weights than males (Glenn 1975a). Spawning occurs from April through June, probably in tributaries of the Great Lakes. Water temperature in the Tennessee and Cumberland Rivers ranged from 46° to 59°F (8° to 15°C) during the spawning period (Wallus and Buchanan 1989). Females are known to deposit from 10,000 to 20,000 eggs in Ontario (Johnson 1951), 5,000 to 9,000 in Manitoba (Glenn and Williams 1976), and 3,000 to 7,800 in the Cumberland and Tennessee river systems (Wallus and Buchanan 1989).

Growth is rapid, particularly from June to mid-August (Glenn 1976), and young attain a length of $4^1/_2$ to $6^1/_2$ inches (11.4 to 16.5 cm) by the first autumn. Mooneyes can reach a length of 17 inches (43.2 cm), and individuals are known to live more than eight years. The mooneye diet consists primarily of insects and other invertebrates (Boesel 1938; Johnson 1951) with occasional crayfish and plant materials (Glenn 1975b).

Conservation/Management: The mooneye's historic status in Michigan's Great Lakes is not well documented, because most records are known to have been confused with other species. The few confirmed specimens show the mooneye to be formerly distributed widely, although current populations are probably restricted to Lakes St. Clair and Erie. The mooneye has declined in the Great Lakes due to increasing water turbidity and pollution (Trautman 1981). Surveys and careful identification of commercial harvests are needed to identify populations and ecological requirements before the mooneye can be considered secure in Michigan waters.

LITERATURE CITED

Boesel, M. W. 1938. The food of nine species of fish from the western end of Lake Erie. *Trans. Am. Fish. Soc.* 67:215–23.

Glenn, C. L. 1975a. Annual growth rates of mooneye, *Hiodon tergisus*, in the Assiniboine River. *J. Fish. Res. Board Can.* 32:407–10.

Glenn, C. L. 1975b. Seasonal diets of mooneye, *Hiodon tergisus*, in the Assiniboine River. *Can. J. Zool.* 53:232–37.

Glenn, C. L. 1976. Seasonal growth rates of mooneye (*Hiodon tergisus*) from the Assiniboine River. *J. Fish. Res. Board Can.* 33:2078–82.

Glenn, C. L., and R. R. G. Williams. 1976. Fecundity of mooneye, *Hiodon tergisus*, in the Assiniboine River. *Can. J. Zool.* 54:156–61.

Johnson, G. H. 1951. An investigation of the mooneye (*Hiodon tergisus*). Abstract, *5th Tech. Session Res.* Council, Ont.

Radforth, I. 1944. Some considerations on the distribution of fishes in Ontario. *Contr. Royal Ont. Mus. Zool.* 25.

Trautman, M. B. 1981. *The fishes of Ohio.* Ohio State Univ. Press, Columbus.

Van Oosten, J. 1961. Records, ages, and growth of the mooneye, *Hiodon tergisus* of the Great Lakes. *Trans. Am. Fish. Soc.* 90:170–74.

Wallace, R., and J. P. Buchanan. 1989. Contribution to the reproductive biology and early life ecology of mooneye in the Tennessee and Cumberland Rivers. *Am. Midl. Nat.* 122:204–7.

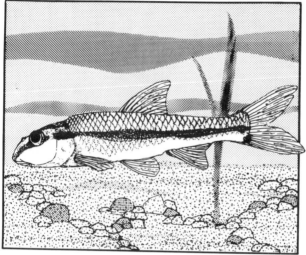

Bigeye Chub

· *Hybopsis amblops* (Rafinesque)
· **State Endangered**

Map 34. Confirmed individuals. ▲ = before 1941 (Michigan DNR). No records known since 1941.

Status Rationale: Although this small minnow may be common in local areas of the southeastern United States, most of its northern populations have been extirpated due to siltation and pollution of streams and rivers. The bigeye chub has not been recorded since 1941 in Michigan and may now be extirpated.

In-hand Identification: This 2¹/₂ to 3¹/₂ inch (0.8 to 1.1 cm) large-eyed, silvery chub has a heavily pigmented lateral band (which is most evident on fish living in clear water). The back typically has a greenish yellow cast. The barbels on each corner of the mouth are usually outwardly visible.

Typical of the minnow family (Cyprinidae), the bigeye chub has one dorsal fin in contrast to the two dorsal fins of darters and other members of the Percidae family. Except for chubs, all other similar Michigan minnows lack barbels. The large eyes, large scales, and ventrally oriented mouth distinguish this species from most other Michigan chubs. One similar species, the silver chub (*Hybopsis storeriana*), may be separated from the bigeye chub by its white streak on the lower edge of the caudal fin. Unlike the bigeye chub, the origin of the dorsal fin in the silver chub is in front of the pelvic fin's origin.

Total Range: The bigeye chub formerly occurred in the Arkansas and Tennessee river drainages north throughout much of the Ohio River basin, including several

Lake Erie tributaries in Michigan, New York, and Pennsylvania. This chub is still common in areas with clear and clean streams, such as in the Ozarks of Missouri (Pflieger 1975). However, many northern populations of the bigeye chub have declined or have been extirpated (Trautman 1981; Smith 1979).

State History and Distribution: The bigeye chub was historically restricted to southeastern Michigan and has been found only in rivers draining into Lake Erie and in the St. Joseph River, Hillsdale County. There are no recent records of this species in Michigan (last record 1941). To illustrate its former abundance, over 80 individuals were found in the Tiffin River, Lenawee County, in 1924. During a 1978 survey of the Raisin River system, no bigeye chubs were found in the Tiffin River or other areas within the drainage (Smith et al. 1981).

Habitat: Optimal habitat conditions are clear, small- to moderate-sized streams with fine gravel-sand bottoms. The extent of accumulated silt in pool and riffle areas indicates the suitability of the habitat. Trautman (1981) found that increased siltation of streams quickly reduced bigeye chub populations. Bigeye chubs can occur in streams with swift or slow gradients, but they prefer moderate currents. This chub also requires some aquatic vegetation. The large historic population in Michigan's Tiffin River was found in clear water with a substrate of sand and stone and alternating pools and riffles.

Ecology and Life History: The biology of this species has not been studied in detail. Females deposit their eggs on vegetation in late spring or early summer (Smith 1979). Protected pools are probably used as spawning sites. The young may reach a length of 2 inches (5.1 cm) by the first summer (Pflieger 1975). Because of its large eyes, most of the chub's feeding activity likely occurs in clear waters during the day. Invertebrates are probably its main diet.

Conservation/Management: The bigeye chub's decline in Michigan is primarily due to increased silt and pollution levels in its stream habitat. This habitat degradation has also been responsible for the disappearance of this species throughout most of its northern range. Michigan's known populations disappeared prior to intensive studies, and recent surveys in historical locations as well as neighboring areas have been unsuccessful in its location (Smith et al. 1981).

Many other Michigan fish species inhabiting clear streams and rivers also have experienced declines due to high silt loads. This limiting factor makes the substrate on the spawning grounds unsuitable, decreases water quality, and reduces prey availability. In addition, the bigeye chub's spawning habits make it susceptible to high silt levels, because the aquatic vegetation needed for egg laying grows best in clear water. Turbid waters generally restrict plant growth. A management technique important in reducing siltation is the retention of riparian vegetation. These areas screen toxins and silt and also reduce erosion of agricultural topsoil. In addition, sand sediment traps will reduce siltation levels (Hansen et al. 1983). With more intensive surveys and restoration of riverine habitats, the bigeye chub may be found once again in Michigan.

LITERATURE CITED

Hansen, E. A., G. R. Alexander, and W. H. Dunn. 1983. Sand sediment in a Michigan trout stream. Part 1. A technique for removing sand bedload from streams. *N. Am. J. Fish. Mgmt.* 3:355–64.

Pflieger, W. L. 1975. *The fishes of Missouri.* Dept. Cons., Jefferson City, MO.

Smith, G. S., J. N. Taylor, and T. W. Grimshaw. 1981. Ecological survey of fishes in the Raisin River drainage, Michigan. *Mich. Acad. Sci.* 13:275–305.

Smith, P. W. 1979. *The fishes of Illinois.* Ill. State Nat. Hist. Surv., Univ. Ill. Press, Urbana.

Trautman, M. B. 1981. *The fishes of Ohio.* Ohio State Univ. Press, Columbus.

Ironcolor Shiner

· *Notropis chalybaeus* (Cope)

· **State Endangered**

Map 35. Confirmed individuals. ▲ = before 1977 (Michigan DNR). No records known since 1977.

Status Rationale: There are no confirmed records of the ironcolor shiner in Michigan since 1940. This shiner historically inhabited streams in extreme southern Michigan, and surveys are needed to evaluate its current status in the state. Stream habitat degradation is probably the major cause of its decline and possible extirpation from the state.

In-hand Identification: The ironcolor shiner is characterized by a sharp upper border on its dark lateral band, which extends from the snout to the tail. Its eight anal and eight dorsal fin rays, a small subterminal and oblique mouth, and black pigmentation on the snout, chin, and inside the mouth are diagnostic. Adults of this shiner are 2 to 2¹/₂ inches (5 to 6 cm) and have a yellowish cast to the back and fins. During the breeding season, the ironcolor shiner has a bright red coloration on the undersides. The lateral line is complete.

Several of Michigan's shiners are similar in appearance to the ironcolor shiner, a member of the minnow family (Cyprinidae). Many lack the solid black lateral band and the dark color inside the mouth. The most similar species, the blackfin and weed shiners (*Notropis heterodon* and *N. texanus*, respectively), differ from the ironcolor shiner by their 7 anal fin rays rather than 8 rays. The blacknose shiner (*Notropis heterolepis*) and young golden shiners (*Notemigonus crysoleucas*) are also similar but are readily identified by their 11 to 13 anal fin rays.

271

Total Range: This species has an extensive eastern range. It is distributed along the coastal lowlands from New Jersey to Florida, west to Texas. Inland, the ironcolor shiner ranges throughout the Mississippi River valley north to Illinois, Indiana, and the St. Joseph River system in Michigan.

State History and Distribution: Five records are known from four counties in extreme south-central Michigan, all within the St. Joseph River drainage. This species was last recorded in June, 1940, in the Rocky River, Cass County. However, in 1989, a population was discovered two miles south of Branch County, in Indiana's Fawn River. Sampling along this river and its tributaries may locate populations in Michigan.

Habitat: The ironcolor shiner is a cyprinid of clear, well-vegetated waterways with sand bottoms (Smith 1979). Sites where this species was collected in Michigan during the early 1900s were larger bodies of water (lakes and rivers) characterized by clear water, depths of 2 to 3 feet (0.6 to 0.9 m), and aquatic vegetation. Water velocities and substrates were variable.

Ecology and Life History: The biology of this species in Michigan has not been studied. In Illinois, the adhesive eggs are deposited on the river or stream bottom in June and July (Smith 1979). Nests are not prepared and eggs are broadcast into areas with minimal currents. The eggs hatch in approximately two days at 60°F. Individuals are probably in breeding condition within one year. The ironcolor shiner typically occurs in small- to medium-sized schools. Its diet primarily consists of small aquatic and terrestrial invertebrates, taken at or near the water surface.

Conservation/Management: The degradation of stream habitat has probably been the main factor for the ironcolor shiner's decline and possible disappearance in Michigan. Although it only occurred in the St. Joseph River system in Michigan, there are several widely distributed records along this drainage. Unlike most of Michigan's other endangered and threatened small fish species that are restricted to extreme southeastern Michigan, this shiner only occurred in south-central Michigan. Recent intensive surveys have been made in southeast Michigan but not in rivers and streams in the south-central region. Sampling of the clear, well-vegetated streams that remain within the St. Joseph River system, particularly along the Fawn River system in Branch and St. Joseph counties, is needed in order to determine if the ironcolor shiner is still surviving in the state.

LITERATURE CITED

Smith, P. W. 1979. *The fishes of Illinois.* Ill. State Nat. Hist. Surv., Univ. Ill. Press, Urbana.

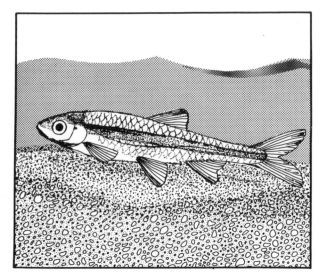

Silver Shiner

· *Notropis photogenis* (Cope)
· **State Threatened**

Map 36. Confirmed individuals. ▲ = before 1977 and
● = 1977 to 1989 (Michigan DNR).

Status Rationale: Michigan's population of this species is currently small but possibly stable. The silver shiner is at the northern fringe of its range in southeastern Michigan and probably was never abundant or widespread. The clear, high-quality river habitat upon which this shiner depends continues to disappear in Michigan.

In-hand Identification: This species has large eyes and shiny, large scales. The silver shiner has a greenish back, a solid predorsal stripe, and a wide lateral band on a silvery background, with a complete lateral line. Adults are large and slender, averaging 4 inches (1.2 cm) in length. A key identifying characteristic for this shiner is the presence of black crescents between the nostrils. The front of the dorsal fin is slightly behind the insertion of the pelvic fin.

Although the silver shiner and many other members of the minnow family (Cyprinidae) appear to be superficially similar, most can be distinguished by the number of scales, the number of rays in the fins, or the number of pharyngeal teeth (found on the fifth gill arch). Two common Michigan species that are particularly similar to the silver shiner are the rosyface shiner (*Notropis rubellus*) and the emerald shiner (*Notropis atherinoides*). The three can be distinguished best from their fin arrangements. Unlike the silver shiner, the front of the dorsal fin in the emerald and rosyface shiners is further behind the pelvic fin insertion (Gruchy et

273

al. 1973). Compared to other shiners, scales of the silver shiner are easier to dislodge under normal handling.

Total Range: This species can be found from the Tennessee River drainage north to the Michigan and Ontario drainages of Lakes Erie and St. Clair (Gruchy et al. 1973; Parker and McKee 1984). Ontario populations are locally abundant and may have increased since the 1970s (Baldwin 1988). The silver shiner's range includes much of the Ohio River basin, from eastern Indiana east to western New York, Pennsylvania, and West Virginia. This shiner is still relatively common in many parts of its range where suitable habitat remains.

State History and Distribution: The silver shiner is at the northern edge of its range in Michigan. In limited stretches of two river systems (Raisin and St. Joseph), silver shiner populations may be stable. Four of the 11 Michigan records of this fish are from the Raisin River in Washtenaw (3) and Monroe (1) counties. However, recent rotenone treatments of stretches of the Raisin River by the state fisheries division may have seriously depleted these populations. Individuals also have been collected in the Huron River in Washtenaw County, Saline River (a tributary of the Raisin River drainage) in Monroe County, and Ore Lake in Livingston County. In addition, a 1982 survey discovered populations in the St. Joseph River, Hillsdale County. Populations also are known on this river system south into Williams County, Ohio.

Habitat: The silver shiner is most prevalent in the swift and turbulent sections of large streams and rivers interrupted with flowing pools. It prefers cobble, gravel, or pebble substrates and relatively clear water. McKee and Parker (1982) found that healthy populations could tolerate temporarily increased levels of siltation from erosion during heavy precipitation. They also found no relationship between the presence of silver shiners and the density of aquatic vegetation; a few large catches were made over concrete aprons below dams. Schools frequently may be located in the deep, swift, silt-free riffles of suitable habitat (Trautman 1981). In the Raisin River, silver shiner populations were found along banks that were usually wooded with an average current of 11 inches (27.9 cm) per second and a maximum depth of nearly 4 feet (1.2 m). The newly discovered St. Joseph River population inhabits a large, wooded stream draining into the principal river. The stream site averages 50 feet (15.2 m) in width and 6¹/₂ feet (2.0 m) in depth, with a current of 20 inches (50.8 cm) per second.

Ecology and Life History: From McKee and Parker's (1982) study in southern Ontario (approximately the same latitude as Michigan's populations), the silver shiner spawns in June, usually at temperatures varying between 63° and 72°F. Spawning grounds are probably confined to deeper regions of riffle areas. Young-of-the-year (by August) average 1⁷/₈ inches (4.8 cm), yearlings 2⁵/₈ inches (6.7 cm), and breeding two-year-olds are 3⁵/₈ inches (9.2 cm) (McKee and Parker 1982). Individuals surviving into a third June breeding season are rare and may reach a maximum length of 4¹/₂ inches (11.4 cm).

This shiner is primarily a surface feeder, surviving on terrestrial and aquatic insects. It occasionally feeds upon benthic (bottom-dwelling) organisms. In Michigan, Smith et al. (1981) found that terrestrial insects dominate this species' diet, with adult Diptera (flies) and Coleoptera (beetles) making up most of the prey. Blackflies (*Chironomus* spp.) were the most frequently eaten aquatic insect in their study. In a Canadian study, 90% of the diet was comprised of insects, with dipterans accounting for more than half the volume (Parker and McKee 1984). This species is reported to occasionally jump out of the water to capture prey (Trautman 1981).

Conservation/Management: The silver shiner may be declining in portions of its range, but, in general, its distribution and abundance are relatively stable. McKee and Parker (1982) studied the two rivers known to harbor this species in Canada (Ontario) and found little evidence suggesting a declining population. A subsequent review found these Canadian populations to be stable and even increasing (Parker and McKee 1984). Trautman (1981) characterized this species as abundant and still widespread, but with numbers decreasing in many locales, especially in the river stretches that have increasing turbidity and siltation (e.g., the Ohio River). The specific relationship between the presence of the silver shiner and quality of river water is unknown.

The known surviving Michigan populations appear to be stable; with further surveys, other populations may be found. McKee and Parker (1982) suggested that the degree of stream gradient, less than 16 feet per mile and greater than 30 feet per mile, restricts silver shiners to a limited number of stream habitats. Thus, damming and channeling rivers and large streams have probably reduced the number of suitable rivers, affecting the survival and expansion of viable populations. Silver shiners can tolerate limited, human-related habitat changes (e.g., agricultural runoff) (Parker and McKee 1984).

Because of the difficulty in identifying shiners, this species may be overlooked or confused with closely related species. With more detailed and intensive surveys of Michigan's southern rivers, protection of existing stream habitats, and monitoring of erosion and pollution, the silver shiner will have a better chance of survival in Michigan.

LITERATURE CITED

Baldwin, M. E. 1988. Updated status of the silver shiner, *Notropis photogenis*, in Canada. *Can. Field-Nat.* 102:147–57.

Gruchy, C. G., R. H. Bowen, and I. M. Gruchy. 1973. First records of the silver shiner, *Notropis photogenis*, from Canada. *J. Fish. Res. Board Can.* 30:1379–82.

McKee, P. M., and B. J. Parker. 1982. The distribution, biology, and status of the fishes *Campostoma anomalum*, *Clinostomus elongatus*, *Notropis photogenis* (Cyprinidae) and *Fundulus notatus* (Cyprinodontidae) in Canada. *Can. J. Zool.* 60:1347–56.

Parker, B., and P. McKee. 1984. Status of the silver shiner, *Notropis photogenis*, in Canada. *Can. Field-Nat.* 98:91–97.

Smith, G. R., J. N. Taylor, and T. W. Grimshaw. 1981. Ecological survey of fishes in the Raisin River Drainage, Michigan. *Mich. Acad. Sci.* 13:275–305.

Trautman, M. B. 1981. *The fishes of Ohio.* Ohio State Univ. Press, Columbus.

Weed Shiner

· *Notropis texanus* (Girard)
· **State Endangered**

Map 37. Confirmed individuals. ▲ = before 1977 (Michigan DNR). No records since 1977.

Status Rationale: Formerly inhabiting the Grand, Kalamazoo, and Saginaw river systems, this shiner has not been found for nearly 40 years. Stream and river habitat degradation probably are responsible for its decline, but surveys of former locales may discover this inconspicuous and little-known species.

In-hand Identification: This typical olive-colored shiner has a dark lateral band, extending to the snout and the tip of the chin. Adults typically range from $1^3/8$ to $2^1/4$ inches (3.3 to 5.4 cm), attaining a maximum length of $2^1/2$ inches (6.3 cm). It has seven anal fin rays and a small caudal fin spot.

The pugnose shiner (*Notropis anogenus*) is the only other black-striped shiner with seven anal fin rays. Its very small mouth is nearly vertical, differing from the larger, less upward-sloping mouth of the weed shiner. The commonly associated, state-endangered ironcolor shiner (*Notropis chalybaeus*) also is similar but has eight or more anal fin rays.

Total Range: The weed shiner is a relatively common southern U.S. species. It occurs in the lowlands of northwestern Florida and southwestern Georgia, west to Texas, and north along the Mississippi River into Missouri. North of this point it is locally distributed. This North-South distinction may represent two differentiated populations (Hubbs and Lagler 1974). Populations in the upper Mississippi River drainage are known in Iowa, Minnesota, Wisconsin, Illinois, and Indiana.

Disjunct populations occur in Lake Michigan tributaries (primarily in the southern end) and in the Saginaw River basin.

State History and Distribution: This is a southern Lower Peninsula species, found in both Lake Michigan and Lake Huron drainages. In the Saginaw River basin, it was recorded in the early 1940s in the Bad River, Saginaw County. Former populations also were known from the Kalamazoo and Grand River systems and Gull Lake in Kalamazoo County. Known records are mostly between the late 1920s and late 1940s. Several were from the Kalamazoo River, its tributaries and bayous, and lakes in central Allegan County. The last confirmed record of the weed shiner in Michigan was in 1953 from the north branch of Rice Creek, Calhoun County.

Habitat: An inhabitant of open, sand-bottomed streams, rivers, and impoundments in the South, the weed shiner occurs in similar areas in the northern part of its range with submerged aquatic vegetation. Most formerly occupied sites in Michigan were in bayous, tributary junctions, and below dams of major rivers. It has been found downstream from the following dams: Allegan on the Kalamazoo River in Allegan County, Dimondale on the Grand River in Eaton County, and Minards Mill on Sandstone Creek in Jackson County.

In Wisconsin, it occurred in lakes of 75° to 82°F (24° to 46°C), with an abundance of aquatic vegetation, and the occupied stream habitat characteristics were 13 to 33 feet (4.0 to 10.0 m) wide, 1 to 3 feet (0.3 to 1.0 m) deep, slight currents, and temperatures between 72° to 77°F (40° to 43°C) (Fago 1984).

Ecology and Life History: Breeding occurs in June. Eggs hatch in a few days, and individuals are in breeding condition within one year. Most live two years, a few live for three years. Its diet is primarily aquatic and terrestrial invertebrates.

Conservation/Management: Michigan, being at the northeast periphery of the weed shiner's range, has few river systems capable of supporting this species. Further, northern populations are rare and locally distributed compared to southern ones. The few extant populations in Michigan probably disappeared in the mid 1900s due to extreme habitat degradation including increased silt loads, turbidity from the destruction of riverine vegetation, and widespread deforestation and wetland alteration. These are the likely reasons for its extirpation (Smith 1979).

Surveys are needed to determine occupied sites, particularly in former locales that remain relatively disturbed. Newly found populations should be monitored and provided complete protection from further habitat degradation. This species' survival in Michigan is questionable, however it deserves both recognition as part of the state's fish fauna and accompanying efforts to ensure its survival.

LITERATURE CITED

Fago, D. 1984. Distribution and relative abundance of fishes in Wisconsin. Part 3. Red Cedar River basin. *Wisc. Dept. Nat. Resour., Tech. Bull.* No. 143.

Hubbs, C. L., and K. F. Lagler. 1974. *Fishes of the Great Lakes Region.* Univ. Mich. Press, Ann Arbor.

Smith, P. W. 1979. *The fishes of Illinois.* Ill. State Nat. Hist. Surv., Univ. Ill. Press, Urbana.

Pugnose Minnow

- *Opsopoeodus emiliae* (Hay)
- **State Threatened**

Map 38. Confirmed individuals. ▲ = before 1977 (Michigan DNR). No records since 1977.

Status Rationale: There are few records of this species in its limited southeastern Michigan range. The pugnose minnow requires clear, slow-moving waterways that are well vegetated and habitat that has not been heavily disrupted by human-related activities.

In-hand Identification: The yellowish silvery pugnose minnow is characterized by a dark lateral band that extends to the small, sharply upturned mouth. This minnow has nine fully developed dorsal fin rays, no caudal spot, and usually has a complete lateral line. Adults attain a length of 2 1/2 inches (6.4 cm) and have dusky, pigmented areas on the dorsal fin.

Together, these identification features distinguish the pugnose minnow from other Cyprinids. All similar minnows, in the genus *Notropis*, have eight fully developed dorsal fin rays. The pugnose shiner (*Notropis anogenus*) is the most similar species but has eight dorsal fin rays, a caudal spot, and a transparent dorsal fin. The bluntnose minnow (*Pimephales notatus*), with its well-developed lateral bands, could be confused with the pugnose minnow.

Total Range: This minnow ranges from the upper Mississippi River valley and southern Great Lakes basin, south through the Ohio River valley to the United States drainages of the Gulf of Mexico. Many populations are rare and disappearing

(Smith 1979; Trautman 1981; Parker et al. 1987; Coffin and Pfannmuller 1988), although recent intensive surveys in Wisconsin have documented a more extensive distribution than once believed (Fago 1983, 1984, and 1986).

State History and Distribution: Historic records for this species are from the shallow waters of Lake St. Clair, the Detroit River, and the western Lake Erie drainage, including the Raisin and Huron rivers. In 1941, three sites were located along the Huron River, Wayne County. This species has not been recorded in Michigan since those occurrences. In the early 1900s, this species was present in western Lake Erie and its tributaries near the Ohio border. However, by 1930 it had dramatically declined (Trautman 1981), and, currently, only a few locales are known in Ohio.

Habitat: The pugnose minnow occurs in streams, oxbows with a fast-flowing current, and in the shallow regions of lakes. Both Trautman (1981) and Smith (1979) comment on this species' habitat requirements of clear, slow-moving waterways and its abundance in areas with well-vegetated bottoms. In Wisconsin, characteristics of stream habitat generally had a sluggish to moderate current, slight to moderate turbidity, and a depth of less than 2 feet (0.6 m)(Fago 1983 and 1986). Substrate composition can be variable, including sand and organic debris, as long as it can support aquatic vegetation. At the Huron River capture sites, the current was typically sluggish. Other river characteristics at this location were variable, indicating that this site may have been only marginally suited for the pugnose minnow.

Ecology and Life History: Little is known about the biology of the pugnose minnow in the lower Great Lakes basin. Spawning probably occurs in late spring and early summer in areas with aquatic vegetation. As suggested by its upturned mouth, the pugnose minnow probably feeds near the water's surface. Blackfly larvae (*Chironomus* spp.) and the minute shrimp (*Cladocera* spp.) are known prey (Cassidy et al. 1930; Gilbert and Bailey 1972).

Conservation/Management: Although the pugnose minnow was probably not widely distributed in Michigan, it may have been a common inhabitant of shallow ecosystems in Lake St. Clair south to western Lake Erie. The clear, sluggish, well-vegetated habitat it requires has been replaced with highly turbid water that, in part, impedes natural plant growth. Trautman (1981) attributed its dramatic decline in the 1930s to excessive siltation. Disturbance through drainage, channelization, and fluctuating water levels have also contributed to its disappearance from many habitats in its northern range.

If populations still exist in the lower Great Lakes or adjacent tributaries, improved water conditions in Michigan waterways may aid in its recovery. Lake St. Clair contains extensive natural, shallow-vegetated habitats and is believed to harbor small populations of this species (Herdendorf et al. 1986). Research in the St. Clair delta to document specific locations inhabited by pugnose minnow populations and their general ecology are needed to retain it as a viable representative of Michigan's fish fauna.

LITERATURE CITED

Cassidy, H., A. Dobkin, and R. Wetzel. 1930. A study of the food of three fish species from Portage Lakes, Ohio. *Ohio J. Sci.* 30:194–98.

Coffin, B., and L. Pfannmuller (eds.). 1988. *Minnesota's endangered flora and fauna.* Univ. Minn. Press, Minneapolis.

Fago, D. 1983. Distribution and relative abundance of fishes in Wisconsin. Part 2. Black, Trempealeau, and Buffalo River basins. *Wisc. Dept. Nat. Resour. Tech. Bull.* No. 140.

Fago, D. 1984. Distribution and relative abundance of fishes in Wisconsin. Part 4. Root, Milwaukee, Des Plaines, and Fox River basins. *Wisc. Dept. Nat. Resour. Tech. Bull.* No. 147.

Fago, D. 1986. Distribution and relative abundance of fishes in Wisconsin. Part 7. St. Croix River basin. *Wisc. Dept. Nat. Resour. Tech. Bull.* No. 159.

Gilbert, C. R., and R. M. Bailey. 1972. Systematics and zoogeography of the American cyprinid fish *Notropis (Opsopoeodus) emiliae. Univ. Mich. Mus. Zool., Occ. Pap.* No. 664.

Herdendorf, C. E., C. N. Raphael, and E. Jawerski. 1986. The ecology of Lake St. Clair wetlands: A community profile. *U.S. Fish Wildl. Serv., Biol. Rept.* 85(7.7).

Parker, B., P. McKee, and R. R. Campbell. 1987. Status of the pugnose minnow, *Notropis emiliae,* in Canada. *Can. Field-Nat.* 101:208–12.

Smith, P. W. 1979. *The fishes of Illinois.* Ill. State Nat. Hist. Surv., Univ. Ill. Press, Urbana.

Trautman, M. B. 1981. *The fishes of Ohio.* Ohio State Univ. Press, Columbus.

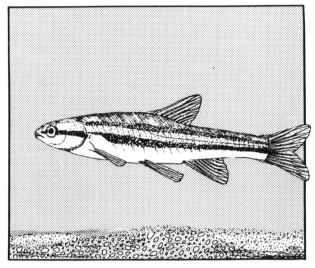

Southern Redbelly Dace

· *Phoxinus erythrogaster* (Rafinesque)
· **State Threatened**

Map 39. Confirmed individuals. ▲ = before 1977 and ● = 1977 to 1989 (Michigan DNR).

Status Rationale: The southern redbelly dace has declined in Michigan; however, recent surveys have shown this species to inhabit the Huron and Raisin river drainages in the southeastern part of the state. Severe declines in adjacent northern Ohio populations due to the degradation of spring-fed stream habitats indicates the need for complete protection in Michigan.

In-hand Identification: The southern redbelly dace, a member of the minnow family (Cyprinidae), averages only 2 to 3 inches (5.1 to 7.6 cm) in length. This species has small scales and lacks a complete lateral line. A series of small black dots are present on the upper back in front of the dorsal fin. The brilliant coloration of this greenish olive species is distinctive. A complete reddish or yellowish side streak separates two dark lateral bands. On breeding males, these colors intensify, the belly turns crimson-colored, and the fins become bright yellow.

The double lateral band is diagnostic for this dace and its close relative, the northern redbelly dace (*Phoxinus eos*). The two species can be readily separated by the mouth length, with the southern redbelly dace having a more pointed snout and a less oblique mouth that does not extend to the eye.

Total Range: This dace occurs from the upper Mississippi River system in Minnesota and Wisconsin south through the Ohio River valley to the Mississippi River lowlands, extending to the lower reaches of the Tennessee and Missouri rivers. Its east-west distribution is from northeastern Oklahoma to southwestern Pennsylvania.

State History and Distribution: As indicated by Trautman (1981) in Ohio, this dace may have been abundant in the wooded streams of extreme southeastern Michigan. However, until recently, the last known record of the southern redbelly dace was in 1930 from a population near the mouth of the Saline River, in the Raisin River drainage, Monroe County. Finally, in 1973, this dace was documented again—in the Huron River drainage, Washtenaw County. Further surveys in southeastern Michigan verified its presence at several sites in the Raisin River drainage, in Lenawee and Washtenaw counties (Smith et al. 1981).

Habitat: Like many of Michigan's rare and declining riverine species, the southern redbelly dace requires permanent streams with clear water. Preferred habitats are spring-fed brooks and clear, cool streams intermixed with small pools within wooded ravines (Settles and Hoyt 1978; Smith 1979). Stream habitats with mud bottoms are frequented by this species. Trautman (1981) considered areas where the water had cut under vegetated banks as crucial habitat during times of low water levels. At two Michigan sites, the following habitat variables were measured: average current was 3 inches (7.6 cm) per second, water transparency was clear, amount of aquatic vegetation was 13%, and the substrate was a broad mix of mud, sand, gravel, and larger rocks (Smith et al. 1981).

Ecology and Life History: The southern redbelly dace spawns in late spring and early summer in headwater streams with shallow gravel riffles. The spawning chronology is similar throughout this species' limited Midwest range (Koff 1961; Phillips 1968; Settles and Hoyt 1978). More than 300 eggs are generally deposited and subsequently abandoned by the adults. Gravel nests constructed by other cyprinids may be used as spawning sites (Hankinson 1932). Growth is fastest within the first year, and individuals are usually sexually mature upon reaching a length of $1^5/8$ inches (4.1 cm)(Settles and Hoyt 1976). Most dace do not live past two years of age.

The southern redbelly dace frequently feeds on algae that it gathers from stones or other objects in the water. Selectiveness of algal food depends on algal abundance (Phillips 1968; Settles and Hoyt 1976). Invertebrates in the larval stage, such as blackflies (*Chironomus* spp.), are also taken.

Conservation/Management: Many fish species suffered precipitous declines in Michigan as stream and streamside habitat were drastically altered by extensive logging. Subsequent agricultural land-use patterns in southern Michigan caused heavy siltation loads and channelization of most small, natural waterways. The headwaters of spring-fed streams also were affected by fluctuating and lower water tables. These conditions and their negative effects on small stream inhabitants, such as the southern redbelly dace, have been documented in Ohio (Trautman

1981). Many streams are now intermittent, turbid, less shaded waterways—conditions extremely unsuitable for this dace.

Michigan surveys in the late 1970s located isolated sites still inhabited by this species (Smith et al. 1981). Protection of these populations is paramount, as are further surveys of the Huron and Raisin river drainages to identify other remaining populations. Manipulation of stream habitats for gamefish introductions by rotenone poisoning needs to be avoided in the few remaining sites where the southern redbelly dace resides. Collection of this species should be carefully monitored, because the southern redbelly dace frequently forms schools in midchannel, making them vulnerable to capture and local population declines. General protection of wooded, spring-fed streams in southeastern Michigan would preserve habitat for this vulnerable species, one of the few aquatic species to inhabit such areas in our state.

Literature Cited

Hankinson, T. L. 1932. Observations on the breeding behavior and habits of fishes in southern Michigan. *Pap. Mich. Acad. Sci., Arts, Ltrs.* 15:411–24.

Koff, W. 1961. Reb-bellied dace *Chrosomus erythrogaster* (Rafinesque). *Prog. Fish-Cult.* 40:60.

Phillips, G. L. 1968. *Chrosomus erythrogaster* and *Chrosomus eos* (*Osteichthyes: Cyprinidae*): *taxonomy, distribution, ecology.* Ph.D. diss., Univ. Minn., Minneapolis.

Settles, W. H., and R. D. Hoyt. 1976. Age structure, growth patterns, and food habits of the southern redbelly dace *Chrosomos erythrogaster* in Kentucky. *Trans. Kentucky Acad. Sci.* 37:1–10.

Settles, W. H., and R. D. Hoyt. 1978. The reproductive biology of the southern redbelly dace *Chrosomus erythrogaster* Rafinesque, in a spring-fed stream in Kentucky. *Am. Midl. Nat.* 99:290–98.

Smith, G. R., J. N. Taylor, and T. W. Grimshaw. 1981. Ecological survey of fishes in the Raisin River drainage, Michigan. *Mich. Acad. Sci.* 13:275–305.

Smith, P. W. 1979. *The fishes of Illinois.* Ill. State Nat. Hist. Surv., Univ. Ill. Press, Urbana.

Trautman, M. B. 1981. *The fishes of Ohio.* Ohio State Univ. Press, Columbus.

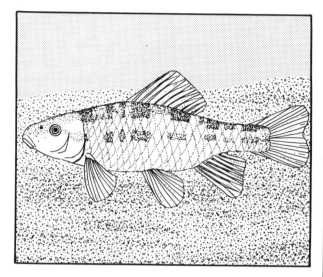

Creek Chubsucker

- *Erimyzon oblongus* (Mitchill)
- **State Threatened**

Map 40. Confirmed individuals. ▲ = before 1977 and ● = 1977 to 1989 (Michigan DNR).

Status Rationale: The creek chubsucker is a peripheral species that ranges into Michigan's extreme southern counties and has been found only in the Raisin River drainage in recent years. Stream habitat degradation is responsible for its decline and its current threatened status in Michigan.

In-hand Identification: The creek chubsucker is a yellowish olive species with a suckerlike mouth that is characteristic of the sucker family (Catostomidae). However, its mouth is less ventral than other members of the sucker family. This is one of the smaller sucker species: adults measure 3 to 6 inches (7.6 to 15.2 cm) but occasionally reach 10 inches (25.4 cm) in length. In adults, pigmentation on the sides forms a series of blotches and an indistinct lateral band. Males have tubercles on the snout and a falcate anal fin. Young-of-the-year have two heavy longitudinal streaks and a red tail. The lateral line is incomplete.

Only one other chubsucker occurs in Michigan, the lake chubsucker (*Erimyzon sucetta*). Separation of these two chubsuckers is reliably made by counting the longitudinal rows of scales. The creek chubsucker has 39 to 41 rows and the lake chubsucker has 36 to 38 rows.

Total Range: This species is distributed over much of the eastern United States, north to the lower Great Lakes basin, but it has disappeared from several regions at the periphery of its range. Two subspecies are recognized: the western creek chubsucker (*E. o. claviformis*), which generally occurs west of the Appalachian Mountains, including Michigan, and the eastern subspecies (*E. o. oblongus*) along the Atlantic coastal plain.

State History and Distribution: This small sucker is known from the St. Joseph, Raisin, and Maumee river systems in seven counties of extreme southern Michigan. There also are undocumented records from Clinton and Ingham counties. All site records are from the early and mid-1900s, except for two locales along the Swamp Raisin Creek, Lenawee County, in 1978 (Smith et al. 1981). This study suggests that the creek chubsucker is not necessarily declining from the southernmost tier of Michigan counties.

Habitat: Clear streams with sand-gravel bottoms and moderate currents are the preferred habitat of this species. In Michigan, stream characteristics at historic sites had depths up to 3 to 5 feet (0.9 to 1.5 m), slight-to-moderate currents, minimal aquatic vegetation, and sand-gravel-mud bottoms. Habitat variables for the most recently discovered sites in Swamp Raisin Creek had depths of less than 3 feet (0.9 m), but were otherwise atypical creek chubsucker habitat since it was extremely turbid (water visibility was 3 inches [7.6 cm]), with no current and a nearly exclusive mud substrate (Smith et al. 1981). In Ohio, the largest populations were historically associated with clear streams and habitats with sand-gravel bottoms (Trautman 1981).

Ecology and Life History: The creek chubsucker probably spawns in May. Adults move to larger streams after spawning or by early summer (Trautman 1981). Like other members of the sucker family, the creek chubsucker is a bottom feeder, feeding on invertebrates and other organisms.

Conservation/Management: Stream habitat degradation is responsible for the decline of the creek chubsucker in Michigan and over much of its range. Siltation and pollution are the primary limiting factors. Trautman (1981) documented marked decreases of this species in turbid water and, on several occasions, found dead individuals with silt packed around the gills. Severe erosion from recently cultivated agricultural lands following periods of rain was responsible for the heavy silt loads.

Recent surveys in the Raisin River have found a few locations that still support this species (Smith et al. 1981). Surprisingly, individuals were collected in stream habitats that were highly disturbed, indicating some tolerance to turbid habitats in Michigan. Surveys in the St. Joseph River drainage are needed to determine its present status in southwestern Michigan. Game fish management using rotenone poisoning and subsequent introduction of exotic fish in streams also must be limited. Research is required to determine the response of self-sustaining populations within highly disturbed stream habitats before currently inhabited sites are secure.

LITERATURE CITED

Smith, G. R., J. N. Taylor, and T. W. Grimshaw. 1981. Ecological survey of fishes in the
Raisin River drainage, Michigan. *Mich. Acad. Sci.* 13:275–305.
Trautman, M. B. 1981. *The fishes of Ohio.* Ohio State Univ. Press, Columbus.

River Redhorse

· *Moxostoma carinatum* (Cope)
· **State Threatened**

Map 41. Confirmed individuals. ▲ = before 1977 and
● = 1977 to 1989 (Michigan DNR).

Status Rationale: The river redhorse reaches the northern limit of its distribution in Michigan. Generally considered an uncommon fish throughout its range, the river redhorse has been documented only three times in Michigan.

In-hand Identification: This member of the sucker family (Catostomidae) is one of the largest species of redhorse, sometimes reaching a length of 30 inches (76.2 cm) and a weight of 10 pounds (4.5 kg). The upper lobe of the caudal fin may be larger than the lower lobe. Like many other redhorse species (also known as mullets), it has striated lips and brassy yellow coloration with a reddish tail.

In Michigan, only three redhorse species exhibit the reddish fin color, thereby separating them from other suckers (the red coloration may disappear shortly after death). The river redhorse can be readily distinguished from all relatives in Michigan by its pharyngeal teeth, which are conspicuously enlarged and molarlike. These teeth are located in the gill region on the fifth supporting arch, which is comparatively larger and is triangular in cross-section.

Total Range: The river redhorse is restricted to eastern North America in the middle and upper Mississippi River drainages, the upper St. Lawrence River, tributaries of Lakes Erie and Michigan, and river systems of the eastern Gulf slope. Populations are generally low and declining throughout its range, except such areas

287

as the Ozark region of Missouri and Arkansas (Pflieger 1975). Recent intensive surveys in Wisconsin found many new locations inhabited by the river redhorse (Fago 1982, 1983, 1984, and 1986). Although the river redhorse is known to survive only in disjunct populations in the lower Great Lakes basin, archaeological finds suggest a continuous distribution in this region (Scott and Crossman 1973).

State History and Distribution: The river redhorse may have existed in widely scattered populations across southern lower Michigan, but its confusion with other species and possible disappearance following the heavy siltation of its river habitat by logging and farming operations have resulted in few confirmed records. Hubbs and Lagler (1974) considered its range to extend north to the Muskegon and Detroit rivers. However, this species is documented only from three river systems in the state: the Muskegon River, Newaygo County, in 1935; the Detroit River in Wayne County; and the Grand River at Grand Haven State Game Area, Ottawa County, in 1978.

Habitat: As its name implies, the river redhorse occurs in rivers, typically medium- to large-sized with strong currents. This species requires unpolluted, clear rivers and is intolerant of turbid conditions (Jenkins 1970). A gravel or rubble bottom is preferred. This substrate and the strong currents minimize siltation levels (Trautman 1981), crucial for its survival. Aquatic vegetation is preferable in isolated patches (Parker and McKee 1984). In Illinois and Missouri, this species was found in small- and medium-sized rivers with gravel or rocky bottoms (Smith 1979; Pflieger 1975).

From a sample of 59 stations inhabited by the river redhorse in Wisconsin, river characteristics included an average stream depth of 1 to 10 feet (0.3 to 3.0 m), a moderate current, and slight to moderate turbidity (Fago 1982, 1983, 1984, and 1986). The Newaygo County individual was found in a clear, quiet dammed lake on the Muskegon River. This occurrence is probably exceptional, since habitats with strong currents are preferred. However, with only three locations for the species in the state, it is difficult to describe its habitat preference in Michigan.

Ecology and Life History: Spawning occurs later in the spring for the river redhorse than for other redhorse species. It apparently migrates upstream and concentrates in shallow areas. The territorial male constructs a redd (nest) on a clean, gravelly substrate within a riffle area. Adult females carry between 6,000 and 23,000 eggs, depending upon their size. These eggs will hatch in a few days with an ambient temperature of approximately 73°F (Parker and McKee 1984). No parental care follows spawning. The river redhorse's growth rate is rapid; by three years of age (about 16 inches [41 cm] long), this species reaches sexual maturity.

The heavy, pharyngeal teeth aid the adults of this unique species in crushing its primary prey of mollusks and crustaceans. From a limited sample, Parker and McKee (1984) found that individuals between 4 and 6 inches (10.1 and 15.2 cm) fed primarily on fly larvae and pupae. Individuals larger than 8 inches (20.3 cm) were partially dependent upon crustaceans, probably reflecting development of the pharyngeal teeth. Prey are located by sight and capture efficiency may depend upon the degree of turbidity.

Conservation/Management: The river redhorse requires clear, fast, undisturbed rivers with a gravelly substrate. Populations of the river redhorse decline as its riverine habitat becomes more polluted and turbid (Trautman 1981). The food sources of this redhorse (e.g., mollusks and insect larvae) also require clear, unpolluted waters and substrates free of silt to survive. Lack of recognition by anglers and researchers are probably a source of unknown harvesting of local populations, which may cause declines (Parker and McKee 1984; Parker 1988). Since detailed historic and recent information on the distribution and abundance of the river redhorse are lacking, surveys of larger river environs need to be made. Shock methods in Wisconsin and Ohio have shown this redhorse to be more widely distributed than previously thought. Efforts to preserve and restore clear waters in larger rivers, and to allow access to upstream spawning grounds (via fish ladders at dams), would help maintain this rare species in Michigan.

LITERATURE CITED

Fago, D. 1982. Distribution and relative abundance of fishes in Wisconsin. Part 1. Greater Rock River basin. *Wisc. Dept. Nat. Resour. Tech. Bull.* No. 136.

Fago, D. 1983. Distribution and relative abundance of fishes in Wisconsin. Part 2. Black, Trempealeau, and Buffalo River basins. *Wisc. Dept. Nat. Resour. Tech. Bull.* No. 140.

Fago, D. 1984. Distribution and relative abundance of fishes in Wisconsin. Part 4. Root, Milwaukee, Des Plaines, and Fox River basins. *Wisc. Dept. Nat. Resour. Tech. Bull.* No. 147.

Fago, D. 1986. Distribution and relative abundance of fishes in Wisconsin. Part 7. St. Croix River basin. *Wisc. Dept. Nat. Resour. Tech. Bull.* No. 159.

Hubbs, C. L., and K. F. Lagler. 1974. *Fishes of the Great Lakes region.* Univ. Mich. Press, Ann Arbor.

Jenkins, R. E. 1970. *Systematic studies of the catostomid fish tribe Moxostomatini.* Ph. D. diss., Cornell Univ., Ithaca, NY.

Parker, B. J. 1988. Updated status of the river redhorse, *Moxostoma carinatum,* in Canada. *Can. Field-Nat.* 102:140–46.

Parker, B., and E. H. McKee. 1984. Status of the river redhorse, *Moxostoma carinatum,* in Canada. *Can. Field-Nat.* 98:110–14.

Pflieger, W. L. 1975. *The fishes of Missouri.* Missouri Dept. Cons. Jefferson City, MO.

Scott, W. B., and E. H. Crossman. 1973. *Freshwater fishes of Canada.* Fish. Res. Board Can., Bull. 184.

Smith, P. W. 1979. *The fishes of Illinois.* Illinois State Nat. Hist. Surv. Univ. Illinois Press, Urbana.

Trautman, M. B. 1981. *The fishes of Ohio.* Ohio State Univ. Press, Columbus.

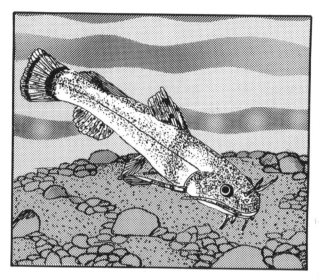

Northern Madtom

· *Noturus stigmosus* Taylor

· **State Endangered**

Map 42. Confirmed individuals. ▲ = before 1977 and ● = 1977 to 1989 (Michigan DNR).

Status Rationale: Generally recognized as an uncommon species throughout its limited range, the northern madtom is experiencing a rangewide decline. In Michigan, this species may be surviving only in a protected section of the Huron River.

In-hand Identification: This small member of the catfish family (Catostomidae) averages 3 to 4 inches (7.6 to 10.2 cm) in length and has an overall chubby body appearance. Its sides and back are golden to tan; the belly is mottled, occasionally with a pinkish yellowish tinge. Like other members of this family, the northern madtom has barbels near its mouth.

Within the catfish family, only madtoms have a long, adipose fin that is continuous with the caudal fin. Unlike the stonecat (*Noturus flavus*), tadpole madtom (*Noturus gyrinus*), and margined madtom (*Noturus insignis*), the northern madtom has serrations on both sides of the curved pectoral fin spine. The most similar madtom in Michigan, the brindled madtom (*Noturus miurus*), differs from the northern madtom by the uninterrupted pigmented band on the adipose fin. The northern madtom's adipose fin band only extends over half the fin.

Total Range and Taxonomic Status: The northern madtom inhabits tributaries of the Mississippi River in western Mississippi and Tennessee, and extends north throughout much of the Ohio River basin, including Kentucky, Indiana, Ohio, and

290

extreme eastern Illinois and western Pennsylvania. It is also found in the western Lake Erie drainage, including Indiana, Ohio, and Michigan. Until 1969, the northern madtom was conspecific with the mountain madtom (*Noturus eleutherus*): known as the Carolina madtom (*Noturus furiosus*). Taylor (1969) convincingly determined the northern madtom's current taxonomic status. Prior to this study, it was presumed that the northern madtom was the female counterpart of the mountain madtom species.

State History and Distribution: In Michigan, this madtom is only known from the Huron and Detroit rivers. In the Huron River, it was recorded nearly 20 times between the Portage Lake outlet and Delhi Rapids, especially in Hudson Mills Metropark, Washtenaw County. The last confirmed specimen was in 1977, when three individuals were collected in the park; however, subsequent sampling has found this population to remain extant. Only two specimens have been found in the Detroit River, Wayne County: one within Detroit's city limits and the other near the junction of Lake St. Clair and the Detroit River. The Huron River population is the only known active site for this species in the state.

Habitat: Typically found in large streams and rivers with sparsely vegetated substrates, the northern madtom avoids extremely silty situations and, as noted in Michigan, a shifting bottom (Taylor 1969). In the Huron River, this madtom is found in clear water with a moderate-to-swift current and a substrate with mixed gravel, sand, or large irregular stones. Some individuals found in the Huron River were utilizing a section of the river that was 5 feet (1.5 m) deep, 70 to 160 feet (21.3 to 48.7 m) in width, and with sparse vegetation.

Ecology and Life History: Little is known about this species' life cycle. In Michigan, spawning activity and egg laying probably occur in midsummer in warm water (74° to 80°F [24° to 27°C]). Redds (nests) containing several dozen to nearly 150 eggs are guarded by the males, as are the young (Taylor 1969). After two to three years (approximately 2 inches [5.1 cm] in length), this madtom is probably sexually mature. The northern madtom is generally most active at night, searching for aquatic invertebrates. During these times, it may leave the deeper and darker waters of the river for shallower regions along river banks.

Conservation/Management: Research is needed on the biology of this species, its habitat requirements, and the effects of human activities. Pollution and siltation levels in the occupied Huron River stretch should be monitored frequently and carefully, because these two factors are the primary cause of this species' decline. Trautman (1981) felt that the loss of stony riffles and river bars may also have been responsible for its disappearance in Ohio rivers. Pollution sources need to be identified and controlled from upstream areas to maintain any remaining populations. The majority of the known northern madtom population is protected within the boundaries of a metropark. Surveys in suitable habitat should be made at night in autumn when this madtom is most active and more likely to be found in shallow areas. Madtoms possess venom glands at the base of their pectoral fin spines and should be handled with care. By maintaining this protected population and initiating surveys along remaining natural areas of the Huron River, the northern madtom's existence in Michigan will be secured.

LITERATURE CITED

Taylor, W. R. 1969. A revision of the catfish genus *Noturus* Rafinesque, with an analysis of higher groups in the Ictaluridae. *Smith. Inst. Press., U.S. Natl. Bull.* No. 282.

Trautman, M. B. 1981. *The fishes of Ohio.* Ohio State Univ. Press, Columbus.

Sauger

- *Stizostedion canadense* (Smith)
- **State Threatened**

Status Rationale: Formerly widespread in the Great Lakes, and particularly common in Lake Erie, sauger populations have experienced large-scale declines, caused by commercial overfishing. Habitat degradation may not be the primary cause of the decline. This situation may permit reestablishment efforts should other environmental limiting factors be minor.

In-hand Identification: The sauger is a member of the family Percidae (perch family), and displays the group's diagnostic features: two distinct dorsal fins and one or two spines in the anal fin (the latter for the sauger). This medium-sized fish has an elongated body with a golden olive back fading to silvery sides. General features include a yellowish dorsal fin, deeply-forked tail, two anal fin rays, and a dark spot at the base of the pectoral fin. By autumn, young-of-the-year measure 3 to 6 inches (7.6 to 15 cm) long; adults average 9 to 10 inches (23 cm) (Trautman 1981), but may reach lengths of 30 inches (76 cm) (Carufel 1963). Average adult weights are 0.5 to 1.0 pound (227 to 454 g), reaching over 3.0 pounds (1361 g) for females with eggs in Ohio (Trautman 1981). The sauger naturally hybridizes; large individuals may be sauger-walleye hybrids (Nelson 1968a; Trautman 1981).

The similar, but larger, walleye (*Stizostedion vitreum*) differs by a dusky blotch at the base of the spinous dorsal fin. The sauger's lower lobe of the tail (caudal) fin generally is not whitish, a feature diagnostic for the walleye. Differences between these two species may be very minor, though, particularly with faded individuals (Eddy and Surber 1960). They are best separated by the number of pyloric caeca (walleye usually have three, sauger have three to nine)(Scott and Crossman 1973). Commercial catches of both species are collectively known as yellow pike. Darters may be mistaken for young-of-the-year saugers, but superficially differ with their smaller mouths and lack of a deeply forked tail.

Total Range: Widely distributed, the sauger occurs across much of central North America. Native populations are found from the western side of the Appalachian Mountains, west to the northern Great Plains and north to the drainages of James Bay. It is an inhabitant of much of the Mississippi River drainage, from its northern tributaries to Arkansas. It occurs throughout the Ohio River valley and parts of the Great Lakes region into the St. Lawrence River–Lake Champlain system. Populations have been established outside this range through introductions, primarily in the Lower Mississippi River basin and in rivers of the southeastern United States.

State History and Distribution: Although the sauger is widely distributed in and around the Great Lakes basin, there are relatively few locations in Michigan. In the mid-1900s, Hubbs and Lagler (1974) considered it "abundant in Lake Erie and Saginaw Bay, less common in Lakes Michigan and Huron and their main tributaries; only locally common in the Superior basin." This distribution is supported by Bailey and Smith (1981). In Ohio waters, Lake Erie commercial catches indicate severe population declines, reaching commerical extinction in the late 1960s (Trautman 1981). Similar trends apparently occurred in Michigan waters. Today, it is rarely found in Michigan waters of the Great Lakes.

Ecology and Life History: Spawning lasts for two weeks between mid-May and mid-June and occurs in turbid, shoal areas with gravel substrate. Males arrive prior to the females, beginning spawning at approximate water temperatures of 40°F (4°C) in waters 2 to 12 feet (0.6 to 3.7 m) deep (Scott and Crossman 1973). Peak spawning occurs at night in water temperatures ranging from 39° to 53°F (4° to 12°C). The number of adhesive eggs produced may range from 9,000 to well over 100,000. The number of eggs depends on individual size and geographic locations, but averages 15,000 to 40,000 eggs per pound of fish. No nest is built, the eggs settle on the gravel substrate.

Eggs hatch after 25 to 29 days at temperatures ranging from 40° to 55°F (5° to 13°C) (Scott and Crossman 1973), and growth rates vary according to environmental conditions. By 2 to 3 years of age, males are sexually mature, measuring an average of 8 to 10^1/3 inches (20 to 26 cm) and females are sexually mature at 4 to 6 years of age, measuring an average of 10^1/3 to 15^3/4 inches (26 to 40 cm) (Hart 1928; Deason 1933; Carlander 1950; Nelson 1968b; Priegel 1969). Northern populations exhibit slower growth rates and live longer. Individuals may live 12 to 13 years in Canada and only 5 to 6 years in more southern U.S. populations (Scott and Crossman 1973).

Young saugers forage over shoals for invertebrate larvae and adults, whereas older sauger take small fish (1.2 to 5.0 cm) and large invertebrates (Priegel 1969).

Conservation/Management: Areas of former abundance in Michigan waters of the Great Lakes (e.g., Lake Erie) now harbor dramatically reduced sauger populations. Commercial fisheries account for this species decline in Ohio; reestablishment efforts are now under way (Trautman 1981). The sauger remains an important commercial fish in Canada, particularly Manitoba.

Its tolerance of turbid waters indicates that its dramatic decline in the Great Lakes region and the larger rivers is probably unrelated to habitat degradation. Furthermore, Scott and Crossman (1973) consider competition with closely related species, such as walleye, to have an insignificant effect on the survival of sauger populations. Others believe that the walleye can outcompete the sauger under less turbid water conditions.

Additional research is needed to determine the limiting factors for self-sustaining sauger populations. This should be followed by reestablishment programs at formerly occupied sites if this species is going to recover and become a viable component of the Great Lakes fish fauna.

LITERATURE CITED

Bailey, R. M., and G. R. Smith. 1981. Origin and geography of the fish fauna of the Laurentian Great Lakes Basin. *Can. J. Fish. Aquat. Sci.* 38:1539–61.

Carlander, K. D. 1950. Growth rate studies of saugers, *Stizostedion canadense canadense* (Smith) and yellow perch, *Perca flavescens* (Mitchill) from Lake of the Woods, Minnesota. *Trans. Am. Fish. Soc.* 79:30–42.

Carufel, L. H. 1963. Life history of saugers in Garrison Reservoir. *J. Wildl. Mgmt.* 27:450–56.

Deason, H. J. 1933. Preliminary report on the growth rate, dominance, and maturity of the pike-perches (Stizostedion) of Lake Erie. *Trans. Am. Fish. Soc.* 63:348–60.

Eddy, S., and T. Surber. 1960. *Northern fishes with special reference to the upper Mississippi Valley.* Charles T. Banford Co., Boston.

Hart, J. L. 1928. Data on the rate of growth of pike perch (*Stizostedion vitreum*) and sauger (*S. canadense*) in Ontario. *Univ. Toronto Publ. Ont. Fish. Res. Lab., Stud. Biol. Ser.* 31, 34:43–55.

Hubbs, C. L., and K. F. Lagler. 1974. *Fishes of the Great Lakes region.* Univ. Mich. Press, Ann Arbor.

Nelson, W. R. 1968a. Reproduction and early life history of sauger, *Stizostedion canadense,* in Lewis and Clark Lake. *Trans. Am. Fish. Soc.* 97:159–66.

Nelson, W. R. 1968b. Embryo and larval characteristics of sauger, walleye, and their reciprocal hybrids. *Trans. Am. Fish. Soc.* 97:167–74.

Priegel, G. R. 1969. The Lake Winnebago sauger: Age, growth, reproduction, food habits and early life history. *Wisc. Dept. Nat. Resour. Tech. Bull.* No. 43.

Scott, W. B., and E. J. Crossman. 1973. *Freshwater fishes of Canada.* Fish. Res. Board Can., Bull. 184.

Trautman, M. B. 1981. *The fishes of Ohio.* Ohio State Univ. Press, Columbus.

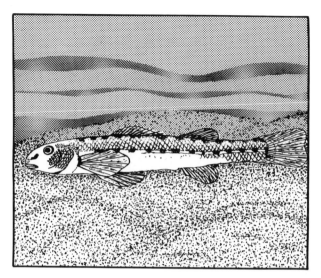

Eastern Sand Darter

· *Ammocrypta pellucida* (Putnam)
· **State Threatened**

Map 43. Confirmed individuals. ▲ = before 1977 and
● = 1977 to 1989 (Michigan DNR).

Status Rationale: The eastern sand darter is considered to be a peripheral species in Michigan, but its protection is crucial because it is experiencing a rangewide decline. Only one known viable population currently exists in the state.

In-hand Identification: The eastern sand darter has an elongated, light yellow body; living individuals are nearly transparent. It averages 2¹/₂ inches (6.4 cm) in length and, as characterized by its genus, has a single-spined anal fin. This darter has a nearly scaleless ventral half, a slightly emarginated caudal fin, and 8 to 13 small, rounded green markings along the sides. The yellow coloration in males intensifies during the breeding season. An identifying behavioral feature is its habit of burrowing head first into sandy substrates with only its eyes protruding above the surface.

A key identification characteristic that separates the eastern sand darter from all other Michigan percid darters except the johnny darter (*Etheostoma nigrum*), is the single spine on the anal fin. To distinguish the johnny darter from the sand darter, look for the darker colored body, W-shaped markings (not rounded spots), and a groove between the upper lip and snout (absent in the eastern sand darter).

Total Range: This darter occurs in much of the Ohio River basin, south to West

Virginia and Kentucky, west to eastern Illinois, east to Ohio, and north to southeastern Michigan and southwestern Ontario. A disjunct population occurs in the St. Lawrence–Lake Champlain drainage. Lake Erie tributaries, shorelines, and island shoals are also frequented by this darter. The eastern sand darter is a steadily declining species in much of its range (Kuehne and Barbour 1983).

State History and Distribution: The eastern sand darter is restricted to the southeastern portion of the Lower Peninsula, primarily in the Huron River drainage of Livingston County. Other systems with historic records include the Detroit, St. Joseph, Raisin, and Rouge rivers. Individuals have been collected from Lake St. Clair, St. Clair County. However, observations were recorded before 1955. Recent studies of the Raisin River drainage did not locate this species, and it may be extirpated in this river system (Smith et al. 1981). Currently, the only known population is restricted to the Huron River near the Island Lake State Recreational Area, Livingston County. Individuals were recently found (in 1985) in the Pine River, St. Clair County, but the status of this population is unknown.

Habitat: The preferred habitats of this darter are the sandy stretches of slow-moving streams and rivers (Lachner et al. 1950) and the wave protected portions of lakes, such as the sandy shoals of islands and beaches (Scott and Crossman 1973). In 1977, a survey of the Huron River in Livingston County (currently containing Michigan's only known viable population) located this darter at five sites each within 5 miles (8.0 km) of one another. This particular stretch of the Huron River was clear and had a moderate current with a substrate composed primarily of sand and some silt and gravel. The river bank was minimally disturbed, with many wooded areas and some aquatic vegetation near the shoreline. Water depths ranged from 2 to over 7 feet (0.6 to 2.1 m).

Ecology and Life History: Little is known about the biology of the eastern sand darter. In the Ohio River basin, the spawning period usually begins during the first part of June and ends in late July (Williams 1975). The Great Lakes population appears to be two to three weeks later. The spawning substrate is probably sand. Blackfly larvae (*Chironomus* spp.) are its main prey (Turner 1921).

Conservation/Management: The eastern sand darter is declining throughout its range. Many populations have been extirpated and those remaining are stressed by disturbed environmental conditions. The primary reason for its decline appears to be heavy siltation of streams, a prevalent problem in highly industrialized and agricultural regions, that has significantly reduced populations of this darter (Trautman 1981; Kuehne and Barbour 1983). Large carp (*Cyprinus carpio*) populations also have affected local sand darter populations because they disturb the river bottom while foraging, thereby increasing water turbidity. Recent changes in farming practices have further affected this darter and other fish species sensitive to water quality changes. The extensive draining of the landscape, cutting of wooded areas, and removing hedgerows (which increases erosion) have all contributed to high silt loads of streams and rivers. Pollution, riparian destruction, and residential development have reduced populations even further.

Protection of riparian zones and lowland areas from clearing or development along darter-inhabited bodies of water is needed. Harmful industrial effluents have been reduced in recent years throughout Michigan, but pollution sources still must be monitored closely. Shoreline protection, monitoring of the populations dynamics, and general ecological studies are necessary to secure Michigan's only known population.

LITERATURE CITED

Kuehne, R. A., and R. W. Barbour. 1983. *The American darters.* Univ. Kentucky Press, Lexington.

Lachner, E. A., E. F. Westlake, and P. S. Handwerkk. 1950. Studies on the biology of some percid fishes from western Pennsylvania. *Am. Midl. Nat.* 43:92–111.

Scott, W. B., and E. H. Crossman. 1973. *Freshwater fishes of Canada.* Fish. Res. Board Can., Bull. No. 184.

Smith, G. R., J. N. Taylor, and T. W. Grimshaw. 1981. Ecological survey of fishes in the Rasin River drainage, Michigan. *Mich. Acad. Sci.* 13:275–305.

Trautman, M. B. 1981. *The fishes of Ohio.* Ohio State Univ. Press, Columbus.

Turner, C. L. 1921. Food of the common Ohio darters. *Ohio J. Sci.* 22: 41–62.

Williams, J. D. 1975. Systematics of the Percid fishes of the subgenus *Ammocrypta,* genus *Ammocrypta,* with descriptions of two new species. *Alabama Mus. Nat. Hist. Bull.* No. 1.

Channel Darter

· *Percina copelandi* (Jordan)
· **State Threatened**

Map 44. Confirmed individuals. ▲ = before 1958 and ● = 1958 to 1989 (Michigan DNR).

Status Rationale: Michigan populations of this species have severely declined over the past two to three decades. Until a 1986 survey discovered populations in the Au Sable and Pine rivers, the channel darter was reported only in small numbers from Monroe County. Reduction from its historic 11-county Michigan range is related to river habitat degradation.

In-hand Identification: This darter has a blunt snout, yellowish olive back, with six to nine indistinct, brown saddle bands and yellowish sides. A series of nine to ten (occasionally more) small, oblong, dark blotches mark the lateral line from the cheek to the tail. A dark vertical teardrop is usually present below the eye. Adults are typically 2$\frac{1}{3}$ to 3 inches (5.9 to 7.6 cm) in length. Breeding males form tubercles and change to a dusky color, with the head, undersides, and pelvic fins turning blackish.

Darters (family Percidae) are easily separated from the single dorsal-finned members of the minnow family (Cyprinidae) by their double dorsal fin structure. Differences are subtle between Michigan's three darter genera. The two anal fin spines distinguish the channel darter from members in the genus *Ammocrypta*, which have single-spined anal fins. Unlike the channel darter, species of the genus *Etheostoma* have unmodified scales between the pelvic fins and scaled bellies. Within the genus *Percina*, the other three Michigan members may be distinguished

299

by a sharper snout and a broad bridge of skin between the snout and lips. The blackside darter (*P. maculata*) also has darker, continuous lateral blotches; the logperch (*P. caprodes*) has an elongate snout and numerous vertical lines on the sides.

Total Range: Channel darters are mainly found in the Ohio River basin, north to the lower Great Lakes basin and upper St. Lawrence River drainage. Disjunct populations occur in the Red and Arkansas rivers west of the Mississippi River drainage, and in several smaller rivers in Louisiana, Mississippi, and Alabama. Although the channel darter has a relatively wide distribution, populations are frequently isolated (Kuehne and Barbour 1983).

State History and Distribution: Eleven counties bordering Lake Huron and Lake Erie comprise the known range of this species in Michigan. Historic populations were found in the Au Sable River, Iosco County; Saginaw Bay region; Pine River, Alcona County; Rifle River, Arenac and Ogemaw counties; and in Thunder Bay, Alpena County. Collected regularly from 1923 to 1957, the channel darter was not found for several decades after this period. However, a sizable population was discovered in the Au Sable and Pine rivers in 1986 (Schultz 1986). Even though historical sites from tributaries flowing into Lake Erie are currently unsuitable for this species, Lake Huron bays and principal tributaries still contain suitable habitat for the channel darter and some local areas may be inhabited by this species.

Habitat: Within rivers, channel darters prefer slow currents, strong enough to create silt-free gravel bottoms (Pflieger 1975), rather than fast-flowing riffles. At the Au Sable and Pine rivers sites, channel darters preferred slow to moderate flows over sand, gravel, and cobble bottoms (Schultz 1986). Sections with rapid currents (e.g., below dams) were avoided; the darters remained in the protected, downstream side of gravel-sand bars in water depths of 2 to 5 feet (0.6 to 1.5 m). Channel darters also inhabit sand or fine gravel beaches of Lake Huron and the gravel bottoms of larger rivers flowing into this lake and Lake Erie. In the Great Lakes, this species usually occurs in water more than 3 feet (0.9 m) deep and ventures into shallower water only at night (Trautman 1981). It is known to overwinter in backwater pools filled with organic debris (Branson 1967).

Ecology and Life History: Adults may reach a total length of $2^1/2$ inches (6.4 cm), but the average size is usually $1^1/2$ to $2^1/6$ inches (3.8 to 5.5 cm). Winn (1953) found that gradual flowing water over the gravel beds and water temperatures ranging from 69° to 72°F (21° to 22°C) are essential for spawning. Usually 350 to 400 eggs (frequently more) are deposited in midsummer on gravel beds (Winn 1958a) and are subsequently abandoned by the adults. Territories, maintained by males and centered by a large rock, are less than 3 feet (0.9 m) in diameter (Winn 1958b). A population studied by Turner (1921), near the Bass islands of western Lake Erie, fed primarily on mayfly and blackfly (*Chironomus* spp.) larvae as well as on algae and bottom debris. This species is one of the few darters that regularly feeds on algae and detritus. Small crustaceans also may serve as important food items.

Conservation/Management: Heavy silt deposits and pollutants in suitable habitat appear to be the primary factors causing the decline of the channel darter (Trautman 1981). It requires clean gravel beaches in Lake Huron, Saginaw Bay, and principal tributaries for its survival in Michigan. The minimally disturbed habitats in the AuSable and Pine rivers attest to this need, while areas that have been severely altered (e.g., Huron and Cass rivers) no longer support channel darter populations.

Sound land-use practices, such as the retention of riparian vegetation along rivers, will significantly reduce erosion and siltation and would increase the likelihood of this species' survival in Michigan. Management tools such as in-stream sand sediment traps can reduce high sand bedloads (Hansen et al. 1983) and can help fish species that require higher stream quality (Alexander and Hansen 1983). Other techniques aimed solely at game fish management, such as the deliberate poisoning of streams, need to be strictly controlled in areas that still retain their native fish fauna.

Protection and management of the known populations are crucial, as are further surveys in Saginaw and Thunder bays, the Rifle River, and other principal Lake Huron tributaries. Suggested sampling methods include using seine nets at depths of 3 to 5 feet (0.9 to 1.5 m) (1) in the Great Lakes from mid- to late spring and (2) in rivers during June and July along the downstream side of gravel-sand bars (Schultz 1986). Low water levels in autumn also allow successful sampling, particularly at night.

LITERATURE CITED

Alexander, G. R., and E. A. Hansen. 1983. Sand sediment in a Michigan trout stream. Part II. Effects of reducing sand bedload on a trout population. N. Am. J. Fish. Mgmt. 3:365–72.

Branson, B. A. 1967. Fishes of the Neosho River System in Oklahoma. Am. Midl. Nat. 78:121–54.

Hansen, E. A., G. R. Alexander, and W. H. Dunn. 1983. Sand sediment in a Michigan trout stream. Part I. A technique for removing sand bedload from streams. N. Am. J. Fish. Mgmt. 3:355–64.

Kuehne, R. A., and R. W. Barbour. 1983. The American darters. Univ. Kentucky Press, Lexington.

Pflieger, W. L. 1975. The fishes of Missouri. Missouri Dept. Conserv., Jefferson City, MO.

Schultz, D. L. 1986. Report of the survey to assess the current status of Percina copelandi and Percina shumardi in Michigan. Mich. Nat. Feat. Invent., Unpubl. Rept.

Trautman, M. B. 1981. The fishes of Ohio. Ohio State Univ. Press, Columbus.

Turner, C. L. 1921. Food of the common Ohio darters. Ohio J. Sci. 22: 41–62.

Winn, H. E. 1953. Breeding habits of the Percid fish Hadropterus copelandi in Michigan. Copeia 1953:26–30.

Winn, H. E. 1958a. Comparative reproductive behavior and ecology of fourteen species of darters (Pisces-Percidae). Ecol. Monogr. 28:155–91.

Winn, H. E. 1958b. Observations on the reproductive habits of darters (Pisces-Percidae). Am. Midl. Nat. 59:190–212.

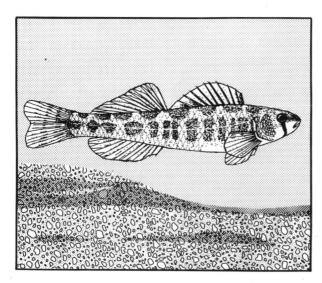

River Darter

· *Percina shumardi* (Girard)

· **State Threatened**

Map 45. Confirmed individuals. ▲ = before 1977 (Michigan DNR). No records since 1977.

Status Rationale: The river darter is distributed widely along the Mississippi River drainage with historic populations in Michigan's rivers along Lake Huron and Lake Erie. This species has not been collected in Michigan since 1941; however, it may still be present in undisturbed Lake Huron tributaries.

In-hand Identification: The river darter is a small fish (adults are 2^1/$_2$ to 3 inches [6.4 to 7.6 cm]) with a terminal mouth, a slender body, five to nine obscure saddle bands, 8 to 15 vertically elongated blotches along the sides, and two distinct dorsal fins. This brownish olive species has a well-developed black vertical marking below each eye. A yellowish brown cast may be observed on the underside. During the breeding season, the male's colors darken, markings intensify, and tubercles develop on the caudal and anal fins.

Many of the state's darters (family Percidae) are similar to this species. There are subtle differences between Michigan's three genera of percids: *Ammocrypta* have single-spined anal fins and *Etheostoma* typically have unmodified scales between the pelvic fins and scaled belly, whereas *Percina* have two anal fin spines and usually a modified row of scales in the middle of a scaleless belly. The other three Michigan *Percina* species differ from the river darter, superficially, with a blunter snout that lacks the bridge of skin between the snout and upper lip (channel darter [*P. copelandi*]); an overall darker color, and usually fewer than eight vertical bars on the sides (blackside darter [*P. maculata*]); and an enlongate snout and numerous vertical lines on the sides (logperch [*P. caprodes*]).

Total Range: This darter has an extensive but irregular north-south distribution. It ranges from southern Manitoba and western Ontario south through Ohio to the Gulf coast. Found throughout much of the Mississippi River basin, the river darter occurs in scattered populations in the Ohio, Tennessee, and Arkansas river systems. Recently, an increase in abundance has been observed in the Mississippi River (Page 1983), and it is the most common darter in the Mississippi River channel (Kuehne and Barbour 1983). Small populations occur within the shorelines of western Lakes Huron and Erie and their tributaries, although the last report along Ohio's shoreline was in 1946 (Van Meter and Trautman 1970).

State History and Distribution: The river darter was distributed historically along the eastern margin of Michigan. Among the Lake Huron tributaries, records exist from the Au Sable and Cass rivers. The island and shoreline shoals of Saginaw Bay also contained populations, and Trautman (1981) found this darter in similar habitat in western Lake Erie. The sole documented Michigan record of the river darter from Lake Erie and its tributaries was in 1941 from the Huron River, Monroe County, but it has not been reported since. A recent study sampled the five known historic sites of the river darter without success (Schultz 1986). Although two of these sites are now considered unsuitable, the Saginaw Bay and Au Sable River appear to offer habitat capable of supporting self-sustaining populations.

Habitat: As its name implies, the river darter is an inhabitant of rivers and large streams. Deeper riffles with gravel substrates and moderate to swift currents are preferred. This darter also occurs in the comparatively shallow regions of the Great Lakes, frequenting areas less than 5 feet (1.5 m) deep. Populations may shift from Lake Huron to the principal tributaries of mid-Michigan during the breeding season (Schultz 1986). This species is moderately tolerant to water turbidity (Pflieger 1975; Trautman 1981), which explains its abundance in parts of the Mississippi River and suggests the potential for its survival in relatively disturbed environments (e.g., Saginaw Bay).

Ecology and Life History: The river darter has not been studied throughout the Great Lakes region, but it is typically associated with the closely related channel darter. For this reason, the well-documented Michigan study on the channel darter (Winn 1953) is used as a reference for the river darter's breeding biology. Spawning probably occurs in midsummer on gravel riffles with water temperatures in the low 70s°F (the low 20s°C). In Canada, the river darter spawned as late as June and July (Scott and Crossman 1973). River darters likely spawn in deeper areas of the river (or lake) than the channel darter. Once deposited, the several hundred eggs are probably abandoned by the adults. Most of the river darter's growth is during the first year (average length is $1^7/8$ inches [4.8 cm]); it occasionally lives three to four years (Thomas 1970). Invertebrates are its primary prey, particularly blackfly (*Chironomus* spp.) and sand caddisfly larvae (Thomas 1970).

Conservation/Management: The river darter was never widespread in the state, but it did have a scattered distribution along the eastern margin of Michigan. The last known record was in 1941; although it is possible that it still occurs in Michigan because systematic collections of small fish in northern Michigan have been few. Unlike most other darters, this species has a tolerance to turbid water. Nevertheless, the river darter has severely declined in the Great Lakes region, primarily due to siltation, channelization (Smith 1979), and pollution (Schultz 1986) of its habitat.

Intensive surveys are needed to identify existing populations. Suggested survey locations are Saginaw Bay and the lower reaches of the AuSable, Pine, and Rifle rivers. Sampling strategies for river darters include the use of seine nets at depths of 3 to 5 feet (0.9 to 1.5 m) in (1) Saginaw Bay from mid- to late spring and in (2) principal tributaries in June and July along gravel riffles (Schultz 1986). Another recommended survey method is to investigate deep, rapid river stretches during lower autumn water levels at night. The 1986 discovery of a self-sustaining channel darter population indicates that the river darter may be more widely distributed, and, if so, a concentrated effort will be needed to maintain it.

LITERATURE CITED

Kuehne, R. A., and R. W. Barbour. 1983. *The American darters.* Univ. Kentucky Press, Lexington.

Page, L. M. 1983. *Handbook of darters.* T.F.H. Publications, Neptune City, NJ.

Pflieger, W. L. 1975. *The fishes of Missouri.* Missouri Dept. Conserv., Jefferson City, MO.

Schultz, D. L. 1986. Report of the survey to assess the current status of *Percina copelandi* and *Percina shumardi* in Michigan. *Mich. Nat. Feat. Invent.,* Unpubl. Rept.

Scott, W. B., and E. H. Crossman. 1973. *Freshwater fishes of Canada.* Fish. Res. Board Can., Bull. No. 184.

Smith, P. W. 1979. *The fishes of Illinois.* Ill. State Nat. Hist. Surv., Univ. Ill. Press, Urbana.

Thomas, D. L. 1970. An ecological study of four darters of the genus Percina (Percidae) in the Kaskaskia River, Illinois. *Ill. Nat. Hist. Surv. Biol. Notes 70.*

Trautman, M. B. 1981. *The fishes of Ohio.* Ohio State Univ. Press, Columbus.

Van Meter, H. O., and M. B. Trautman. 1970. An annotated list of the fishes of Lake Erie and its tributary waters exclusive of the Detroit River. *Ohio. J. Sci.* 70:65–78.

Winn, L. M. 1953. Breeding habits of the percid fish *Hadropterus copelandi* in Michigan. *Copeia* 1953:26–30.

Redside Dace

· *Clinostomus elongatus* (Kirtland)
· **State Threatened**

Map 46. Confirmed individuals. ▲ = before 1977 and ● = 1977 to 1989 (Michigan DNR).

Status Rationale: Although the redside dace has few Michigan records, its special concern status over the remainder of its range warrants its protection in the state. This species occurs naturally in discontinuous populations and, therefore, should not be regarded as a peripheral species.

In-hand Identification: This small, 3-inch (7.6-cm), red-sided fish has a distinctively large mouth that extends below the front half of the eyes. Its lower jaw projects beyond the upper jaw, producing a somewhat triangular head appearance. Its body is laterally compressed, producing a "slab-sided" appearance. Adult redside dace are brightly colored, particularly the breeding males with their reflections of iridescent blue, green, and purple body scales. The back is usually blue-green or steel blue, and a yellowish band separates the back from the reddish orange sides. Nuptial tubercles are small and widely distributed over the body in both sexes. The lateral line is complete and slightly decurved.

Similar species are the northern and southern redbelly dace (*Phoxinus eos* and *P. erythrogaster*, respectively) and the finescale dace (*P. neogaeus*). Compared to the redside dace, both the northern and southern redbelly dace species have smaller mouths, blunter heads, and red on the sides, which may continue to the belly. Also, the finescale dace's yellow or red lateral stripes are the reverse of the redside dace. The finescale dace has a smaller mouth, blunter head, and an incomplete lateral line.

305

Total Range: The redside dace occurs in discontinuous and widely disjunct populations throughout its limited eastern range. It is primarily found in the upper Susquehanna River drainage of New York and Pennsylvania, the upper Ohio River basin, the drainages of Lakes Erie and Ontario in southeastern Michigan, Ontario, Ohio, Pennsylvania, and New York, and in the upper Mississippi River basin of Wisconsin and southeastern Minnesota. Although declining in Ontario (Parker et al. 1988), recent intensive surveys in Wisconsin found redside dace at more drainages, and at sites within them, than previously recorded (Fago 1982 and 1983).

State History and Distribution: The redside dace occurs naturally in distinctly separate populations. In Michigan, only seven records have been confirmed, including the River Rouge drainage in Oakland and Wayne counties, and the Huron River drainage in Washtenaw County. A rediscovered population from a tributary of the River Rouge, Wayne County, in 1988 is threatened by urbanization; 80 individuals have been translocated to a more protected stream habitat in Washtenaw County.

Habitat: As with many endangered and threatened fish species, the redside dace is declining because it requires clear water, usually in the headwater regions of small- to medium-sized streams. These clear, cool streams, with overhanging woody vegetation shading the sun, have pebble or gravel bottoms and a moderate to fast current (Trautman 1981). Streams meeting these physical characteristics also need riffles for spawning grounds and pools for the nonbreeding season to support redside dace populations. McKee and Parker (1982) found this species most prevalent under the shade of overhanging vegetation adjacent to banks that had been undercut by the current. Submerged woody debris also may enhance habitat by providing added shelter.

Ecology and Life History: Unlike most small, nongame fish, the redside dace's life history has been intensively studied (Koster 1939; Schwartz and Norvell 1958; Trautman 1981; McKee and Parker 1982). Koster's (1939) study of a population in New York was at the approximate latitude as Michigan's occurrences, thereby allowing comparisons. Spawning occurs in late May with water temperatures at least 65°F (18°C). During this time, adults move from pools to gravel riffle areas. The male redside dace defends a small, nonbinding territory until spawning begins. When a female arrives in the area, two or more males follow her to a chosen nest site. Females lay from 400 to 2,000 eggs (Koster 1939; McKee and Parker 1982), which are then fertilized by males.

Young-of-the-year (by October) average a length of 1 1/8 inches (2.9 cm), yearlings average 1 5/8 inches (4.1 cm), two-year-olds average 2 1/4 inches (5.7 cm), and redside dace of three or more years average 2 3/4 inches (7 cm) (McKee and Parker 1982). Maximum total length recorded is 4 1/2 inches (11.4 cm). Females are normally larger than males. Although two-year-olds occasionally are in breeding condition, individuals that are three years of age or older make up the majority of a breeding population.

Insects are the major prey of the redside dace. The morphology of its large mouth accounts for the unique feeding habits, such as jumping out of the water to catch flying and other terrestrial insects. A study in Pennsylvania found that more than 75% of this species' diet consisted of terrestrial insects (Schwartz and Norvell 1958).

Conservation/Management: The redside dace is declining throughout its limited range due to its stringent habitat requirements. It needs clear, cool, small streams with riffles, pools, and overhanging woody vegetation for shade. Substrate characteristics are not as crucial, but gravel or pebble bottoms are preferred during the spawning season. It appears that relatively stable water temperatures and dissolved oxygen levels also are important. Trautman (1981) attributed decreasing populations to the effects of turbidity, which usually occurs when agricultural practices or development alter the riparian zones.

In Michigan, areas with suitable habitat need to be surveyed and protected. Protection of riparian habitat adjacent to small streams would provide two critical components—shaded water and reduced siltation. Pollution and channelization also need to be monitored near areas potentially harboring populations. If redside dace populations are located, they should receive immediate and complete protection.

LITERATURE CITED

Fago, D. 1982. Distribution and relative abundance of fishes in Wisconsin. Part 1. Greater Rock River basin. *Wisc. Dept. Nat. Resour., Tech. Bull.* No. 136.

Fago, D. 1983. Distribution and relative abundance of fishes in Wisconsin. Part 2. Black, Trempealeau, and Buffalo River basins. *Wisc. Dept. Nat. Resour., Tech. Bull.* No. 140.

Koster, W. J. 1939. Some phases of the life history and relationships of the cyprinid, *Clinostomus elongatus* (Kirtland). *Copeia* 1939:201–8.

McKee, P. M., and B. J. Parker. 1982. The distribution, biology, and status of the fishes *Campostoma anomalum, Clinostomus elongatus, Notropis photogenis* (Cyprinidae), and *Fundulus notatus* (Cyprinodontidae) in Canada. *Can. J. Zool.* 60:1347–56.

Parker, B. J., P. McKee, and R. R. Campbell. 1988. Status of the redside dace, *Clinostomus elongatus*, in Canada. *Can. Field-Nat.* 102:163–69.

Schwartz, J., and J. Norvell. 1958. Food, growth, and sexual dimorphism of the redside dace, *Clinostomus elongatus* (Kirtland) in Linesville Creek, Crawford County, Pennsylvania. *Ohio J. Sci.* 58:311–16.

Trautman, M. B. 1981. *The fishes of Ohio.* Ohio State Univ. Press, Columbus.

The Great Lakes Coregonids

David C. Evers

Ten species of coregonids inhabited the Great Lakes, six of which are considered chubs or deepwater ciscoes. These six include, in descending order of average length, the blackfin (*Coregonus nigripinnis*), deepwater (*C. johannae*), shortjaw (*C. zenithicus*), shortnose (*C. reighardi*), kiyi (*C. kiyi*), and the bloater (*C. hoyi*). Beside the shortjaw cisco and extinct blue pike (*Stizostedion vitreum glaucum*), these ciscoes are the only fish endemic to the Great Lakes basin. The two other species in the genus *Coregonus* are the lake whitefish (*C. clupeaformis*) and the state-threatened lake herring (*C. artedii*) or shallow-water cisco. In 1961, these ciscoes were placed in the genus *Coregonus*, formerly the genus *Leucichthys* (Norden 1961). The longjaw cisco (*C. alpenae*) is now considered conspecific with the shortjaw cisco (Todd et al. 1981).

Taxonomic radiation is extreme within the genus *Coregonus*, with 40 species and subspecies commonly recognized in the Great Lakes basin (Koelz 1929 and 1931). Most of the subspecies (e.g., 24 in the lake herring species) are allopatric-small inland-lake populations. As expected, the number of species and subspecies recognized varies among authorities; one hypothesis raises the possibility that lake herring, chubs, and ciscoes in the Great Lakes basin are not completely reproductively isolated (Bailey and Smith 1981). The number and size of gill rakers, mouth size and shape, body shape, and color are the main features allowing species separation, although, genetic recognition may be a potential method (Todd 1981).

Currently, however, the combination of morphological and behavioral characteristics may provide the best method for correct identification of a coregonid (Todd 1981). The difficulty in recognizing a particular coregonid is due to the ability of populations to quickly respond specifically to their unique environment and to adapt morphologically (e.g., shape and size) and biologically (e.g., spawning periods) (Hile 1937; Todd et al. 1981). The result is a tremendous variation within this genus. Recently, attempts to accurately identify a coregonid have been further confused due to the hybridization of rarer species (Smith 1968). Originally, populations were maintained by their tendency to spawn in specific areas and times (Koelz 1929; Smith 1964). Many coregonid populations were reduced as a result of human-related alterations of the Great Lakes ecosystem.

Great Lakes coregonids were formerly a dominant deepwater component of the four deeper lakes (Superior, Michigan, Huron, and Ontario) and less of a component in the relatively shallow Lake Erie. The chubs and ciscoes are very similar in appearance, all probably evolving from a recent ancestor related to the lake herring (Smith and Todd 1984). Until the last century, they managed to maintain distinct populations utilizing different niches and breeding strategies. Variations in size, spawning periods, and habitat (e.g., water depth) strengthened species isolation. However, the coregonid species were unable to withstand intense pressures from commercial fisheries, the immigration of sea lamprey (*Petromyzon marinus*) and other exotic fish into the Great Lakes basin, and a complicated process of changes in species' composition altering the entire balance of the Great Lakes ecosystem. Various forms of contamination also may have affected local populations.

The first major threat to the Great Lakes coregonids began in the mid-1800s, when a commercial fishery was formed in Lake Michigan. The larger coregonids were the initial major target during the early years of abundance, including the two largest species—the blackfin and deepwater. Following this successful market, the commercial fisheries subsequently turned to the larger species of Lakes Huron, Ontario, and Erie; by the late 1800s, Lake Superior stocks also were being exploited. The effects of harvesting became apparent by the early 1930s, when the blackfin and deepwater cisco were scarcely represented in index catches throughout their range (Smith 1964). As these large coregonids declined, pressure was directed toward the intermediate size species, including the shortjaw, shortnose, kiyi, and lake herring.

In the 1940s, when the blackfin and deepwater cisco were near the brink of extinction and the populations of the intermediate-sized species were stressed, an additional limiting factor was introduced—the sea lamprey. This parasitic fish entered the Great Lakes system shortly after the completion of the Welland Canal in 1825 and became established in the upper Great Lakes in the late 1940s (Applegate 1950). The sea lamprey encountered an unfilled niche with few predators and an abundant prey source (e.g., coregonids and lake trout [*Salvelinus namaycush*]) allowing it to expand and increase without apparent natural control. It was finally artificially controlled in the upper Great Lakes by the early1970s. However, it did affect the coregonids, particularly the larger species (Lawrie and Rahrer 1972). Due to sea lamprey predation, the lake trout commercial fishery disappeared. Even though the larger coregonids were scarce and most others were relatively uncommon, the Great Lakes commercial fisheries shifted their harvest in the late 1940s to the remaining stocks (Hile and Buettner 1955; Moffett 1957). By the 1960s, the blackfin and deepwater ciscoes were extinct, and the shortjaw and shortnose cisco populations were severely depleted, causing a collapse in the commercial fisheries. Exotic fish, primarily the abundant alewife (*Alosa pseudoharengus*) and rainbow smelt (*Osmerus mordax*), significantly influenced coregonids through competition for resources beginning in the 1950s in the upper Great Lakes (Berst and Spangler 1972; Christie 1972; Crowder and Crawford 1984).

The mass reduction in diversity and abundance of the coregonids by the mid-1950s had an unexpected effect on fish composition within the Great Lakes. The smallest coregonid, the bloater, increased rapidly and invaded the niches previously occupied by the other coregonids. The domination of closely related coregonids by the bloater (and lake herring) caused genetic swamping of the rarer chub species (Smith 1968). Today, the shortjaw and shortnose ciscoes are endangered. The kiyi survives in self-sustaining populations in western Lake Superior, while the bloater remains abundant in Lakes Superior, Michigan, and Huron. The remaining two *Coregonus*, the lake whitefish and lake herring, are still found throughout the Great Lakes (as well as many inland lakes), although the latter is rare or declining except in Lake Superior.

In summary, three of the ten Great Lakes coregonids are now extinct, two species are endangered, and one species has been extirpated in three of the four Great Lakes that it occupied. The relative number of coregonid species that are extinct in each of the Great Lakes is as follows: Lake Ontario, one-half; Lake Erie, one-third; Lake Huron, one-half; and Lake Michigan, one-half. Lake Superior still retains its full complement of five coregonid species, although at depressed levels. Identification of limiting factors and the management and protection of remaining coregonid populations are paramount to the natural stability of the Great Lakes ecosystem and need to be addressed immediately.

LITERATURE CITED

Applegate, V. C. 1950. Natural history of the sea lamprey (*Petromyzon marinus*) in Michigan. *U.S. Fish Wildl. Serv., Spec. Sci. Rept. Fish.* No. 55.

Bailey, R. M., and G. R. Smith. 1981. Origin and geography of the fish fauna of the Laurentian Great Lakes Basin. *Can. J. Fish. Aquat. Sci.* 38:1539–61.

Berst, A. H., and G. R. Spangler. 1972. Lake Huron: Effect of exploitation, introductions and eutrophication on the salmonid community. *J. Fish. Res. Board Can.* 29:877–87.

Christie, W. J. 1972. Lake Ontario: Effects of exploitation, introductions and eutrophication on the salmonid community. *J. Fish. Res. Board Can.* 29:975–83.

Crowder, L. B., and H. L. Crawford. 1984. Ecological shifts in resource use by bloaters in Lake Michigan. *Trans. Am. Fish. Soc.* 113:694–700.

Hile, R. 1937. Morphometry of the cisco, *Leucichthys artedi* (LeSueur) in the lakes of northeastern highlands, Wisconsin. *Lepzig* 37:57–130.

Hile, R., and H. J. Buettner. 1955. Commercial fishery for chubs (ciscoes) in Lake Michigan through 1953. *U.S. Fish Wildl. Serv., Spec. Sci. Rept. Fish.* No. 163.

Koelz, W. 1929. Coregonid fishes of the Great Lakes. *Bull. U.S. Bur. Fish.* 43:297–643.

Koelz, W. 1931. The coregonid fishes of the northeastern America. *Pap. Mich. Acad. Sci., Arts, Ltrs.* 8:303–432.

Lawrie, A. H., and J. F. Rahrer. 1972. Lake Superior: Effects of exploitation and introductions on the salmonid community. *J. Fish Res. Board Can.* 29:765–76.

Moffett, J. W. 1957. Recent changes in the deep-water fish populations of Lake Michigan. *Trans. Am. Fish. Soc.* 86:393–408.

Norden, C. R. 1961. Comparative osteology of representative salmonid fishes, with particular reference to the grayling (*Thymallus arcticus*) and its phylogeny. *J. Fish. Res. Board Can.* 18:679–791.

Smith, G. R., and T. N. Todd. 1984. Evolution of species flocks of fishes in north temperate lakes. Pp. 45–68 *in* A. A. Echelle and I. Kornfield (eds.), *Evolution of fish species flocks*. Univ. Maine Orono Press, Orono.

Smith, S. H. 1964. Status of the deepwater cisco population of Lake Michigan. *Trans. Am. Fish. Soc.* 69:63–84.

Smith, S. H. 1968. Species succession and fishery exploitation in the Great Lakes. *J. Fish. Res. Board Can.* 25:667–93.

Todd, T. N. 1981. Allelic variability in species and stocks of Lake Superior ciscoes (Coregoninae). *Can. J. Fish. Aquat. Sci.* 38:1808–13.

Todd, T. N., G. R. Smith, and L. E. Cable. 1981. Environmental and genetic contributions to morphological differentiation in ciscoes (Coregoninae) of the Great Lakes. *Can. J. Fish. Aquat. Sci.* 38:59–67.

Lake Herring

- *Coregonus artedii* LeSueur
- **State Threatened**

Map 47. Confirmed individuals. Historically widespread in Great Lakes and large inland lakes (Michigan DNR); ● = 1989 inland lake population and occurs in all Great Lakes (Michigan DNR).

Status Rationale: Because of its extensive distribution, the lake herring has been spared the nearly complete decimation experienced by most of its Great Lakes relatives. Throughout much of the southern portion of its range, however, populations have been depleted severely. In Michigan's Great Lakes, it is now abundant only in Lake Superior.

In-hand Identification: The lake herring is a silvery, streamlined fish that ranges from 8 to 15 inches (20.3 to 38.1 cm) in length. As with all members of the salmon family (Salmonidae), the lake herring has an adipose fin and an axillary process at the base of its pelvic fin. This species varies considerably in color and form, among isolated populations and within populations, thus causing taxonomic complications. The variability is expressed in the number of gill rakers, which may range from 38 to 64. They number 41 to 54 in Lake Ontario, 38 to 53 in Lake Erie, 40 to 53 in Lake Huron, 38 to 53 in Lake Superior, and 38 to 64 in many inland lakes (Scott and Crossman 1973). The number of gill rakers is generally the most reliable method for identifying many coregonid species. Genetic recognition may soon be possible (Todd 1981).

Six of the Great Lakes coregonids (also known as chubs or deepwater ciscoes) closely resemble the lake herring. However, species identification can be determined through knowledge of their current distribution, average size, breeding period, water depth preferences, and number of gill rakers. One other coregonid, the lake whitefish (*Coregonus clupeaformis*), differs by its larger size (up to 20 inches [50.8 cm]) and blunt, overhanging snout (the lake herring has a pointed, terminal snout).

Total Range and Taxonomic Status: This coregonid has a widespread distribution in north-central North America, ranging from the Great Lakes basin north to Hudson Bay and the Mackenzie River drainage in Canada's Northwest Territories. Unlike most other members of the genus *Coregonus*, the lake herring is still common over much of its range. It is part of a taxonomically perplexing genus that is recognized by some authorities as containing nine species and 31 subspecies (24 are considered lake herring subspecies) (Koelz 1929 and 1931). Hubbs and Lagler (1964) recognize 22 lake herring subspecies. However, the variability and plasticity of the coregonids is so great that several taxa may be unrecognized or combined. It is not known if the variable characteristics are genetic and/or environmentally induced.

State History and Distribution: The lake herring was an abundant species prior to unregulated harvest in the early and mid-1900s. It inhabited all five Great Lakes and, unlike other coregonids, this species also resided in Michigan's deeper inland lakes. Except for the large fisheries in Lake Erie (Van Oosten 1930), the lake herring was usually disregarded by commercial fisherman for the larger coregonids prior to the mid-1900s. In the mid-1950s, commercial persecution of the lake herring intensified due to the sharp decline in numbers of lake trout (*Salvelinus namaycush*) and lake whitefish (*Coregonus clupeaformis*) caused by fishing pressures and parasitism by the sea lamprey (*Petromyzon marinus*) (Moffett 1957).

Today, many of the shallow water stocks have been severely depleted in the four lower Great Lakes. Each of the Great Lakes still maintains a self-sustaining population; however, lake herring populations are low in Lake Superior and are declining in Lake Huron. It is commercially extinct in Lakes Erie, Michigan, and Ontario. Historic and current spawning grounds are well known for the Great Lakes (Goodyear et al. 1982). Along Michigan's Lake Superior shoreline, spawning grounds are known in five locations: Isle Royale, Ontonogan County, Keweenaw Peninsula, Marquette County, and Whitefish Bay. Spawning occurs in the lower St. Marys River (e.g., Lake Nicolet) and on Lake Michigan in Big Bay de Noc, Grand Traverse Bay, and other scattered locales. In Lake Huron, historical spawning areas are known along much of Michigan's shoreline; Saginaw Bay formerly served as an important spawning and nursery area until the early 1950s. Isolated shoals still contain breeding populations but are now locally distributed. Populations in Lake St. Clair and western Lake Erie were extirpated by the early 1900s.

Habitat: In Michigan's Great Lakes, the pelagic lake herring resides in both shallow and deep waters. Spawning activity may occur in shallow waters of 3 feet (0.9 m) (Smith 1956) to depths of over 200 feet (61 m) (Dryer and Beil 1964). Throughout the year, Lake Superior lake herring were found to inhabit depths ranging from

60 to 174 feet (18 to 53 m) (Dryer 1966). Inland lake populations typically occur in lakes that are relatively deep.

Ecology and Life History: Like many coregonids, the lake herring forms large aggregations between late fall and early winter to spawn. Egg laying is primarily temperature dependent. Optimum water temperatures during the spawning period range from 36° to 46°F (2° to 8°C) (Colby and Brooke in Lindsey and Woods 1970), although females delayed from spawning will use slightly warmer waters (John 1956). The spawning substrate is variable; in deep water, eggs are distributed in the open water. Western Lake Superior populations are bottom spawners, while eastern populations are deep-water spawners (Goodyear et al. 1982).

Thousands of eggs are deposited and abandoned. The number of eggs is directly proportional to the size of the female. Since the eggs are released in cold water, development is slow. In experimental conditions, the optimal incubation temperature is 42°F (Colby and Brooke in Lindsey and Woods 1970). John and Hasler (1956) found that the young hatched only after ice breakup in the spring. After one year, young lake herring are approximately 6 inches (15 cm) long (Scott and Crossman 1973). In three to four years this species usually attains maturity and typically measures nearly 12 inches (30 cm). Lake herring one to three years old are easily aged using the scale method (Van Oosten 1929). Lake herring feed primarily on plankton, small crustaceans, and larval aquatic insects (Langford 1938; Dryer and Beil 1964).

Conservation/Management: Commercial exploitation is the primary reason for the decline of the lake herring. Populations are low in four of the five Great Lakes, but, with proper management, this downward trend could be reversed. Competition for food with the recently introduced American smelt (*Osmerus mordax*) may also hamper lake herring recovery, although Selgeby et al. (1978) have shown predation by smelt on young lake herring does not significantly suppress healthy populations. Ideally, management of lake herring populations in the lower Great Lakes and the establishment of a refuge system in Lake Superior should be implemented to provide the greatest recovery in the shortest amount of time. By monitoring and controlling the commercial harvest in Lake Superior, restricting fishing of depleted stocks, and conducting studies to determine recruitment and interspecific relationships, the lake herring can return to its former Michigan abundance.

LITERATURE CITED

Dryer, W. R. 1966. Bathymetric distribution of fish in the Apostle Islands region, Lake Superior. *Trans. Am. Fish. Soc.* 95:248–59.

Dryer, W. R., and J. Beil. 1964. Life history of lake herring in Lake Superior. *U. S. Fish Wildl. Serv. Fish. Bull.* 63:248–59.

Goodyear, C. D., T. A. Edsall, D. M. Ormsby Dempsey, G. D. Moss, P. E. Polanski. 1982. Atlas of the spawning and nursery areas of Great Lake fishes (vol. 1–9). *U.S. Fish Wildl. Serv.*, FWS/OBS-82/52.

Hubbs, C. L., and K. F. Lagler. 1964. *Fishes of the Great Lakes region*. Univ. Mich. Press, Ann Arbor.

John, K. R. 1956. Onset of spawning activities of the shallow water cisco, *Leucichthys artedi* (sic) (LeSueur), in Lake Mendota, Wisconsin, relative to water temperatures. *Copeia* 1956:116–18.

John, K. R., and A. D. Hasler. 1956. Observations on some factors affecting the hatching of eggs and the survival of young shallowwater cisco, *Leucichthys artedi* LeSueur, in Lake Mendota, Wisconsin. *Limnol. Oceanogr.* 1:176–94.

Koelz, W. 1929. Coregonid fishes of the Great Lakes. *Bull. U.S. Bur. Fish.* 43:297–643.

Koelz, W. 1931. The coregonid fishes of the northeastern America. *Pap. Mich. Acad. Sci., Arts, Ltrs.* 8:303–432.

Langford, R. R. 1938. The food of the Lake Nipissing cisco *Leucichthys artedi* (LeSueur), with special reference to the utilization of the limnetic crustacea. *Publ. Ont. Fish. Res. Lab.* 57:143–90.

Lindsey, C. C., and C. S. Woods (eds.). 1970. *Biology of coregonid fishes*. Univ. Manitoba Press, Winnipeg.

Moffett, J. W. 1957. Recent changes in the deep-water fish populations of Lake Michigan. *Trans. Am. Fish. Soc.* 86:393–408.

Scott, W. B., and E. J. Crossman. 1973. *Freshwater fishes of Canada*. Fish. Res. Board Can., Bull. 184.

Selgeby, S. H., W. R. MacCallum, and D. V. Swedberg. 1978. Predation by rainbow smelt (*Osmerus mordax*) on lake herring (*Coregonus artedii*) in western Lake Superior. *J. Fish. Res. Board Can.* 35:1457–63.

Smith, S. H. 1956. Life history of lake herring of Green Bay, Lake Michigan. *U.S. Fish Wildl. Serv. Fish. Bull.* No. 109, 57:87–138.

Todd, T. N. 1981. Allelic variability in species and stocks of Lake Superior ciscoes (Coregoninae). *Can. J. Fish. Aquat. Sci.* 38:1808–13.

Van Oosten, J. 1929. Life history of the lake herring (*Leucichthys artedi* LeSueur) of Lake Huron as revealed by its scales, with a critique of the scale method. *Bull. U.S. Bur. Fish.* 44:265–428.

Van Oosten, J. 1930. The disappearance of the Lake Erie cisco—a preliminary report. *Trans. Am. Fish. Soc.* 60:204–14.

Shortnose Cisco

· *Coregonus reighardi* (Koelz)

· **State Endangered**

Formerly widespread in Lakes Huron, Michigan, and Ontario, the endemic shortnose cisco is currently restricted to a small but self-sustaining population in Canada's Georgian Bay. Until recently, the shortnose cisco was thought to occur in Lakes Superior and Nipigon. These northern populations have since been documented as the similar shortjaw cisco (*Coregonus zenithicus*) (Todd and Smith 1980). The shortnose cisco was a valuable commercial species, but its population could not withstand commercial fishing pressure and was quickly extirpated in Lake Ontario (by 1964) and in Lake Michigan (by 1974). The last Lake Michigan (1974) occurrence was near Grand Haven in Ottawa County. In Georgian Bay, the northeastern part of Lake Huron, the shortnose cisco's depleted and declining population precariously persists (Parker 1988). The last known Michigan (and U.S.) specimen was captured in 1982.

This pelagic species is silvery with a bluish cast on its upper body and averages a length of around 10 inches (25.4 cm). The shortnose cisco has 30 to 43 gill rakers (Scott and Crossman 1973). Unlike the other Great Lakes ciscoes, this species spawns in the spring, from May to early June, in waters of 12 to 470 feet (3.7 to 143 m) deep with a fine substrate (e.g., sand and clay) and ambient temperatures at approximately 39° to 40°F (Jobes 1943). Pritchard (1931) found shortnose ciscoes in western Lake Ontario to spawn between April and mid-May in approximately 250 feet (76 m) of water. The shortnose cisco prefers small crustaceans (e.g., *Mysis relicta*), but also may eat other invertebrates (Pritchard 1931).

LITERATURE CITED

Jobes, F. W. 1943. The age, growth, and bathymetric distribution of Reighard's chub, *Leucich-thys reighardi* Koelz, in Lake Michigan. *Trans. Am. Fish. Soc.* 72:108–35.

Parker, B. J. 1988. Status of the shortnose cisco, *Coregonus reighardi*, in Canada. *Can. Field-Nat.* 102:92–96.

Pritchard, A. L. 1931. Taxonomic and life history studies of the ciscoes of Lake Ontario. *Publ. Ont. Fish. Res. Lab.* No. 41.

Scott, W. B., and E. H. Crossman. 1973. *Freshwater fishes of Canada.* Fish. Res. Board Can., Bull. 184.

Todd, T. N., and G. R. Smith. 1980. Differentiation in *Coregonus zenithicus* in Lake Superior. *Can. J. Fish. Aquat. Sci.* 37:2228–35.

Shortjaw Cisco

- *Coregonus zenithicus*
  (Jordan and Everman)
- State Endangered

The shortjaw cisco formerly occurred in abundance throughout the three largest Great Lakes. Due to intensive commercial fishing and exotic fish introduction, this species was commercially extinct by 1950; it completely disappeared from Lake Michigan in 1975 and from Lake Huron in 1982. This most recent occurrence was at Au Sable Point in Iosco County (Houston 1988).

Today, the shortjaw cisco only occurs in Lake Superior, Lake Nipigon, and in a few scattered lakes north and west of the Great Lakes to Great Slave Lake in the Northwest Territories, Canada. Lake Superior populations of the blackfin cisco (*C. nigripinnis cyanopterus*), shortnose cisco (*C. reighardi dymondi*), and longjaw cisco (*C. alpenae*) are now considered to be variations of this species (Todd and Smith 1980; Todd and Smith 1981; Smith and Todd 1984).

This cisco, also known as the pale-back tullibee, is a silvery fish with a dark greenish back, averaging around 11 inches (28.0 cm) in length. In Lakes Superior and Nipigon, this cisco has 32 to 46 gill rakers, and between 34 and 44 gill rakers from its former residence in Lakes Huron and Michigan (Scott and Crossman 1973). The shortjaw cisco is a pelagic species and inhabits the deeper waters of large lakes (typically 200 to 400 feet [61 to 122 m]). During the spawning season (between late November and early December), this cisco may be found over a clay substrate at a depth of 120 to 240 feet (37 to 73 m) (Koelz 1929; Van Oosten 1937). Primary food items include small crustaceans (e.g., *Mysis relicta* and *Pontoporeia hoyi*) and occasionally aquatic insect larvae (Scott and Crossman 1973).

LITERATURE CITED

Houston, J. J. 1988. Status of the shortjaw cisco, *Coregonus zenithicus*, in Canada. *Can. Field-Nat.* 102:97–102.

Koelz, W. 1929. Coregonid fishes of the Great Lakes. *Bull. U.S. Bur. Fish.* 43:297–643.

Scott, W. B., and E. H. Crossman. 1973. *Freshwater fishes of Canada*. Fish. Res. Board Can., Bull. 184.

Smith, G. R., and T. N. Todd. 1984. Evolution of species flocks of fishes in north temperate lakes. Pp. 45–68 *in* A. A. Echelle and I. Kornfield (eds.), *Evolution of fish species flocks*. Univ. Maine Orono Press, Orono.

Todd, T. N., and G. R. Smith. 1980. Differentiation in *Coregonus zenithicus* in Lake Superior. *Can. J. Fish. Aquat. Sci.* 37:2228–35.

Todd, T. N., and G. R. Smith. 1981. Environmental and genetic contributions to morphological differentiation in ciscoes (Coregoninae) of the Great Lakes. *Can. J. Fish. Aquat. Sci.* 38:59–67.

Van Oosten, J. 1937. The age, growth, and sex ratio of the Lake Superior longjaw, *Leucichthys zenithicus* (Jordan and Everman). *Pap. Mich. Acad. Sci., Arts, Ltrs.* 22:691–711.

Longjaw Cisco

· *Coregonus alpenae* (Koelz)

· **Extinct**

Due to overexploitation from commercial fisheries and competition with exotic fish, the longjaw cisco is now extinct. This historically common species was last recorded in Lake Erie in 1957, in Lake Michigan's Grand Traverse Bay in 1967, and in Lake Huron's Georgian Bay in 1975 (Campbell 1987). The small Lake Erie population was discovered after the species had, more than likely, become locally extinct (Scott and Smith 1962). Scattered populations also were affected by sea lamprey (*Petromyzon marinus*) and hybridization with other ciscoes.

This silvery fish averaged a length of 10 to 11 inches (25 to 28 cm), with a gill raker range of 30 to 39 in Lake Erie, 31 to 44 in Lake Huron, and 33 to 46 in Lake Michigan (Scott and Crossman 1973). Before its extinction, longjaw ciscoes apparently spawned in November (Jobes 1949). This deepwater species typically inhabited depths of approximately 200 to 350 feet (61 to 107 m) and moved into shallower waters during the spawning period. Koelz (1929) considered the species to spawn in depths of 60 to 150 feet (18 to 46 m). Small crustaceans (e.g., *Mysis relicta*) were the preferred food of the longjaw cisco (Koelz 1929; Bersamin 1958).

LITERATURE CITED

Bersamin, S. V. 1958. A preliminary study of the nutritional ecology and food habits of the chubs (*Leucichthys* spp.) and their relation to the ecology of Lake Michigan. *Pap. Mich. Acad. Sci., Arts, Ltrs.* 43:107–18.

Campbell, R. R. 1987. Status of the Longjaw Cisco, *Coregonus alpenae*, in Canada. *Can. Field-Nat.* 101:241–44.

Jobes, F. W. 1949. The age, growth, and distribution of the longjaw cisco, *Leucichthys alpenae* (Koelz) in Lake Michigan. *Trans. Am. Fish. Soc.* 76:215–47.

Koelz, W. 1929. Coregonid fishes of the Great Lakes. *Bull. U. S. Bur. Fish.* 43:297–643.

Scott, W. B., and E. H. Crossman. 1973. *Freshwater fishes of Canada.* Fish. Res. Board Can., Bull. 184.

Scott, W. B., and S. H. Smith. 1962. The occurrence of the longjaw cisco, *Leucichthys alpenae*, in Lake Erie. *J. Fish. Res. Board Can.* 19:1013–23.

Blackfin Cisco

· *Coregonus nigripinnis* (Gill)

· **Extinct**

The large, endemic blackfin cisco formerly occupied Lakes Huron and Michigan and is now believed to be extinct. Formerly confused with other coregonids, the blackfin cisco was thought to have a much larger distribution. Todd and Smith (1980) showed that the alleged blackfin cisco populations in Lake Superior and Nipigon were actually a form of large shortjaw ciscoes (*Coregonus zenithicus*). Dark-finned lake herring (*Coregonus artedii*) of Lake Superior and northern Canada also were originally and incorrectly identified as blackfin ciscoes. The geographic narrowing of the blackfin cisco's range to Lakes Huron and Michigan changed its status considerably, from possibly "holding on" to probably extinct. By 1907, this chub approached extinction because of its inability to recover from the early overfishing and later because of the parasitic sea lamprey (*Petromyzon marinus*) (Moffett 1957). The last known blackfin cisco from Lake Huron was in 1923 and from Lake Michigan in 1969 at Marinette, Wisconsin.

This species, also known as the black-backed tullibee, mooneye cisco, or bluefin, is silvery, like all other ciscoes, but its back is heavily pigmented. This deep-bodied chub averaged around 13 inches (33 cm) and was known to reach 20 inches (51 cm) in length. The number of gill rakers varied from 41 to 52 in Lake Michigan and 40 to 52 in Lake Huron (Scott and Crossman 1973). The blackfin cisco formerly gathered in relatively shallow spawning areas between late fall and midwinter, and later shifted to the deepest regions of the two Great Lakes. The blackfin cisco was primarily dependent on small crustaceans (*Mysis relicta*) (Koelz 1929).

LITERATURE CITED

Koelz, W. 1929. Coregonid fishes of the Great Lakes. *Bull. U.S. Bur. Fish.* 43:297–643.
Moffett, J. W. 1957. Recent changes in the deepwater fish populations of Lake Michigan. *Trans. Amer. Fish. Soc.* 86:393–408.
Scott, W. B., and E. H. Crossman. 1973. *Freshwater fishes of Canada.* Fish. Res. Board Can., Bull. 184.
Todd, T. N., and G. R. Smith. 1980. Differentiation in *Coregonus zenithicus* in Lake Superior. *Can. J. Fish. Aquat. Sci.* 37:2228–35.

Deepwater Cisco

- *Coregonus johannae* (Wagner)
- **Extinct**

One of the largest ciscoes, the deepwater cisco was the first species to be intensively pressured by commercial fisheries and, as a result, was the first to become extinct. Already existing in stressed populations in the mid-1900s, it declined further due to sea lamprey (*Petromyzon marinus*) parasitism beginning in the 1940s. Formerly occurring in the deepest portions of Lakes Huron and Michigan, typically 300 to 500 feet (91 to 152 m), the deepwater cisco was last confirmed in Lake Michigan's Grand Traverse Bay in 1951 (Moffett 1957) and Ontario's Lake Huron in 1952.

This silvery cisco averaged a length of around 11 to 12 inches (28 to 30 cm), with a gill raker range of 25 to 35 in Lake Huron and 26 to 36 in Lake Michigan (Scott and Crossman 1973). Its life history is not well known because it was quickly exterminated. It spawned in August and September in Lakes Huron and Michigan. Unlike the other, smaller ciscoes, the deepwater cisco may have spawned every other year (Koelz 1929). Spawning grounds were never located, but it probably frequented the shallower waters. The diet consisted mainly of small crustaceans (e.g., *Mysis relicta*) (Koelz 1929).

LITERATURE CITED

Koelz, W. 1929. Coregonid fishes of the Great Lakes. *Bull. U.S. Bur. Fish.* 43:297–643.
Moffett, J. W. 1957. Recent changes in the deepwater fish populations of Lake Michigan. *Trans. Am. Fish. Soc.* 86:393–408.
Scott, W. B., and E. H. Crossman. 1973. *Freshwater fishes of Canada.* Fish. Res. Board Can., Bull. 184.

Paddlefish

- *Polyodon spathula* (Walbaum)
- **State Extirpated**

The ancient and unique paddlefish was formerly abundant throughout much of the Mississippi River valley extending west along the Missouri River to Montana and east through Lake Erie and into New York. Its original distribution in the Great Lakes basin is not fully known, and it may have reached some areas through canals (Hubbs and Lagler 1964). Confirmed records exist for Lakes Erie and Huron. Since the turn of the century, paddlefish populations have declined dramatically due to overharvesting and habitat destruction. Today, this species has retreated from the extreme parts of its range; it is currently surviving in scattered populations in the larger rivers of the United States (e.g., Mississippi and Missouri). The last Great Lakes report of a paddlefish was more than 70 years ago.

The paddlefish inhabits open, large bodies of water, particularly reservoirs and slow-flowing rivers. Summer backwaters of large rivers are used extensively by paddlefish as feeding areas (Southall and Hubert 1984). Spawning is the most critical period of this species' life cycle, during which time it requires an accessible shallow site consisting of extensive, silt-free gravel bars (Pflieger 1975), preferably with running water (Stockard 1907).

This species and a paddlefish (*Psephurus gladius*) of the Yangtze valley in China are the only representatives of their family, Polyodontidae. The most distinctive feature of the paddlefish (also known as spoonbill) is the long, paddle-shaped snout. Although this snout may be used as a sensory organ for locating prey concentrations, its function is still not fully understood (Pflieger 1975).

This is one of the largest North American freshwater fish, occasionally attaining a length of 7 feet (2.1 m) and a weight of 160 pounds (73 kg) or more. In addition to these characteristics, the paddlefish has a heterocercal (sharklike) tail, a tough, nearly scaleless outer skin, an extremely large, toothless mouth, and numerous filamentous gill rakers for planktivorous feeding. Paddlefish are very mobile throughout the year and have been known to travel over 360 miles (nearly 600 km) in one year (Rosen and Hales 1982).

Spawning occurs on shallow gravel bars in spring. The eggs hatch if inundated by spring floods after nine days at an ambient temperature of 57°F (14°C) (Purkett 1961 and 1963). Under optimal conditions, young may reach an average length of over 28 inches (71 cm) in one year (Houser and Bross 1959). Female paddlefish are sexually mature in seven to eight years (Adams 1942), but may not breed every year (Houser and Bross 1959). Paddlefish feed by continuously swimming with the mouth open, filtering plankton, crustacea, and insect larvae from the water (Eddy and Simer 1929). This fish also will take small fish (Fitz 1966).

LITERATURE CITED

Adams, L. A. 1942. Age determination and rate of growth in *Polyodon spathula*, by means of growth rings of the otoliths and dentary bone. *Am. Midl. Nat.* 28:617–30.

Eddy, S., and P. H. Simer. 1929. Notes on the food of the paddlefish and the plankton of its habitat. *Trans. Ill. State Acad. Sci.* 21:59–68.

Fitz, R. B. 1966. Unusual food of a paddlefish (*Polyodon spathula*) in Tennessee. *Copeia* 1966:356.

Houser, A., and M. G. Bross. 1959. Observations on growth and reproduction of the paddlefish. *Trans. Am. Fish. Soc.* 88:50–52.

Hubbs, C. L., and K. F. Lagler. 1964. *Fishes of the Great Lakes region.* Univ. Mich. Press, Ann Arbor.

Pflieger, W. L. 1975. *The fishes of Missouri.* Missouri Dept. Conserv., Jefferson City, MO.

Purkett, C. A. 1961. Reproduction and early development of the paddlefish. *Trans. Am. Fish. Soc.* 90:125–29.

Purkett, C. A. 1963. The paddlefish fishery of the Osage River and the Lake of the Ozarks, Missouri. *Trans. Am. Fish. Soc.* 92:239–44.

Rosen, R. A., and D. C. Hales. 1982. Biology and exploitation of paddlefish in the Missouri River below Gavins Point Dam. *Trans. Am. Fish. Soc.* 11:216–22.

Southall, P. D., and W. A. Hubert. 1984. Habitat use by adult paddlefish in the upper Mississippi River. *Trans. Am. Fish. Soc.* 113:125–31.

Stockard, C. R. 1907. Observations on the natural history of Polyodon spathula. *Am. Nat.* 41:753–66.

Arctic Grayling

· *Thymallus arcticus* (Richardson)
· State Extirpated

This holarctic salmonid occurs in North America across much of north and central Canada, west of Hudson Bay. Disjunct native populations once occurred in Michigan and Montana. Michigan's population is now extinct, but Montana populations survived and several introductions throughout the northern Rocky Mountains have been made. The four distinct historical populations were once considered separate species, including *Thymallus tricolor* in Michigan, but are now lumped into a single species (Walters 1955).

The historical Michigan population was isolated from the primary contiguous range by nearly 1,000 miles (approximately 1,600 km). In the Lower Peninsula, the grayling occurred in the Lake Michigan drainages from the Jordan to the Muskegon river, and in Lake Huron drainages from the Cheboygan to the Rifle river; in the Upper Peninsula, populations were known in the Otter and Little Carp rivers in the Lake Superior drainage (Taylor 1954; Hubbs and Lagler 1974).

Favored habitats were cool, clear streams and rivers, particularly in areas with swift waters. Its abundance in these rivers and habitats was staggering. Mather (in Mershon 1923) commented in the late 1870s of the slaughter by two large camps on the Au Sable River: "... killed five thousand fish, not going beyond five miles of the mouth of the north branch."

By the turn of the century, deforestation caused streams to warm and become silty. This habitat degradation, coupled with river log drives, severely depleted suitable spawning areas. Anglers and illegal market fishing also served to stress and extirpate populations. However, the introduction of exotic salmonids, such as brown and rainbow trout (*Salmo trutta* and *S. gairdnerii*, respectively) probably was most responsible for the extinction of Michigan's grayling population. The grayling was last known from the Otter River in 1936 (Creaser and Creaser 1935; Hubbs and Lagler 1974).

Since the early 1900s, reintroductions in various Michigan lakes were unsuccessful. In the late 1980s, further attempts were made—releases of tens of thousands of grayling in streams, rivers, and lakes. Eggs were hatched in the Michigan Department of Natural Resources Wolf Lake Fish Hatchery and at Lake Superior State University in Sault Ste. Marie and reared to yearlings before release.

Fish plants in lakes included Alcona and Crawford counties in the northern Lower Peninsula, and Alger and Luce counties in the Upper Peninsula. In rivers, grayling were released in the Manistee, Au Sable, Cedar, and others. These transplants have been unsuccessful.

The grayling averages a length of 12 to 15 inches (30 to 38 cm). Its most characteristic feature is its saillike dorsal fin and blue-black coloring with a pink-ish iridescence. It spawns soon after ice-off in spring. No nest is prepared; the female lays several thousand eggs. Young feed on zooplankton and adults on a variety of aquatic and terrestrial invertebrates.

LITERATURE CITED

Creaser, C., and E. Creaser. 1935. The grayling of Michigan. *Pap. Mich. Acad. Sci., Arts Ltrs.* 29:599–611.

Hubbs, C. L., and K. F. Lagler. 1974. *Fishes of the Great Lakes region.* Univ. Mich. Press, Ann Arbor.

Mershon, W. B. 1923. *Recollections of my fifty years hunting and fishing.* Stratford Co., Boston.

Taylor, W. R. 1954. Records of fishes in the John N. Lowe collection from the Upper Peninsula of Michigan. *Univ. Mich. Mus. Zool. Misc. Publ.* No. 87.

Walters, V. 1955. Fishes of the western Arctic American and eastern Arctic Siberia: Taxonomy and zoogeography. *Bull. Am. Mus. Nat. Hist.* No. 106, art. 5:259–368.

Blue Pike

· *Stizostedion vitreum gluacum* (Hubbs)

· Extinct

Initially described as a distinct species (Hubbs 1926), the documented intergrades with the yellow walleye (*S. v. vitreum*) led to its subspecific status. The blue pike differed from the yellow walleye in color as well as size, growth rate, spawning time and location, and habitat (Campbell 1987).

Found only in Lakes Erie and Ontario and the lower Niagara River, it was relatively common and an important component of the commercial fish industry. Unregulated overexploitation was directly responsible for the decline of this subspecies. Habitat degradation by pollution, eutrophication, competition with such exotic fish as rainbow smelt (*Osmerus mordax*) (Reiger et al. 1969), and eventually genetic swamping by the yellow walleye contributed to its extinction.

This medium-sized fish had a blueish cast to the back and rarely exceeded lengths of 13.4 inches (34 cm) or weights of 700 g. It preferred deep, cool waters, except in May while spawning in shallow areas over gravel substrate. Tens of thousands of eggs are laid. At two to three years, males were sexually mature, females averaged one year later (Deason 1933).

It was abundant until the 1950s in Lake Erie and was less common in Lake Ontario. The last confirmed individual was from 1965 in Lake Erie (Campbell 1987). Few populations ranged into western Lake Erie and it is questionable if the blue pike occurred in Michigan waters.

LITERATURE CITED

Campbell, R. R. 1987. Status of the Blue Walleye, *Stizostedion vitreum glaucum*, in Canada. *Can. Field-Nat.* 101:245–52.
Deason, H. J. 1933. Preliminary report on the growth rate, dominance and maturity of the pike-perches (Stizostedion) of Lake Erie. *Trans. Am. Fish. Soc.* 63:348–60.
Hubbs, C. L. 1926. A checklist of the fishes of the Great Lakes and tributary waters with nomenclature notes and analytical keys. *Univ. Mich. Mus. Zool. Misc. Publ.* 15.
Reiger, N. A., V. C. Applegate, and R. A. Ryder. 1969. The ecology and management of the walleye in western Lake Erie. *Great Lakes Fish. Comm. Tech. Rept.* 15.

Endangered and Threatened

Pipe vine swallowtail (*Battus philenor*)
Mitchell's satyr (*Neonympha mitchellii*)
Regal fritillary (*Speyeria idalia*)
Frosted elfin (*Incisalia irus*)
Northern blue (*Lycaeides idas nabokovi*)
Karner blue (*Lycaeides melissa samuelis*)
Ottoe skipper (*Hesperia ottoe*)
Powesheik skipper (*Oarisma powesheik*)
Dukes' skipper (*Euphyes dukesi*)
Dusted skipper (*Atrytonopsis hianna*)
Persius dusky wing (*Erynnis persius*)
Three-staff underwing (*Catocala amestris*)
Leadplant moth (*Schinia lucens*)
Phlox moth (*Schinia indiana*)
Silphium borer moth (*Papaipema silphii*)
Greyback (*Tachopteryx thoreyi*)
Lake Huron locust (*Trimerotropis huroniana*)
Great Plains spittlebug (*Lepyronia gibbosa*)
Hungerford's crawling water beetle (*Brychius hungerfordi*)
American burying beetle (*Nicrophorus americanus*)

An Introduction to Insects

Mogens C. Nielsen
Michigan State University

Michigan's Endangered and Threatened Species (ETS) program began in 1970 with the adoption of Act 210, that provided protection for federally endangered species. It did not include plants and invertebrates. In 1974, Act 203 broadened the program by including both of these, but, more important, it provided authorization for the Michigan Department of Natural Resources to carry out scientific investigations for the protection and enhancement of endangered and threatened plants and animals. Originally, no species of insects were state listed. Twelve years later, in 1986, 8 species of insects, 2 beetles, and 6 lepidoptera were listed as threatened in Michigan. Currently, 20 species are listed, 15 of which are lepidoptera.

The ETS program is now gaining greater public support for the inclusion of such invertebrate animals as beetles, butterflies, dragonflies, and grasshoppers. Insects and their roles as important constituents of the state's natural fauna are now recognized. Invertebrates, because of their numbers and the diversity of species may, in the long term, be the best barometer of the state's natural environment.

Before this century, the state's insect fauna was not actively investigated. Pest species, such as the coddling moth, armyworm, and various stored-grain insects that affected the agriculture and forest industry, were first studied at Michigan State University (MSU). Consequently, pesticides were developed and vigorously used into the 1960s, until research and public opinion halted the use of many, and severely limited the use of other chemicals on the land. Since that time, the use of pest-specific chemicals and the introduction of biological controls have dramatically altered the impact of chemicals used in fighting pest species. The recent use of "BT," *Bacillus thuringiensis*, in an attempt to control nuisance populations of the gypsy moth (*Lymantria dispar*) in Michigan, shows promise in pest management. Much research and field study still remains to be done as we question the effect of pest management on nontarget insect species, some of which are now threatened with extinction.

Since 1900, entomologists have published several local and state lists of insects. Some of these lists were compiled largely from data contained in private and institutional collections (e.g., the collections of Sherman Moore, John H. Newman, University of Michigan [UM], and MSU). However, no special efforts were made to collect in unique habitats or in previously uncollected areas of the state. These authors apparently made little effort to systematically survey a county or collect both early- and late-emerging insect species. Some of our scarce Cuculliinae moths (e.g., *Lithophane lepida* and *Platypolia anceps*) have recently been recorded for the first time in Michigan, mainly because of searches in April and October. In the early years, much of the interest in collecting insects, particularly Coleoptera and

329

Lepidoptera, came from collectors living in the southern Lower Peninsula. One notable exception was the geological expeditions to Isle Royale in the early 1900s (Adams 1909), which included brief accounts of insects collected on that remote northern island.

The Detroit Entomological Society (DES), organized in 1942, and forerunner to the Michigan Entomological Society (1954), helped promote interest in and the study of Michigan insects. Most of the collecting was done in the Lower Peninsula, especially in the southeastern counties. Sherman Moore, a charter DES member, published several papers on lepidoptera from northern areas, including Beaver Island. He later published an annotated list of moths (1955) and a revised annotated list of butterflies (1960). These are still in use and considered the most comprehensive lists of Michigan macrolepidoptera.

Between 1904 and 1920, the University of Michigan, on behalf of their Zoological Museum, sponsored biological expeditions to Alger, Berrien, Chippewa, and Schoolcraft counties to gather specimens, including invertebrates. Several local lists of insects were published as a result of these and other investigations (e.g., McAlpine 1918; Andrews 1929). In recent years, other specialists have published annotated lists of various orders and families of Michigan insects (Snider 1967; Cantrall 1968; Hanna 1970; McPherson 1970; Gosling 1973; O'Brien 1989). The late Robert R. Dreisbach of Midland attempted to publish a comprehensive list of all Michigan insects, comparable to that written by Leonard (1928) for the state of New York. Unfortunately, this project ended with his death in 1964. Along with the late John H. Newman, I collected lepidoptera in several huge sphagnum bogs and other boreal habitats in the Upper Peninsula—in areas previously overlooked by Moore. These efforts produced an addendum (Newman and Nielsen 1973) that added 154 new species to Moore's moth lists exlusive of "micromoths."An addendum to Moore's butterfly list (Nielsen 1970) brought the recorded Michigan butterfly fauna to 147 species.

Dr. Edward G. Voss has produced lists of lepidoptera found in the Douglas Lake Region, Emmet and Cheboygan counties (Voss 1954, 1969, 1981, 1983, and 1991). His efforts continue with his planned treatment of the micromoths. This unique baseline data will be of great value to future lepidopterists and ecologists in determining distribution shifts. It is hoped that future interest and funds can be directed to similar evaluations of insect fauna in other unique and sensitive areas of the state, such as coastal dunes, tallgrass prairies, and various wetland areas.

Overview of Listed Insects

Before European settlement, insect populations were probably stable with only minor changes due to wildfires and the development of local trading centers. As more land was cleared of its pine and hardwood forests, destructive fires raged across the land. This allowed recolonization by extirpated insect populations and the invasion of many new species. Dramatic vegetation changes occurred after fire suppression policies and reforestation efforts under the Civilian Conservation

Corps (CCC) program were initiated in the 1930s. These efforts had a remarkable impact on insect populations, allowing some species to increase and allowing other species from neighboring states to establish resident populations. It appears that the diversity of insect species reached its peak earlier in this century. Now, as more land is cultivated and as communities expand, insect populations and species will be pushed farther into smaller areas—some ultimately toward extinction.

Fire suppression has had a severe impact on several insects, such as the karner blue (*Lycaeides melissa samuelis*). Michigan once contained the largest population of this butterfly in the eastern United States. The butterfly's larval foodplant, lupine (*Lupinus perennis*), grows best in sandy oak openings and edges in the southern and western counties of the Lower Peninsula. Unfortunately, oak and other woody successional plants have grown and expanded to shade and eliminate lupine, causing the decline of this butterfly. Two other threatened lepidopteran species also utilize lupine as their only larval host plant in Michigan, the Persius dusky wing (*Erynnis persius*) and frosted elfin (*Incisalia irus*). The application of controlled burning in these oak barrens could greatly enhance the regeneration of the foodplant and thereby maintain and increase populations of the karner blue and its companion species.

Intensive use of our former tallgrass prairies in southwestern Michigan by agricultural and forest interests had a negative impact on many native insect species, especially Great Plains spittlebug (*Lepyronia gibbosa*), phlox moth (*Schinia indiana*), three-staff underwing (*Catocala amestris*), Ottoe's skipper (*Hesperia ottoe*), and regal fritillary (*Speyeria idalia*). In addition, the elimination of various wetland types, wooded seeps, sedge marshes, and fens in southern counties have further restricted the distribution of the greyback (*Tachopteryx thoreyi*), Powesheik's skipper (*Oarisma powesheik*), Dukes' skipper (*Euphyes dukesi*), and Mitchell's satyr (*Neonympha mitchellii*). Michigan's Wetland Protection Act now protects these sensitive wetlands and their inhabitants, especially threatened species. The Lake Huron locust (*Trimerotropis huroniana*) occurs primarily on Great Lakes coastal dunes and is now rarely seen because of declining undisturbed dunes habitat. The new dunes protection law should help to protect some of the remaining dunes environment for this and other species.

Several other lepidoptera are also threatened, such as the northern blue (*Lycaeides idas nabokovi*). This butterfly was discovered in 1979, and has been found only in Alger, Dickinson, and Marquette counties and on Isle Royale, Keweenaw County. The dwarf bilberry (*Vaccinium cespitosum*) recently was found to be the blue's larval foodplant (Nielsen and Ferge 1982); this low-lying shrub also is state threatened. Some of Michigan's other, rarer lepidoptera, such as the pipevine swallowtail (*Battus philenor*) and a boreal hawk-moth (*Proserpinus flavofasciata*) are actually peripheral in range; their distributions in North America extend only a short distance into Michigan's southern and northern counties, respectively. These species may develop breeding populations in years of favorable weather conditions.

It is nearly impossible to estimate the total number of insect species in the state. Certain insect orders (e.g., Coleoptera, Lepidoptera, Odonata) have been the focus of many amateur collectors over the years and estimates have been made of their diversity. The state's butterfly fauna currently has been recorded at 158 species, of which 135 species are known to be breeding residents. In contrast, the "macromoth" fauna is estimated at approximately 1,300 species with new species reported frequently. Despite the continuing loss of habitats in the more populated areas of the state, there are still many species of insects, including lepidoptera, to be discovered. The vastness of the Upper Peninsula also provides a haven for many unrecorded insect species, especially members of the "microlepidoptera" and the lesser known insect orders. The use of ultraviolet lights, (especially portable, battery-operated units) during the past 30 years to attract nocturnal species has greatly expanded the knowledge of our insect fauna.

Collecting activities by amateurs have been the main source for the distributional and biological knowledge of many of the state's insect species, especially those currently listed. Collecting insects as a recreational pastime by amateurs has not caused the demise of any species, but may become a limiting factor should habitat destruction continue, simply because shrinking populations become more vulnerable to all forms of mortality.

It is hoped that we can maintain the diversity of Michigan's insect fauna with the current land-use statutes enacted during the past several years, the ETS law, and the creation of new habitats and enhancement of others. The creation of more prairies and wetlands shows promise in promoting larger populations and restoring extirpated insects inhabiting these areas. Tallgrass prairies now being developed by Adrian College and the Fernwood and Seven Ponds Nature Centers show promise that new management techniques can create productive plant and animal communities. New control-burn practices and methods have helped to maintain and enhance existing prairies without eliminating such insect populations as the Ottoe's skipper and various *Papaipema* moth species.

The purchase of remnant prairies (and other habitats sensitive to human development) by the Michigan Nature Association (MNA) and the Nature Conservancy has saved many acres from development and helps ensure viable environments for many invertebrate species. The MNA now has acquired 6,000 acres (2,400 ha) statewide, including 117 sanctuaries inhabited by nearly 200 listed plant and animal species, including most of the listed insects. Unless larger areas of unique habitat, with connecting corridors to similar habitat, can be protected or purchased, many of our rarer species will be further stressed and pushed toward extinction.

The ETS program has now focused public attention on the need to protect Michigan's rare insect species; likewise, similar programs are now in place in many other states. If we are to succeed in protecting our unique insect fauna, we must place greater emphasis on field studies. This cannot be accomplished without more funding support and coordinated efforts to enlist the assistance of both amateur and professional researchers.

LITERATURE CITED

Adams, C. C. 1909. Annotations on certain Isle Royale invertebrates. Pp. 249–77 *in* An ecological survey of Isle Royale, Lake Superior, *Ann. Rept. Mich. Geol. Surv.* (1908), Lansing.

Andrews, A. W. 1929. List of some of the insects found at Huron Mountain (Marquette, Michigan). Pp. 116–52 *in* B. H. Christy (ed.), *The book of Huron Mountain*. Chicago.

Cantrall, I. J. 1968. An annotated list of the Dermaptera, Dictyoptera, Phasmatoptera and Orthoptera of Michigan. *Mich. Entom.* 1 (9): 299–345.

Gosling, D. C. L. 1973. An annotated list of the Cerambycidae of Michigan (Coleoptera), Part 1. *Great Lakes Entom.* 6 (3): 65–85.

Hanna, M. 1970. An annotated list of the spittlebugs of Michigan (Homoptera: Ceropidae). *Mich. Entom.* 3 (1): 2–16.

Leonard, M. D. (ed.). 1928. A list of the insects of New York, with a list of the spiders and certain other allied groups. *Cornell Univ. Agric. Exp. Sta. Mem.* No. 101.

McAlpine, W. S. 1918. A collection of lepidoptera from Whitefish Point, Michigan. *Univ. Mich. Mus. Zool. Occ. Pap.* No. 54.

McPherson, J. E. 1970. An annotated list of the Scutelleroidea of Michigan (Hemiptera). *Mich. Entom.* 3 (2): 34–63.

Moore, S. 1955. An annotated list of the moths of Michigan exclusive of Tineoidea (Lepidoptera). *Univ. Mich. Mus. Zool. Misc. Publ.* No. 88.

Moore, S. 1960. A revised annotated list of the butterflies of Michigan. *Univ. Mich. Mus. Zool. Occ. Pap.* No. 617.

Newman, J. H., and M. C. Nielsen. 1973. Moth species new to Michigan. *Great Lakes Entom.* 6 (2): 33–45.

Nielsen, M. C. 1970. New Michigan butterfly records. *J. Lepid. Soc.* 24:42–47.

Nielsen, M. C., and L. A. Ferge. 1982. Observations of *Lycaeides argyrognomon nabokovi* in the Great Lakes Region (Lycaenidae). *J. Lepid. Soc.* 36:233–34.

O'Brien, M. F. 1989. Distribution and biology of the Sphecinae wasps of Michigan (Hymenoptera: Sphecidae: Sphecinae). *Great Lakes Entom.* 22 (4): 199–217.

Snider, R. J. 1967. An annotated list of the Collembola (springtails) of Michigan. *Mich. Entom.* 1 (6): 179–234.

Voss, E. G. 1954. The butterflies of Emmet and Cheboygan counties, Michigan, with notes on northern Michigan butterflies. *Am. Midl. Nat.* 51:87–104.

Voss, E. G. 1969. Moths of the Douglas Lake region (Emmet and Cheboygan counties), Michigan: I. Sphingidae-Ctenuchidae (Lepidoptera). *Mich. Entom.* 2 (3–4): 48–54.

Voss, E. G. 1981. Notes of the Douglas Lake region (Emmet and Cheboygan counties), Michigan. Part 2. Noctuidae (Lepidoptera). *Great Lakes Entom.* 14 (2): 87–101.

Voss, E. G. 1983. Moths of the Douglas Lake region (Emmet and Cheboygan counties), Michigan. Part 3. Thyatiridae, Drepanidae, Lasiocampidae, Notodontidae, Lymantridae (Lepidoptera). *Great Lakes Entom.* 16 (4): 131–37.

Voss, E. G. 1991. Moths of the Douglas Lake Region (Emmet and Cheboygan counties), Michigan. Part 4. Geometridae (Lepidoptera). *Great Lakes Entom.* 24 (6): 187–201.

Pipe Vine Swallowtail

· **Battus philenor** (Linnaeus)
· **State Threatened**

Map 48. Confirmed individuals. ▲ = before 1980 and
● = 1980 to 1989 (Michigan DNR; Michigan State University).

Status Rationale: Restricted primarily to the southern three tiers of counties, this species is known from only a few sites in recent years. The larval host plant in Michigan, Virginia snakeroot (*Aristolochia serpentaria*), is also quite rare. It is listed as state threatened and has been reported from only a dozen locations in the past 25 years.

Field Identification: The pipe vine swallowtail (Lepidoptera: Papilionidae) is a large (2.75–3.40 in. [7.0–8.6 cm]) wingspan (Pyle 1981), predominantly black butterfly that may serve as the distasteful model for four palatable mimicking butterflies found over parts of its range (Opler and Krizek 1984). It is characterized by a black dorsal surface washed with metallic blue-green on the hindwings that is brighter on males. The hindwing margins have a row of white or cream spots that are more distinct in females; the tails are pronounced. Below, the forewings are dull gray; the hindwings have a row of large, bright orange spots in a field of iridescent blue and there is a row of white marginal spots (Pyle 1981; Opler and Krizek 1984). The four Michigan species that may mimic the pipe vine swallowtail are the spicebush swallowtail (*Papilio troilus*), the dark female form of the tiger swallowtail (*Papilio glaucus*), the female black swallowtail (*Papilio polyxenes asterius*), and the red-spotted purple (*Limenitis arthemis astyanax*). The pipe vine swallowtail can be easily distinguished from its most similar mimics (female

334

spicebush, female black, and dark female tiger) by the lack of an orange spot on the dorsal hindwing (Pyle 1981).

The eggs are rust colored, and the larva (1.9–2.1 in. [4.8–5.4 cm]) at maturity (Pyle 1981) is black or rusty black with rows of black or red tubercles along its body. When disturbed, it everts a brightly colored, forked organ on the first thoracic segment that emits a pungent odor and acts as an antipredator device (Opler and Krizek 1984). The chrysalis, 1.1 inch (2.8 cm) (Pyle 1981) is lavender, green, or brown with yellow splotches and is attached in an upright position with a silk pad at the base and silk strands at the middle (Pyle 1981; Opler and Krizek 1984). Flight dates are early May through the end of September (Moore 1960), though there should be a break between the two broods sometime in June.

Total Range: The pipe vine swallowtail is one of only two species in the genus occurring in the United States. Other species in the genus occur in the tropical Americas, but all members of the genus feed on pipe vines (*Aristolochia* spp.) (Opler and Krizek 1984). The pipe vine swallowtail ranges from southern Ontario and New England to Florida and westward through Nebraska, Texas, and Colorado into Mexico (Pyle 1981; Opler and Krizek 1984). It has been reported from Costa Rica, though not recently; a widely distributed but isolated population occurs in California (Opler and Krizek 1984).

State History and Distribution: Reported in Moore (1960) from Ingham, Lenawee, Washtenaw, and Wayne counties; recent records are from Lenawee, Barry, Wayne, and St. Joseph counties.

Habitat: In Michigan the pipe vine swallowtail has been found in association with oak-hickory woods of varying condition. It is reported to be most abundant in the deciduous forests of the Appalachian Mountains, but may be found in a variety of habitats from woods to open fields and floodplains (Opler and Krizek 1984).

Ecology and Life History: There are no literature reports of the life history of this species in Michigan. Observations from elsewhere indicate it should have two generations per year, with adults emerging in early May and early July (Opler and Krizek 1984). Adults prefer pink to purple flowers as nectar sources (Fee 1979). Males patrol for females during warm daylight hours near host plants and may be seen flying rapidly along a creek or trail or visiting mud puddles; females are more secretive and tend to be found within a woods (Fee 1979).

Ovipositing females tend to lay eggs in small clusters on the lower surfaces of sunny leaves of the host plant; the number of eggs in the cluster varies with the size of the plant (Opler and Krizek 1984). Eggs hatch in about a week (Rausher 1980), after which the young larvae are gregarious, feeding in tightly clustered groups. Later they become solitary (Opler and Krizek 1984), due to the large quantities of food they require. A larva will consume all the edible tissue on a plant before moving to the next one. Before pupating, it may consume as many as 25 plants (Rausher 1979), sometimes producing a loss of over 40% of the annual production of the host plant population (Rausher 1978).

During their development, larva sequester toxic and bad-tasting alkaloid compounds from the host plant, which are responsible for their distastefulness to some vertebrate predators (e.g., birds) as adults. The pipe vine swallowtail is not only distasteful, but it has a tough body that allows it to survive after being tasted by a predator (Scott 1986). The effectiveness of the pipe vine swallowtail mimetic system to deter bird predation has been demonstrated both in the lab (Brower 1958) and the field (Jeffords et al. 1979).

Conservation/Management: Management for healthy populations of the host plant and its habitat are, of course, a necessity and should be conducted at several locations across the limited natural range of the butterfly in Michigan. Additional surveys should be conducted for new populations of both the host plant and the butterfly. Populations of *Aristolochia macrophylla* in Washtenaw County, apparently escaped cultivars (Voss 1985), should also be checked because this species has been reported as a food plant elsewhere (Scott 1986).

LITERATURE CITED

Brower, J. V. Z. 1958. Experimental studies of mimicry in some North American butterflies. Part 2. *Battus philenor* and *Papilio troilus, P. polyxenes* and *P. glaucus. Evolution* 13 (2): 123–36.

Fee, F. D. 1979. Notes on the biology of *Battus philenor* (Papilionidae in Centre County, Pennsylvania). *J. Lepid. Soc.* 33 (4): 267–68.

Jeffords, M. R., J. G. Sternburg, and G. P. Waldbauer. 1979. Batesian mimicry: Field demonstration of the survival value of pipe vine swallowtail and monarch color patterns. *Evolution* 33 (1): 275–86.

Moore, S. 1960. A revised annotated list of the butterflies of Michigan. *Univ. Mich. Mus. Zool. Occ. Pap.* No. 617.

Opler, P. A., and G. O. Krizek. 1984. *Butterflies east of the Great Plains.* Johns Hopkins Univ. Press, Baltimore.

Pyle, R. M. 1981. *The Audubon Society guide to North American butterflies.* Alfred A. Knopf, New York.

Rausher, M. D. 1978. Search image for leaf shape in a butterfly. *Science* 200:1071–73.

Rausher, M. D. 1979. Egg recognition: Its advantage to a butterfly. *Anim. Behav.* 27:1034–40.

Rausher, M. D. 1980. Host abundance, juvenile survival, and oviposition preference in *Battus philenor. Evolution* 34 (2): 342–55.

Scott, J. A. 1986. *The butterflies of North America: A natural history and field guide.* Stanford Univ. Press, Stanford, CA.

Voss, E. G. 1985. *Michigan Flora. Part 2. Dicots (Saururaceae—Cornaceae).* Cranbrook Inst. Science, Bloomfield Hills, MI.

Mitchell's Satyr

· *Neonympha mitchellii* French
· **Federally and State Endangered**

Map 49. Confirmed individuals. ▲ = before 1980 and ● = 1980 to 1989 (Michigan DNR and Michigan State University).

Status Rationale: Originally known from 23 populations in Michigan, only 12 are known to still be extant. Many of the populations are known from only a few individuals and, in most cases, the habitat requires extensive management for the population to remain viable. Michigan is the remaining stronghold for this species, with 4 or fewer populations known outside the state, one of which has recently been identified as a different subspecies (Parshall and Kral 1989).

Field Identification: Mitchell's satyr (Lepidoptera: Nymphalidae) is a medium-sized (38–44 mm wingspan) (Pyle 1981), rich brown butterfly with a distinctive series of submarginal, yellow-ringed black eyespots on the ventral surface of both pairs of wings. The eyespots are accented distally by two orange bands near the wing margins. This species is most easily confused with the little wood satyr (*Megisto cymela*), the northern eyed brown (*Satyrodes eurydice*), and the appalachian eyed brown (*S. appalachia*). Mitchell's satyr is smaller, darker, tends to fly more slowly and erratically, and typically stays closer to the tops of the grasses and sedges than do the other species. The eyespots on the ventral surface of Mitchell's satyr's wings are round to slightly oval, relatively uniform in size, but with the largest eyespots in the hindwing series being the third and fourth. The eyespots of the other species are more variable in size and are not accented distally by two orange bands at the wing margin. Mitchell's satyr larvae have a lime green

337

head capsule and body with pale longitudinal lines on the body and a bifurcate tail (McAlpine et al. 1960). The flight period is late June through mid-July, peaking around July 4.

Total Range: Historically, Mitchell's satyr was known from southern Michigan (Badger 1958; McAlpine et al. 1960), northern Indiana (Badger 1958), northern Ohio (Pallister 1927), and northern New Jersey (Rutkowski 1966). Intensive searches in Pennsylvania have failed to discover any colonies there. A newly described subspecies (*Neonympha mitchellii francisci*) was recently discovered in North Carolina (Parshall and Kral 1989). Outside of North Carolina, this species is found today only in Michigan and Indiana. The New Jersey populations have been lost to habitat changes and overcollecting and the Ohio population to habitat loss.

State History and Distribution: Originally known from 23 sites statewide, extending as far north as Kent County, Mitchell's satyr is now known to be extant at only 12 sites in seven southwestern counties. The decline of the Michigan populations appears to have been caused by habitat loss due to succession and alteration by human activity.

Habitat: Mitchell's satyr is restricted to calcareous wetlands known as prairie fens. In Michigan, this habitat type is characterized by scattered tamaracks (*Larix laricina*), poison sumac (*Toxicodendron vernix*), and dogwood (*Cornus* spp.) clones with a ground cover of sedges (*Carex* spp.), shrubby cinquefoil (*Potentilla fruticosa*), and a variety of herbaceous species with prairie affinities. In Michigan, the butterfly is typically found in very restricted portions of the fen characterized by scattered tamaracks and tall sedges, although much of the remainder of the habitat may appear suitable.

Ecology and Life History: Adults of Mitchell's satyr are active two to three weeks each summer, with males emerging before females. Only one published report (McAlpine et al. 1960) discusses the early life history of this species, and most of the observations were made on captive individuals. Females lay eggs in early to mid-July and the first instar larvae hatched 7 to 11 days later. Larval behavior and the duration of the instars indicated that the larvae hibernate as a fourth instar attached near the bottom of a sedge leaf. In May they become active again to complete their growth through the fifth and sixth instars before pupating in mid-June. The larval food plant is thought to be several species of sedge.

Conservation/Management: Adequate protection of this species will require protection of its habitat, protection from collection, and well-designed habitat management. Protection of Mitchell's satyr habitat precludes development, off-road vehicles, and any changes to the fen or the surrounding area that would affect the hydrology of the fen. Acquisition of the highest quality sites should be pursued by private conservation organizations and public agencies. Protection from collection is a critical component of a conservation plan for this species, which is very easy to capture in flight. In the past, collectors travelled great distances to add a series of Mitchell's satyr specimens to their collections. Although collecting has declined as the critical status of Mitchell's satyr has become known, efforts to discourage collection should be continued through lepidopterist organizations and, when necessary, through law enforcement.

Habitat management is necessary at most Mitchell's satyr sites where habitat succession due to fire suppression and altered hydrological regimes are causing the fen to become overgrown with shrubs. Controlling shrubs through selective cutting and prescribed burns will be required. Burns must be carefully conducted in subunits of the fen so that only part of the satyr population is subjected to the burn in any year. Fires may be stressful to many fen species that exhibit no adaptations for surviving fires.

LITERATURE CITED

Badger, F. S., Jr. 1958. *Euptychia mitchellii* (Satyridae) in Michigan and Indiana tamarack bogs. *Lepid. News* 13:41–46.

McAlpine, W. S., S. P. Hubbell, and T. E. Pliske. 1960. The distribution, habits, and life history of *Euptychia mitchellii* (Satyridae). *J. Lepid. Soc.* 14 (4): 209–25.

Pallister, J. C. 1927. *Cissia mitchellii* (French) found in Ohio, with notes on its habits. Lepidoptera—Satyridae. *Ohio J. Sci.* 27:203–4.

Parshall, D. K., and T. W. Kral. 1989. A new subspecies of *Neonympha mitchellii* (French) (Satyridae) from North Carolina. *J. Lepid. Soc.* 43 (2): 114–19.

Pyle, R. M. 1981. *The Audubon Society guide to North American butterflies.* Alfred A. Knopf, New York.

Rutkowski, F. 1966. Rediscovery of *Euptychia mitchellii* (Satyridae) in New Jersey. *J. Lepid. Soc.* 20 (1): 43–44.

Regal Fritillary

· *Speyeria idalia* (Drury)
· **State Endangered**

Map 50. Confirmed individuals. ▲ = before 1980 and ● = 1980 to 1989 (Michigan DNR and Michigan State University).

Status Rationale: Formerly known from 16 counties, this species has not been reported since 1978, despite intensive surveys. The decline in Michigan is consistent with a precipitous decline over much of the eastern and southeastern parts of its range.

Field Identification: The regal fritillary (Lepidoptera: Nymphalidae) is a very large (2.6–3.6 in. [66–92 mm] wingspan) (Pyle 1981) red-orange butterfly; females are larger than males. The upper surface of the forewing of both sexes is red-orange with blue-black spots and a black margin; females have a broader black margin with white spots. The upper surface of the hindwing is dark brown-black to blue-black, shading to orange-brown near the base with two rows of light spots. Both rows are cream-white in females; the outer row is orange in males. The lower surface of the hindwing is rich olive-brown with several rows of large, silvery spots. The hindwing is the key to distinguishing the regal fritillary from other fritillaries. When viewed from above, the regal fritillary has dark hindwings that contrast sharply with the lighter and brighter forewings. Other fritillaries appear darkest at the base of the fore- and hindwings close to the body, with the outer areas of the wings being lighter (Bliss and Schweitzer 1987; Pyle 1981). The mature larva reaches 1.75 in (4.4 cm) and is velvety black with yellow to red-orange stripes. It has six rows of tapering, fleshy spines, the dorsal two rows are silver

340

with black tips and the lateral rows are orange (Bliss and Schweitzer 1987). Flight dates are late June through early September, with males emerging first.

Total Range: Historically, this species ranged from the southern parts of Maine, New Hampshire, Vermont, and New York southward along the Appalachian Mountains to North Carolina and possibly northern Georgia (Bliss and Schweitzer 1987), westward from Kentucky to Colorado and from Ohio and Michigan through Wisconsin and Minnesota to Montana and Wyoming (Opler and Krizek 1984; Bliss and Schweitzer 1987). It is now extirpated from New England except for several populations on the off-shore islands of Massachusetts and Rhode Island. It is also extirpated from the mainland of New York, New Jersey, Pennsylvania, Delaware, Virginia, and southward (Bliss and Schweitzer 1987).

State History and Distribution: Early Michigan records date back to Wolcott (1893), who found the regal fritillary in Kent County. More recent records are reported from at least 16 counties in the southern Lower Peninsula, ranging as far north as Newaygo County. The last known occurrence was in St. Joseph County, where it has not been seen since 1980. This species may now be extirpated in Michigan, but additional surveys should be conducted, particularly at the lesser known historical locations, before that is concluded.

Habitat: In the western part of its range, the regal fritillary inhabits the tallgrass prairie landscape complex, which is an open habitat matrix of dry prairies and more mesic habitats, including riparian areas, in a rolling landscape. Farther east it is also found in prairies and other open habitats, such as meadows, pastures, and old fields, and along streams (Bliss and Schweitzer 1987; Masters 1975 and 1979).

Conversion of pastures and hay meadows to row crops and the fragmentation and conversion of former prairie and barrens landscape complexes by agriculture, silviculture, and development have reduced the habitat available to it. Fire suppression has undoubtedly played a significant role in habitat reduction by permitting the succession of barrens-prairie complexes to forests and the inhibition of violets by accumulated thatch cover in some of the remaining open areas (Hammond and McCorkle 1983). Different habitat units would also be differentially susceptible to fires of different intensities and frequencies.

Ecology and Life History: The regal fritillary has a single generation each year. Males begin emerging in late June, with females following a week or two later. Some females may mate soon after emerging, though the preponderance of mating takes place later in the summer (Clark 1932). Males are vigorous fliers that wander widely; females fly near the ground or the tops of the ground cover and tend not to stray as far as males do. Both sexes avoid single trees and woodlots, usually going around them rather than over (Clark 1932). Both sexes are avid nectarers, usually seeking tall, isolated flowers such as milkweed (*Asclepias* spp.), though they will also use red clover and alfalfa.

Late in the flight season, females become more active, flying farther afield. Females drop to the ground every 100 feet or so to deposit eggs near, but not necessarily on, violets (*Viola* spp.) (Clark 1932). Larvae hatch in three to four weeks and soon go into diapause beneath leaves near the surface of the ground. In early spring the larvae begin feeding on violets at night and remain inactive during the day (Bliss and Schweitzer 1987). Larvae reared in captivity have accepted a diversity of viola species and the northeastern most population sites lacked *Viola pedata*, which was once thought to be the sole food plant of the regal fritillary in the east. There are five larval molts and the pupal stage lasts about 17 days (Bliss and Schweitzer 1987).

Conservation/Management: The reasons for the decline of the regal fritillary in the eastern part of its range are not well understood and it is likely that no single factor is fully responsible. Habitat destruction has undoubtedly played a significant role (Leahy 1982; Bliss and Schweitzer 1987) though apparently suitable habitat remains unoccupied in many eastern states (Bliss and Schweitzer 1987). Other factors that may have aided its decline, at least in some locations, include pesticide spraying for gypsy moth control (Bliss and Schweitzer 1987) and habitat degradation resulting in the loss of larval and adult food sources.

If an active regal fritillary population is found in Michigan, the site should be investigated to determine the population status and habitat use pattern of the potential of the site to be managed for the butterfly. A well-designed preserve would be nearly treeless, include both dry and mesic habitats, have dense field/wood interface populations of violets and nectar sources, and be large enough to contain several management units. The management plan should adequately address both temporal and spatial aspects of the resource needs of the butterfly. The regal fritillary may be sensitive to fire, so prescribed burns and all management activities should be conducted in units on a rotating schedule that provides larval and adult food resources in surrounding areas, as well as sources of colonizers to the managed area.

LITERATURE CITED

Bliss, P., and D. F. Schweitzer. 1987. *Speyeria idalia* element stewardship abstract. *The Nature Conservancy*, Arlington, VA.

Clark, A. H. 1932. The butterflies of the District of Columbia and vicinity. *U.S. Natl. Mus. Bull.* 157:1–337.

Hammond, P. C., and D. V. McCorkle. 1983. The decline and extinction of *Speyeria* populations resulting from human environmental disturbances (Nymphalidae: Argynninae). *J. Res. Lepid.* 22 (4): 217–24.

Leahy, C. W. 1982. Spineless wonders. *Sanctuary* 21:5.

Masters, J. H. 1975. Occurrence of *Speyeria idalia* (Nymphalidae) on remnant prairie in northwest Wisconsin. *J. Lepid. Soc.* 29 (1): 76–77.

Masters, J. H. 1979. A recent record of *Speyeria idalia* (Nymphalidae) from Manitoba. *J. Lepid. Soc.* 33 (4): 265.

Opler, P. A., and G. O. Krizek. 1984. *Butterflies east of the Great Plains.* Johns Hopkins Univ. Press, Baltimore.

Pyle, R. M. 1981. *The Audubon Society guide to North American butterflies.* Alfred A. Knopf, New York.

Wolcott, R. H. 1893. The butterflies of Grand Rapids, Michigan. *Can. Entom.* 25:98–107.

Frosted Elfin

- *Incisalia irus* Godart
- **State Threatened**

Map 51. Confirmed individuals. ▲ = before 1980 and ● = 1980 to 1989 (Michigan DNR and Michigan State University).

Status Rationale: Formerly known from numerous, scattered populations throughout the southern half of the Lower Peninsula, the frosted elfin has undergone a significant decline since the 1960s. It now occurs in small populations at several locations in the southwestern Lower Peninsula, many of which are degraded.

Field Identification: The frosted elfin (Lepidoptera: Lycaenidae) is a small (0.87–1.30 in. [22–32 mm] wingspan) (Pyle 1981), gray-brown butterfly that is most easily confused with Henry's elfin (*Incisalia henrici*) and the hoary elfin (*Incisalia polios*). All three species occur in the same habitat and can be found flying together. Henry's elfin can be distinguished from the frosted elfin by the sharp border on the chocolate-brown basal portion of the ventral hindwing in Henry's elfin; in the frosted elfin the border is obscured by gray scaling that may have a greenish cast in newly emerged individuals. Both species have short "tails" on the hindwing, but the frosted elfin has a small, dark spot at the anal (nearest the body) angle of the hindwing, which Henry's elfin lacks. Male Henry's elfins also lack the long forewing stigma of male frosted elfins (Opler and Krizek 1984). The hoary elfin has a more extensive gray scaling than the frosted elfin and it lacks the hindwing tails, although the posterior margin of the hindwing appears slightly scalloped. Female frosted elfins tend to be reddish overall (Pyle 1981). The pale, bluish green larvae are marked with three faint white lines dorsally, oblique white dashes subdorsally,

343

and a whitish lateral line (Scott 1986). Flight dates are the first week of May to the first week of June, with a peak in the latter half of May.

Total Range: The frosted elfin occurs from southern Maine to New York and south along the Atlantic seaboard and the Appalachians to Georgia and Alabama. In the Midwest, it occurs in southern Michigan, central Wisconsin, and in the Great Lakes plain of extreme southeastern Wisconsin, Illinois, Indiana, and western Ohio. To the south it is found from western Arkansas and Louisiana to eastern Texas (Opler and Krizek 1984).

State History and Distribution: Historical records indicate that this species was once more widespread and abundant in Michigan than it is today, with records from 15 counties. Today it is found principally in Allegan and Montcalm counties, with a few additional populations in Newaygo, Kalamazoo, Barry, Bay, and Monroe counties. The localized populations found in Michigan are typical of this species throughout its range (Opler and Krizek 1984). The loss of barrens communities and large, healthy populations of wild blue lupine (*Lupinus perennis*) to succession and development appears to be the most important factor in the decline of this species. A similar pattern of decline has been observed in the karner blue butterfly (*Lycaeides melissa*), which also requires lupine as a larva.

Habitat: The frosted elfin is restricted to open oak and oak-pine barrens containing large populations of blue lupine, the larval food plant. Today, with the closing of the canopy in most former oak barrens, it is found in transitional areas between the closed canopy oak woods and the scattered openings of sand prairie.

Ecology and Life History: The frosted elfin has a single generation per year. Adults tend to be sedentary and remain near patches of the larval food plant. Males may perch near or on the ground in small, grassy openings. Adults are more active in the morning than in the afternoon. Cook's (1906) observations provide basic life history information. Females lay eggs on the opening buds of lupine, placing only a single egg on each plant. Plants selected as oviposition sites tend to be exposed to the midday sun. Eggs hatch in three to five days, and the newly hatched larvae bore through petals to feed on the center of blossoms; older larvae enter the flowers between the petals. Developing seed pods replace flowers as the food source when the flowers wither. Larvae undergo three molts before pupating in a loose cocoon constructed from leaves and silk (Cook 1906) at the base of the host plant (Opler and Krizek 1984). All larvae reared by Cook (1906) pupated by July 2. Overwintering occurs in the pupal stage (Scott 1986). In Michigan, the host plant is blue lupine, though the frosted elfin also is reported to feed on wild indigo (*Baptisia tinctoria*) (Opler and Krizek 1984).

Conservation/Management: As with many barrens species, protection of existing populations and habitat management are the most important factors in maintaining the frosted elfin in Michigan. Off-road vehicles should be excluded from all known population sites and those areas where habitat restoration holds good potential. Conversion of habitat for silvicultural, agricultural, or other uses should be stopped. Sites in the early stages of pine plantations or row-cropping should be allowed to revert to a barrens and sand prairie landscape complex, which may require removal of the plantations. Opening the closed canopy of the former barrens will allow the expansion of lupine populations. While fire will be an important tool in this process, the frosted elfin is probably sensitive to fire in all life stages and management plans should be developed accordingly. Refer to the treatment of *Hesperia ottoe* for a more complete discussion of barrens management. Populations of *Baptisia tinctoria* should be surveyed for populations of the frosted elfin.

LITERATURE CITED

Cook, J. H. 1906. Studies in the genus *Incisalia. Can. Entom.* 38: (5): 140–44.

Opler, P. A., and G. O. Krizek. 1984. *Butterflies east of the Great Plains.* Johns Hopkins Univ. Press, Baltimore.

Pyle, R. M. 1981. *The Audubon Society guide to North American butterflies.* Alfred A. Knopf, New York.

Scott, J. A. 1986. *The butterflies of North America: A natural history and field guide.* Stanford Univ. Press, Stanford, CA.

Northern Blue

· *Lycaeides idas nabokovi* Masters
· **State Threatened**

Map 52. Confirmed individuals. ▲ = before 1980 and
● = 1980 to 1989 (Michigan DNR and Michigan State
University).

Status Rationale: Known from five locations in the Upper Peninsula and on Isle Royale, including one that appears to be extirpated and two others that may still exist but cannot be located. The host plant, dwarf bilberry (*Vaccinium cespitosum*), is also state threatened.

Field Identification: The northern blue (Lepidoptera: Lycaenidae) is a small (0.90–1.25 in. [22–32 mm] wingspan) (Pyle 1981) butterfly. The dorsal surface is silvery blue in males with a narrow, dark border and white fringe; in females it is gray-brown near the anterior and outer edges of the wings with areas of blue toward the bases and posterior edges of the wings. The hindwing of the female has a row of dark spots, sometimes with orange, along the outer edge. The ventral surface of both sexes is pearly gray to white with several rows of small black spots on the inner portions of both wings and a row of metallic blue-green, orange, and black spots just inside the outer margin of both wings, becoming less pronounced in the forewing. The thin, black marginal line is inflated into triangles at the ends of the veins. Several other blues resemble the northern blue, but none has the combination of being tailless with orange spots on the underside border of the hindwing. Neither the silvery blue (*Glaucopsyche lygdamus*) nor the spring azure (*Celastrina ladon*) has orange on any wing surface, and the eastern tailed blue (*Everes comyntas*) has a similar pattern and coloration, but both sexes have tails. The karner blue (*Lycaeides melissa samuelis*) occurs only in the Lower Peninsula. Flight dates are

346

variable, centering around mid-June to mid-July with up to two weeks variation on either side (Mueller 1987 and 1988).

Total Range: This subspecies is one of 11 Nearctic subspecies of *Lycaeides idas* (Masters 1972) and occurs in northern Minnesota, northern Wisconsin, and Michigan's Upper Peninsula. *L. idas* was formerly confused with the European species *L. argyrognomon* (Higgins 1985).

State History and Distribution: First discovered in Michigan in 1979 in Dickinson County (Nielsen and Ferge 1982), it is now also known from Alger, Marquette, and Keweenaw (Isle Royale) counties. Recent surveys have failed to confirm three populations, one that is likely extirpated and the other two that probably still exist somewhere in the area.

Habitat: The northern blue is found in open sandy or rocky habitats, such as in patches of open habitat in spruce (*Picea glauca*) forests, along rights-of-way, and near rock outcroppings that support the larval host plant, dwarf bilberry (*Vaccinium cespitosum*). The best known population occurs in Alger County on a nearly flat, sandy plain adjacent to a railroad track. There are scattered shrubs (e.g., *Salix*) and the ground cover is typical of open, dry, sandy areas and includes bracken fern (*Pteridium aquilinum*), hair grass (*Deschampsia flexuosa*), and reindeer moss (*Cladina rangiferina*), as well as a variety of native and nonnative flowering plants that are used as nectar sources, such as blackberry (*Rubus* sp.), oxeye daisy (*Chrysanthemum leucanthemum*), hawkweed (*Hieracium* spp.), spreading dogbane (*Apocynnum androsaemifolium*), and wild basil (*Satureja vulgaris*) (Mueller 1987). The dwarf bilberry occurs in large, circular patches representing different clones, the predominant mode of reproduction in this population; flowering and fruiting are sporadic with virtually none occurring in some years.

Ecology and Life History: The northern blue has a single generation per year, which is markedly different from the two generations of the karner blue. Mueller (1987) found that adults emerge in mid-June, though they have been found as early as the first week of June, often basking on the ground or nectaring. Males typically fly within a meter of the ground patrolling for females and, depending upon the site characteristics, may or may not be found in close association with the dwarf bilberry (Mueller 1987; Nielsen and Ferge 1982). This species overwinters in the egg stage (Nielsen and Ferge 1982) and completes its life cycle the next spring. Northern blue larvae at one site were tended by at least two species of ants (Mueller 1988), but the importance of this interaction to larval survival is unknown.

Conservation/Management: Protection of existing populations and their habitat should be the first conservation priority. Habitat enhancement by maintaining existing openings and creating additional openings in a matrix on the landscape would maintain patches in different successional stages, providing patches of new habitat and ready sources of colonizers.

Historically, fire was important to maintaining open habitat patches. To some extent, railroads assisted that process for many years, through both accidental fires and prescribed burns along rights-of-way, which may account for the vigor and persistence of the Alger County site adjacent to a railroad. The northern blue is fire sensitive at all stages in its life history, so any burn management program should include areas that are managed on a rotating basis, always leaving significant portions of the habitat, including both dwarf bilberry and nectar sources, unburned.

Refer to the *Hesperia ottoe* account for more information on fire management. Any management program should be carefully monitored to determine its effect on both the butterfly and dwarf bilberry. Additional surveys are needed to locate new populations and to develop management plans for existing populations. In addition, research on the life histories of both the butterfly and dwarf bilberry are needed to provide guidance to the management program.

LITERATURE CITED

Higgins, L. G. 1985. The correct name for what has been called *Lycaeides argyrognomon* in North America. *J. Lepid. Soc.* 39 (2): 145–46.

Masters, J. H. 1972. A new subspecies of *Lycaeides argyrognomon* (Lycaenidae) from the eastern Canadian forest zone. *J. Lepid. Soc.* 26 (3): 150–54.

Mueller, S. J. 1987. Biogeographical survey of *Lycaeides idas nabokovi* (Lepidoptera, Lycaenidae) in Michigan. *Mich. Dept. Nat. Resour.*, Unpubl. Rept.

Mueller, S. J. 1988. Biogeography of the northern blue butterfly, *Lycaeides idas nabokovi*. *Mich. Dept. Nat. Resour.*, Unpubl. Rept.

Nielsen, M. C., and L. A. Ferge. 1982. Observations of *Lycaeides argyrognomon nabokovi* in the Great Lakes region (Lycaenidae). *J. Lepid. Soc.* 36 (3): 233–34.

Pyle, R. M. 1981. *The Audubon Society guide to North American butterflies.* Alfred A. Knopf, New York.

Karner Blue

· **Lycaeides melissa samuelis Nabakov**

· **Federally Endangered and State Threatened**

Map 53. Confirmed individuals. ▲ = before 1980 and ● = 1980 to 1989 (Michigan DNR and Michigan State University).

Status Rationale: Known historically from oak and pine-oak savannas in ten counties, recent surveys have found many historical locations to no longer be suitable due to habitat degradation and loss. The principle factors affecting habitat conditions are fire suppression, agriculture, silviculture, and off-road vehicles.

Field Identification: The karner blue (Lepidoptera: Lycaenidae) is a small (0.90–1.25 in. [22–32 mm] wingspan) (Pyle 1981) silvery blue butterfly. The dorsal surface is silvery blue in males with a narrow, dark border and white fringe; in females it is slate gray-brown near the anterior and outer edges of the wings with areas of blue toward the bases and posterior edges of the wings. The hindwing of the female has a row of dark spots with orange along the outer edge. The ventral surface of both sexes is pearly gray to white with several rows of small black spots on the inner portions of both wings and a row of metallic blue-green, orange, and black spots just inside the outer margin of both wings, becoming less pronounced in the forewing. The black marginal line is not distinctly inflated into triangles at the ends of the veins.

Several other blues resemble the karner blue, but none has the combination of being tailless with orange spots on the underside border of the hindwing. Neither the silvery blue (*Glaucopsyche lygdamus*) nor the spring azure (*Celastrina ladon*) has orange on any wing surface, and the eastern tailed blue (*Everes comyn-*

tas) has a similar pattern and coloration, but both sexes have tails. The northern blue occurs only in the Upper Peninsula. Larvae are green or whitish green, covered with white hairs, with a cream lateral stripe (Scott 1986); the head is small and dark. Peak flight dates are late May through early June and mid-July through early August, with stragglers found between peak dates.

Total Range: The karner blue has a disjunct range, occurring historically in eastern Minnesota, northwestern and central Wisconsin (Bleser 1990), southwestern Michigan and northern Indiana, extreme southeastern Michigan and northeastern Ohio, central Ontario near the southern Lake Huron shoreline, in the pine barrens near Albany, New York, and at a few localized sites elsewhere in New York and in New Hampshire, Massachusetts, Pennsylvania, and Illinois. This species is now extirpated in Illinois, Ohio, and Massachusetts, and probably extirpated in Minnesota (Schweitzer 1989b). The drought of 1988 followed by a cold, rainy June in 1989 saw the demise of the Ontario population and the Albany pinebush population is now down to just a few hundred individuals.

State History and Distribution: The karner blue has declined in Michigan due to habitat loss from silviculture (pine plantations), agriculture, and off-road vehicles, as well as to the fire-suppressed succession of open-canopied oak and pine-oak barrens to closed woods. It still occurs in most of the counties from which it was known historically except Monroe County, where it has likely been extirpated by habitat succession and degradation. However, existing populations are much reduced over earlier levels and have become highly fragmented, with expanses of unsuitable habitat between the remaining subpopulations. Schweitzer's (1989b) recent population studies in Allegan State Game Area (Lawrence and Cook 1989) indicate that the karner blue population there has declined to a dangerously low level, similar to that experienced by other populations before becoming extirpated.

Habitat: The karner blue is a species of open-canopied barrens communities, including oak and pine-oak barrens of the Midwest and the "pine bush" of the east. Its larval host plant, blue lupine (*Lupinus perennis*), has declined markedly within its range due to shading from closed canopies, competition from other plants, and extreme soil scarification following farming or intensive logging and burning regimes. Lupine, an early successional species, can become abundant after appropriate disturbances, fire or logging, for example. It relies on periodic disturbances to reset natural succession to an early stage at which it is competitive. At present, the Allegan State Game Area and portions of the southern Manistee National Forest provide the best opportunities for enhancing karner blue habitat and populations.

Ecology and Life History: The life history of the karner blue is generally well understood, though some of the details, particularly some of those important to good management, are unknown. Schweitzer's (1989a) summary of karner blue life history and ecology provides the most concise overview of the species.

The karner blue has two generations per year, with the later (summer) generation typically having three to four times the number of adults as the earlier (spring) brood. Males emerge earlier than females and some may disperse for a short time after emergence. Soon, they become somewhat sedentary, perching on tall grass stems or flowering plants, or engaging in short chases of other karner blues, but generally not ranging far. Adults are active most of the day, decreasing activity during midday and during cool, rainy weather. Females can live up to two weeks in the field, but adults typically live an average of five days.

Females lay eggs on or near lupine and the summer brood larvae hatch in about a week. The larvae grow rapidly, feeding on the upper surfaces of the lupine leaves. Pupation occurs in the litter near the host plant or occasionally on the plant. The summer brood mates and lays eggs that overwinter; the spring brood larvae hatch in April. Karner blue larvae are frequently attended by ants (Packer 1987; Savignano 1987) that feed on the sweet secretions they produce. The ants may confer some protection on the larvae against parasitism or predation, but the results of recent experiments are inconclusive. Adults require adequate nectar resources and will utilize a wide variety of native and introduced species. In Michigan, they frequently nectar on dewberry (*Rubus* spp.) during the first brood and *Monarda punctata*, *Asclepias tuberosa*, *Centaurea maculosa*, and *Liatris aspera* during the second brood (Ewert and Ballard 1990). All life stages are fire sensitive.

Conservation/Management: The loss of habitat has been the most significant cause of the decline of the karner blue in Michigan. The decline and loss of lupine populations is the greatest, though often reversible, threat to the remaining karner blue populations. Shading from overstory closing and competition from sod-forming grasses and sedges have excluded lupine from many former barrens and prairies where it once was common (Bess et al. 1989). Well-planned fire management is the single most important tool for rehabilitating karner blue habitat. The frequency of fire management should be tailored to each management unit, taking into consideration the desired final community matrix, current community conditions, site characteristics, and the life histories of the fire-sensitive species known to occur there. The final product should be a landscape complex of barrens, prairies, and forested and unforested wetlands at different stages in succession inhabited by thriving populations of habitat-sensitive species and their host species. In this setting, semi-isolated karner blue populations within the landscape complex would wax and wane as lupine populations changed and, over short distances, would provide colonizers to new sites opened by fire or lost to localized extinction events.

Of particular importance is research to determine the relationship of fluctuation in the butterfly population to that in the size, phenology, and habitat distribution of the lupine population. The dispersal capabilities of the butterfly must also be determined to ensure proper design and spacing of habitat patches within the landscape complex. Finally, the impact of ant species on the reproductive success of the butterflies and the effects of management activities on the ants must be determined.

LITERATURE CITED

Bess, J. A., R. M. Strand, and L. A. Wilsmann. 1989. Status of the karner blue butterfly, *Lycaeides melissa samuelis* Nabokov, in the Manistee National Forest. *Huron-Manistee Nat. For.*, Unpubl. Rept.

Bleser, C. A. 1990. The Wisconsin 1990 status survey for the karner blue butterfly (*Lycaeides melissa samuelis*). *U.S. Fish Wildl. Serv.*, Unpubl. Rept.

Ewert, D. N., and H. E. Ballard, Jr. 1990. Distribution and management of an indicator species of oak barrens, the karner blue butterfly (*Lycaeides melissa samuelis*): Pre-treatment population studies. *Huron-Manistee Nat. For.*, Unpubl. Rept.

Lawrence, W. S., and A. C. Cook. 1989. The status and management of karner blue (*Lycaeides melissa samuelis*) populations in the Allegan State Game Area, Michigan. *Mich. Chapter, The Nature Conservancy*, Unpubl. Rept.

Packer, L. 1987. Status report on the karner blue butterfly, *Lycaeides melissa samuelis* Nabokov, in Ontario. *Unpubl. Rept. to the Comm. on the Status of Endangered Wildlife of Canada* (COSEWICO).

Pyle, R. M. 1981. *The Audubon Society guide to North American butterflies*. Alfred A. Knopf, New York.

Savignano, D. A. 1987. The association of *Lycaeides melissa samuelis* (Lepidoptera: Lycaenidae), the karner blue butterfly, with attendant ants. *New York Dept. Conserv.*, Unpubl. Rept.

Schweitzer, D. F. 1989a. Fact sheet for the karner blue butterfly with special reference to New York. Ms.

Schweitzer, D. F. 1989b. More bad news: An update on the continuing rangewide decline of the karner blue butterfly, with revised global and state ranks. Memorandum dated 18 December 1989 and global rank form of *The Nature Conservancy*.

Scott, J. S. 1986. *The butterflies of North America: A natural history and field guide*. Stanford Univ. Press, Stanford, CA.

Ottoe Skipper

· *Hesperia ottoe* Edwards
· **State Threatened**

Map 54. Confirmed individuals. ▲ = before 1980 and
● = 1980 to 1989 (Michigan DNR and Michigan State
University).

Status Rationale: The populations of the ottoe skipper have declined over the last 20 years as the amount and quality of available habitat have declined. Formerly known from Newaygo, Montcalm, St. Joseph, Kent, Barry, and Allegan counties, the ottoe skipper is currently known from only a small area in Allegan County.

Field Identification: The ottoe skipper (Lepidoptera: Hesperiidae) is a small (1.25–1.60 in. [32–41 mm] wingspan) (Pyle 1981) butterfly, with darkly bordered, orange-yellow wings that appear finely toothed on the margins. Females are tawnier than males ventrally and have broader, dark borders (Pyle 1981). This species is most similar to the Delaware skipper, *Atrytone delaware,* and the European skipper, *Thymelicus lineola.* These species are both orange dorsally, but lack the broad, gray terminal area found on the forewing of the male ottoe skipper. Behaviorally, the ottoe skipper is a much warier and faster flyer. The newly hatched caterpillar is yellowish white with a dark brown head and black collar (prothorax); mature larvae are light greenish brown with a dark brown head and black collar (Nielsen 1958). The flight period is late June to early August, with peak abundances in early July.

Total Range: It ranges from eastern Montana to southwestern Michigan and south to Texas and Colorado.

State History and Distribution: Known historically from southwestern Michigan extending as far north as Newaygo County, this species has recently been found only in very localized areas in Allegan County.

Habitat: The ottoe skipper is a habitat-restricted species (Panzer 1988) that is found in tallgrass prairie in the western part of its range, often on hills or slopes (Opler and Krizek 1984). In Michigan, it is found in high-quality dry sand prairies and in old fields with a strong prairie flora component.

Ecology and Life History: The ottoe skipper has a single generation each year, with adult males emerging before females in late June and July; females may be found as late as early August. Males typically perch on flowers or, occasionally, on low plants or the ground; they sometimes patrol. Adults readily nectar on *Opuntia, Centaurea,* and *Asclepias* and are quite wary, flying swiftly and low to the ground when disturbed (Nielsen 1958).

Eggs are typically laid near the base of the host plant. Females oviposit on fall witchgrass (*Leptoloma cognatum*) in Michigan (Nielsen 1958) and on little bluestem (*Andropogon scoparius*) in Iowa (Opler and Krizek 1984). The larvae emerge in 8–12 days, feeding first upon their eggshell and then upon grasses, retreating to their silken nests at the base of the grass to consume their food. The ottoe skipper probably feeds upon a diversity of grass species, as is typical for the genus (Scott 1986). The fourth or fifth instar larva overwinters in its nest, where it likely pupates (Scott 1986).

Conservation/Management: Habitat protection and enhancement are essential to the conservation and long-term management of the ottoe skipper in Michigan. Habitat destruction from off-road vehicles, agricultural and silvicultural practices, and development continues to threaten this species. Immediate action should be taken to protect existing populations from further habitat degradation and loss. Fire suppression has encouraged the closing of formerly open-canopied oak and pine-oak barrens and reduced the size of adjoining prairies. Managing the prairie and barrens communities, especially through carefully controlled, prescribed burns, is critical to the long-term survival of the ottoe skipper and other prairie and barrens plant and animal species.

Prior to beginning a burn management program, the location and extent of habitat use of populations of the ottoe skipper and other rare plant and animal species should be determined. Burn management units should be established with special attention to microgeographic variation in the distribution of rare species and their host plants (Opler 1981). Summer burns that mimic natural, lightning-strike fires are more desirable than fall or spring burns, which would rarely occur without human intervention.

Prescribed burns should cover no more than 25–50% (Panzer 1988) of the existing habitat of a known population each year. Return times for fire in a subunit should be no more frequent than four years (Schweitzer 1989) and fire-sensitive and rare species should be carefully monitored as part of the management program. Backfires (burning against the wind) should be avoided because hotter temperatures are produced for longer periods of time near the ground (Panzer 1988). Unburned patches should be left unburned; these likely have provided a safe haven for some species that are unable to survive fire without them. The reintroduction of native nectar sources should be considered in areas where needed.

LITERATURE CITED

Nielsen, M. C. 1958. Observations of *Hesperia ottoe* in Michigan. *Lepid. News* 12 (2): 37–40.

Opler, P. A. 1981. Management of prairie habitats for insect conservation. *Natural Areas J.* 1 (4): 3–6.

Opler, P. A., and G. O. Krizek. 1984. *Butterflies east of the Great Plains.* Johns Hopkins Univ. Press, Baltimore.

Panzer, R. 1988. Managing prairie remnants for insect conservation. *Natural Areas J.* 8 (2): 83–90.

Pyle, R. M. 1981. *The Audubon Society guide to North American butterflies.* Alfred A. Knopf, New York.

Schweitzer, D. F. 1989. A review of Category 2 insects in U.S.F.W.S. Regions 3, 4, and 5. *U.S. Fish Wildl. Serv.,* Unpubl. Rept.

Scott, J. A. 1986. *The butterflies of North America: A natural history and field guide.* Stanford Univ. Press, Stanford, CA.

Powesheik Skipper

· *Oarisma powesheik* (Parker)
· **State Threatened**

Map 55. Confirmed individuals. ▲ = before 1980 and
● = 1980 to 1989 (Michigan DNR and Michigan State
University).

Status Rationale: Currently known from seven sites in Michigan, the powesheik skipper is found only in high-quality fens or as remnant populations in fens of declining quality. Only one population is fully protected on a nature reserve, and several others are receiving partial protection.

Field Identification: The powesheik skipper (Lepidoptera: Hesperiidae) is a small (1.00–1.25 in. [25–32 mm] wingspan) (Pyle 1981), distinctive butterfly not easily confused with any other species in Michigan. The dorsal wing surfaces are dark brown with a burnt orange patch along the anterior margins of the forewings. The ventral forewing is blackish brown with orange on the anterior edge and the ventral hindwing is blackish with the veins distinctively outlined in white on the anterior two-thirds of the wing. The ventral surface of the body is also white. The species in Michigan most similar to the powesheik skipper is an introduced species, the European skipper, *Thymelicus lineola*, a light yellowish brown butterfly with the edges of the wings and veins black (McAlpine 1972). The powesheik skipper is most often observed nectaring on flowers and its rapid flight makes it difficult to track on the wing. The larva is dark green with six longitudinal, cream-colored stripes that give it a yellowish green color on its sides in contrast to the broad, dark green mid-dorsal stripe; the head capsule is light green (McAlpine 1972). Flight dates are late June through the first three weeks of July.

Total Range: Historically found from southeastern Michigan around the south end of Lake Michigan, then northwest through Iowa, western Minnesota, and the eastern Dakotas (Opler and Krizek 1984) in relatively undisturbed, remnant tallgrass prairies (Opler and Krizek 1984) and fens. Although it can be locally common, this species has declined throughout its range.

State History and Distribution: The powesheik skipper was first reported from Michigan in 1893 by Wolcott, who discovered it at Lamberton Lake fen in Kent County (Holzman 1972). Today it is known from seven populations in five southeastern counties. Several populations have been lost to habitat succession and urban development, including the original Kent County site.

Habitat: In Michigan, the powesheik skipper inhabits alkaline wetlands known as fens. This habitat is characterized by scattered tamaracks (*Latrix laricina*), poison sumac (*Toxicodendron vernix*), and dogwood (*Cornus*) clones with a ground cover of sedges (*Carex*), shrubby cinquefoil (*Potentilla fruticosa*), black-eyed Susans (*Rudbeckia*), and other herbaceous species. Many fens have been ditched and drained for agriculture or development. Many others are rapidly being lost to habitat succession due to fire suppression and hydrological changes.

Ecology and Life History: The powesheik skipper has a single generation each year. Holzman (1972) observed a female lay an egg on a blade of spike-rush, *Eleocharis elliptica*, and found additional similar eggs on nearby spike-rushes. McAlpine (1972) reported on larval development from individuals reared in captivity on lawn grass. First instar larvae hatched eight to ten days after the eggs were laid in early July. The caterpillars generally remained on the underside of the blade of grass in a head down position and fed only occasionally. Fifth instar larvae hibernated through the winter on the underside of the blade of grass on which they had been feeding. In early April when they resumed feeding, McAlpine offered them fresh grass. By the seventh instar, the one remaining caterpillar was nearly 24 mm long. The total number of instars is not known, but McAlpine predicted it would be eight or nine. Based upon McAlpine's work, it is likely that the larvae are not restricted to spike-rush as a host food plant, but probably use other sedges and rushes as well.

Conservation/Management: Protection of the known populations and management of their habitat are the top conservation priorities for this species. Landowners should be notified and advised to protect the habitat from degradation and to protect the insect from collection and pesticides. Several of the best sites, with an intact fen community, should be acquired by the state or private conservation organizations. Surveys to identify additional populations of this species are also necessary. High-quality fens and those on state lands should be surveyed first. All prescribed burns should be conducted in subunits and there should be a year or two between burns at the same site to allow the population to recover.

LITERATURE CITED

Holzman, R. W. 1972. Eastern range extension for *Oarisma powesheik* Parker (Lepidoptera: Hesperiidae). *Great Lakes Entom.* 5 (4): 111–14.

McAlpine, W. S. 1972. Observations on life history of *Oarisma powesheik* (Parker) 1870. *J. Res. Lepidoptera* 11 (2): 83–93.

Opler, P. A., and G. O. Krizek. 1984. *Butterflies east of the Great Plains.* Johns Hopkins Univ. Press, Baltimore.

Pyle, R. M. 1981. *The Audubon Society guide to North American butterflies.* Alfred A. Knopf, New York.

Dukes' Skipper

· *Euphyes dukesi* (Lindsey)
· **State Threatened**

Map 56. Confirmed individuals. ▲ = before 1980 and ● = 1980 to 1989 (Michigan DNR and Michigan State University).

Status Rationale: Formerly recorded from four southeastern counties, this species is now known from only five very small, isolated sites in Monroe and Lenawee counties. It is associated with the *Carex lacustris*–dominated wet meadows found on the poorly drained lake plain, a community type that is now highly fragmented and threatened with continuing urbanization.

Field Identification: Dukes' skipper (Lepidoptera: Hesperiidae) is a small (1.25–1.50 in. [32–38 mm] wingspan) butterfly with short, broad, rounded wings (Pyle 1981). The dorsal surface of the male is dark brown with a black stigma near the center of the forewing. The female is similar, but with a band of two or three pale spots near the middle of the forewing. The ventral surface of both sexes is dark brown on the forewing and somewhat paler on the hindwing with a pale streak in the center. Dukes' skipper is most similar to the dion skipper, *Euphyes dion dion*, but can be distinguished from it by its morphology and flight date. Dukes' skipper flies later than the dion skipper, though their flight periods overlap somewhat. The underside of the forewing of Dukes' skipper is almost entirely dark; that of the dion skipper is orangish with three to four yellowish spots in the postmedian area. Dukes' skipper lacks the small light spots found near the costal margin just below the apex of the forewing on several similar species, including the dion skipper (Klots 1979). Flight dates are mid-July to early August.

Total Range: First described from Mobile County, Alabama (Lindsey 1923), this species has a disjunct range extending along the Atlantic coast from southeastern Virginia to northeastern North Carolina, along the Gulf coast from Alabama to Louisiana, up the Mississippi valley to southern Illinois and western Kentucky, and in the Great Lakes region from southeastern Michigan to northeastern Indiana and northwestern Ohio (Covell et al. 1979; Irwin 1969 and 1972; Mather 1963; Price and Shull 1969; Scott 1986).

State History and Distribution: Formerly known from Washtenaw, Wayne, Lenawee, and Monroe counties, this species is now known only from several small, isolated locations in Monroe and Lenawee counties. Habitat degradation and loss from draining and development have caused the decline of this species in Michigan. Dukes' skipper is typically thought of as occurring only on the lake plain of former Lake Maumee. However its occurrence in Ann Arbor, Washtenaw County (Pliske 1957), indicates that it does occur beyond Lake Maumee in the Lake Huron watershed.

The host plant of Dukes' skipper, *Carex lacustris*, is found throughout the southern Lower Peninsula (Voss 1972). The availability of *Carex lacustris* and known concentrations of coastal disjunct species suggests that new populations of Dukes' skipper should be sought in southwestern Michigan.

Habitat: Dukes' skipper is found in Michigan in dense stands of the wide-leaf lake sedge, *Carex lacustris*, usually in association with a swamp or floodplain forest. The patches of sedges are often quite small, frequently not more than 33–66 ft. (10–20 m) across. Throughout its range it is reported from similar habitat of dense sedges in association with wet woods ranging from thin woods to deep swamp (Mather 1963), though some sites are more open (Irwin 1972).

Ecology and Life History: Little is known of the life history of Dukes' skipper. Collection dates indicate that it is single brooded in Michigan, though there may be two broods in Virginia and three in Louisiana (Opler and Krizek 1984). *Carex lacustris* is reported to be the larval food plant in the North (Pliske 1957) and *Carex hyalinolepis* in the South (Irwin 1972). *Carex hyalinolepis* is also known from Monroe, Wayne, Macomb, and St. Clair counties in southeastern Michigan (Voss 1972) and should be investigated for Dukes' skipper populations. Adults are found flying through the sedges or nectaring at nearby flowers, typically swamp milkweed (*Asclepias incarnata*) and buttonbush (*Cephalanthus occidentalis*). Larvae overwinter in the fourth instar (Opler and Krizek 1984).

Conservation/Management: Protection of known populations, both the habitat and the skippers themselves, is critical in Michigan. This would include protection from physical destruction of the habitat, as well as from mowing and spraying of roadside *Carex* patches. In addition, sites should be managed to enhance the quality and extent of the habitat. This can be accomplished by restoring natural hydrological conditions to previously disturbed areas and by selecting restoration sites that will connect already existing, but isolated, populations with suitable habitat.

Additional surveys should be conducted to determine whether Duke's skipper occurs elsewhere in Michigan in areas rich in coastal plain disjunct species.

LITERATURE CITED

Covell, C. V., Jr., L. D. Gibson, R. A. Henderson, and M. L. McInnis. 1979. Six new butterfly records from Kentucky. *J. Lepid. Soc.* 33 (3): 189–91.

Irwin, R. R. 1969. *Euphyes dukesi* and other Illinois Hesperiidae. *J. Res. Lepid.* 8 (2): 183–86.

Irwin, R. R. 1972. Further notes on *Euphyes dukesi* (Hesperiidae). *J. Res. Lepid.* 10 (2): 185–88.

Klots, A. B. 1979. *A field guide to the butterflies of eastern North America.* Houghton Mifflin Co., Boston.

Lindsey, A. W. 1923. New North American Hesperiidae. *Entom. News* 24:209–10.

Mather, B. 1963. *Euphyes dukesi:* A review of knowledge of its distribution in time and space and its habitat. *J. Res. Lepid.* 2 (2): 161–69.

Opler, P. A., and G. O. Krizek. 1984. *Butterflies east of the Great Plains.* Johns Hopkins Univ. Press, Baltimore.

Pliske, T. 1957. Notes on *Atrytone dukesi*, a rare species new to southern Michigan (Hesperiidae). *Lepid. News* 11:42.

Price, H. F., and E. M. Shull. 1969. Uncommon butterflies of northeastern Indiana. *J. Lepid. Soc.* 23 (3): 186–88.

Pyle, R. M. 1981. *The Audubon Society guide to North American butterflies.* Alfred A. Knopf, New York.

Scott, J. A. 1986. *The butterflies of North America: A natural history and field guide.* Stanford Univ. Press, Stanford, CA.

Voss, E. G. 1972. *Michigan Flora. Part I. Gymnosperms and monocots.* Cranbrook Inst. Science, Bloomfield Hills, MI.

Dusted Skipper

· *Atrytonopsis hianna* Scudder
· **State Threatened**

Map 57. Confirmed individuals. ▲ = before 1980 and ● = 1980 to 1989 (Michigan DNR and Michigan State University).

Status Rationale: Recorded historically from 13 counties, this species has become rarer as prairies and prairie-like areas have been degraded or destroyed.

Field Identification: The dusted skipper is a small (1.24–1.50 in. [32–35 mm]), dark brown butterfly (Pyle 1981). The forewings are pointed with several small, translucent spots near the tip both above and below. The hindwings have a small translucent spot near their base. The underside of the wings are brown at the base and violet to gray toward the edge. The club at the end of the antenna is gently curved (Pyle 1981). Similar species are the northern cloudywing (*Thorybes pylades*) and southern cloudywing (*Thorybes bathyllus*), which tend to be larger, at 1.24–1.70 in. (32–44 mm) and 1.24–1.60 in. (32–41 mm), respectively. The cloudywings also have more numerous translucent markings on the wings, but lack the hind-wing spot, and have a distinctly bent antennal club. The final (seventh) instar larva is 1.15–1.30 in. (29–32 mm) long with a deep reddish purple head. The abdomen is pale pinkish lavender on top, grayish white beneath. The first thoracic segment is gray with a dark brown shield, the anal segment is glossy brown and depressed, and the entire body is covered with long yellowish white hair (Heitzman and Heitzman 1974). Flight dates are late May to early or mid-June.

Total Range: The dusted skipper has a discontinuous range that extends from southern New Hampshire to southern Manitoba and southeastern Saskatchewan,

362

south to Florida, eastern Wyoming, central Colorado and north-central New Mexico, but absent from the upper Ohio River and Missouri River valleys (Scott 1986). In the Great Lakes region, it occurs in limited areas at the western end of Lake Erie, around the southern end of Lake Michigan, in western Wisconsin, and in central Minnesota (Opler and Krizek 1984).

State History and Distribution: In Michigan, the dusted skipper has a disjunct distribution and is known from 13 counties in the Lower Peninsula, including the sandy pine barrens in the north and the prairies and oak barrens of the southeast and southwest. It has been reported to be rare in the prairie areas of Newaygo County in spite of habitat degradation that has resulted in the extirpation of at least one other species. Recent surveys of suitable habitat during the flight season in Allegan, Newaygo, Montcalm, and Oceana counties failed to record this species.

Habitat: The dusted skipper is found in dry prairies and prairielike openings in pine and oak barrens in Michigan. This is similar to reported habitats elsewhere in its range (Opler and Krizek 1984; Shapiro 1965). Shapiro (1965) reports that, in Pennsylvania and New Jersey, the dusted skipper and the cobweb skipper, *Hesperia metea*, are found in close association at sites with dense populations of *Andropogon scoparius*, the larval food plant of both species. The skippers are lost from these sites as the *Andropogon* declines from competition with other grasses and from shading as trees become established and mature.

Ecology and Life History: The dusted skipper has a single generation per year and adult population numbers may fluctuate annually (Heitzman and Heitzman 1974). Adults emerge in early spring for a short flight period of about three weeks. Adult males are very active; they perch on tall plants or on the ground and often pursue passing insects that resemble females of the species (Heitzman and Heitzman 1974). Females fly low to the ground and remain within about a foot of the ground even when disturbed. Both sexes may be found on such nectar plants as wild strawberry (*Fragaria*), raspberry (*Rubus*), and clover (*Trifolium*) (Shapiro 1965).

Both the dusted and cobweb skippers utilize the same larval food plant, little bluestem (*Andropogon scoparius*), but they partition the resource by using different feeding and resting positions. In the later instars, the dusted skipper larva lives in a tent near the top of the host plant, while the cobweb skipper larva constructs its tent at the base of the grass clump. Not until midfall, when it stops feeding and prepares to overwinter, does the dusted skipper larva move to the base of the plant. There it overwinters in the last instar, sealed in a case of silk and grass leaves and pupates in early spring (Heitzman and Heitzman 1974).

Conservation/Management: As with many species that rely on early successional plants, the dusted skipper probably evolved in a landscape system where a patchwork of early successional habitats was maintained by such natural processes as fire. Throughout its range, the dusted skipper occupies droughty habitats where, prior to European settlement, natural fires would have been frequent enough to maintain habitat patches with healthy populations of little bluestem, an early successional species. The population dynamics of the skipper across the landscape would likely have reflected the dynamics of the little bluestem patches.

Recent prescribed burns in Allegan State Game Area have demonstrated that little bluestem responds markedly to fire. Open sites in which little bluestem was sparse before the fire were carpeted with it after the burn. Habitat management on a landscape scale resulting in patches of various successional ages would benefit not only the dusted skipper but also other barrens and prairie species, such as the ottoe skipper and the karner blue. In the northern Lower Peninsula, the potential for Kirtland's warbler management to enhance dusted skipper populations, as well as those of other rare plants and animals of the jack pine barrens, should be investigated.

LITERATURE CITED

Heitzman, J. R., and R. L. Heitzman. 1974. *Atrytonopsis hianna* biology and life history in the Ozarks. *J. Res. Lepid.* 13:239–45.

Opler, P. A., and G. O. Krizek. 1984. *Butterflies east of the Great Plains.* Johns Hopkins Univ. Press, Baltimore.

Pyle, R. M. 1981. *The Audubon Society guide to North American butterflies.* Alfred A. Knopf, New York.

Scott, J. A. 1986. *The butterflies of North America: A natural history and field guide.* Stanford Univ. Press, Stanford, CA.

Shapiro, A. M. 1965. Ecological and behavioral notes on *Hesperia metea* and *Atrytonopsis hianna* (Hesperiidae). *J. Lepid. Soc.* 19 (4): 215–21.

Persius Dusky Wing

- *Erynnis persius* Scudder
- **State Threatened**

Map 58. Confirmed individuals. ▲ = before 1980 and ● = 1980 to 1989 (Michigan DNR and Michigan State University).

Status Rationale: Intensive surveys of the barrens and prairie insect fauna of western Michigan over the past three years by several lepidopterists have failed to find any populations of the persius dusky wing, indicating its scarcity in what once would have been prime habitat. Many of these areas have now been destroyed or degraded beyond suitability for many barrens butterfly species; lupine populations, in particular, have suffered from fire suppression and habitat destruction.

Field Identification: The persius dusky wing (Lepidoptera: Hesperiidae) is a small (1.0–1.4 in. [25–35 mm] wingspan) (Pyle 1981), inconspicuous, brown butterfly. The upper forewing is brown with small, translucent dots that are obscured by long, gray, hairlike scales on the male; the hindwing is brown with rows of dark and light markings (Opler and Krizek 1984; Pyle 1981). Females are lighter and more marked (Pyle 1981). The persius dusky wing is very similar in appearance to both the wild indigo dusky wing, *E. baptisiae*, and the columbine dusky wing, *E. lucilius*, which are named for their respective host plants. These three similar species have frequently been confused and misidentified, even by Scudder in his initial description of the species (Burns 1964). They are best distinguished by their association with their respective larval food plants. Mature larvae of persius dusky wing are hairy, light green with white specks. There is a dark green mid-dorsal line, and yellowish dorso-lateral line on the body; the head is red, yellow, or orange

365

with pale vertical streaks (Pyle 1981; Scott 1986). The pupa is a dull olive-green with a pinkish brown abdomen and pale mottling (Scott 1986). Flight dates are early to late May.

Total Range: The persius dusky wing is a northern species absent from the eastern Great Plains. It ranges in the west from Alaska and northwestern Canada southward into the mountains of California, Arizona, and New Mexico. In the east it occurs from eastern Minnesota to Quebec and Maine, extending southward in the Appalachian Mountains as far south as Tennessee (Scott 1986). The eastern populations are disjunct, occurring in the Great Lakes region and limited areas of New England, Connecticut to New Jersey, and the Appalachians (Opler and Krizek 1984; Pyle 1981). It is uncommon and sparsely distributed in the east (Pyle 1981). It is now less common than it was previously and many historical records are false due to misidentification (Schweitzer 1986); it is now thought to be extirpated from the Albany pinebush.

State History and Distribution: The persius dusky wing has been found in 17 counties, ranging from Monroe County in the southeast to Oceana County in the northwest, with an outlying record from Crawford County. Nielsen (pers. comm.) has reviewed Michigan specimens to verify the range of the persius dusky wing. He found records for several counties, including Chippewa, Dickinson, Menominee, Iosco, and Oscoda to be in error, as were the late summer records from counties with verified persius specimens.

Habitat: The eastern subspecies, *E. persius persius,* found east of the Great Plains (Schweitzer 1986) requires wild blue lupine, *Lupinus perennis,* or possibly wild indigo, *Baptisia tinctoria,* as a larval food plant. In Michigan, the persius dusky wing is found in open, sandy habitats that support these larval host plants. Large, healthy populations of lupine have declined in the state with the loss and degradation of oak and pine barrens and their associated prairies. Succession from barrens to closed-canopy oak or pine forests has resulted from fire suppression. Agriculture, silviculture, recreation, and development have destroyed other areas and fragmented the landscape.

Ecology and Life History: Little is known of the persius dusky wing's life history in the east. It has a single generation in early May, and it does not fly with the indigo and columbine dusky wings late in the summer when the second broods of these species are on the wing. Males observed in Colorado perched on or near the ground on ridges or hilltops during most of the day awaiting females (Opler and Krizek 1984; Scott 1986). A passing female is pursued by a male and, after hovering together, mating takes place on the ground. The pale yellow-green or yellow eggs are deposited singly on the undersides of leaves of the host plant, and they turn red as they mature. The larvae eat the leaves and live in nests of rolled or tied leaves on the host plant. The mature larvae overwinter in their nests and pupate the following spring (Opler and Krizek 1984; Scott 1986). Adults require nectar sources and all stages of the life history of the persius dusky wing are fire sensitive.

Conservation/Management: Management for the persius dusky wing is the same as for many other species found in the barrens and prairie landscape complex. Fire management would be an important tool for opening up closed forests and prairies with dense thatch cover, reduced nectar sources, and poor lupine populations (refer to the ottoe skipper account for more details on prairie and barrens management). Any management plan should include monitoring to determine the effects of management on the species. Research is also needed to determine whether wild indigo is an acceptable food source in Michigan and what the habitat requirements are for the persius dusky wing, including patch sizes, food abundance and distribution, and topography.

LITERATURE CITED

Burns, J. M. 1964. Evolution in skipper butterflies of the genus *Erynnis. Univ. Calif. Publ. Entom.* 37.
Opler, P. A., and G. O. Krizek. 1984. *Butterflies east of the Great Plains.* Johns Hopkins Univ. Press, Baltimore.
Pyle, R. M. 1981. *The Audubon Society guide to North American butterflies.* Alfred A. Knopf, New York.
Schweitzer, D. F. 1986. The Nature Conservancy global rank form for *Erynnis persius persius.*
Scott, J. A. 1986. *The butterflies of North America: A natural history and field guide.* Stanford Univ. Press, Stanford, CA.

Three-staff Underwing

· *Catocala amestris* Strecker
· **State Endangered**

Map 59. Confirmed individuals. ▲ = before 1980 and
● = 1980 to 1989 (Michigan DNR and Michigan State
University).

Status Rationale: This moth is known from only one site in Michigan and that site is subject to heavy collection pressure. The larvae feed exclusively on leadplant, *Amorpha canescens*, a special concern plant that is usually found only in small colonies that may not be suitable to sustain the underwing.

Field Identification: The three-staff underwing (Lepidoptera: Noctuidae) is a medium-sized (1.6–1.8 in. [40–45 mm] wingspan) (Covell 1984), nocturnally active moth. It is similar to the other *Amorpha*-feeding *Catocala*, but has stronger lines and shadings on the light gray-brown forewings, including double lines perpendicular to the costal (front) margin of the forewings (Sargent 1976). The hindwings are yellow-orange with two black bands, the outer of which is broken near the anal angle (nearest the body) (Sargent 1976, plate 7 no. 4; Covell 1984, plate 32 no. 12). The larvae are bluish white with an orange stigmatal band and seven thin, lateral black stripes on each side; the head capsule is bluish with several black lines in front and spotted behind (Forbes 1954). Larvae may be found on the food plant, *Amorpha canescens*, in late May and June, where they may seem to be poorly camouflaged. Adults are taken at blacklights in late July and early August.

Total Range: Florida to North Carolina, westward to Michigan, Wisconsin, South Dakota, and Texas. It is reported to be locally common except in the northern part

368

of its range by Covell (1984), though Sargent (1976) reports it as common in Texas and rare and sporadic elsewhere.

State History and Distribution: This species is known from only a single location in Barry County and has not been found at other locations of its larval food plant.

Habitat: This species is associated with dry-mesic sand prairies and silt-loam prairies containing large populations of the food plant. In Michigan, these communities are typically restricted to railroad rights-of-way remnants that have received numerous disturbances in the past and today are subject to herbicides and mechanical disturbances. The single extant colony is in a power line right-of-way.

Ecology and Life History: Underwing moths produce only one generation each year and overwinter in the egg stage (Sargent 1976). The larvae of the three-staff underwing hatch in late spring and may be found feeding on their sole food plant, leadplant, from late May through June. In other states, larvae have been observed feeding on locust (*Robinia*), as well as *Amorpha* (Sargent 1976), but leadplant is the only known host plant in Michigan. Leadplant, a species of special concern in Michigan, is restricted to small populations usually found in degraded prairies and rights-of-way in the southern part of the state. Pupation probably takes place in the litter below the leadplant, with adults emerging three to four weeks later. Standard techniques for attracting adult underwings are "sugaring" (sugar syrup bait painted on trees) and light trapping.

Conservation/Management: Protection of the only existing colony in Michigan is the most immediate action needed. Collectors must be made aware of the sensitivity of this species and the effects of collecting on a small, localized population. Habitat management is a necessity at the extant site, where autumn olive has become a problem.

LITERATURE CITED

Covell, C. V., Jr. 1984. *A field guide to the moths of eastern North America.* Houghton Mifflin Co., Boston.
Forbes, W. T. M. 1954. Lepidoptera of New York and neighboring states. Part 3. Noctuidae. *Cornell Univ. Agric. Exp. Sta. Mem. 329.*
Sargent, T. D. 1976. *Legion of the night: The underwing moths.* Univ. Mass. Press, Amherst, MA.

Leadplant Moth

- *Schinia lucens* (Morrison)
- **State Endangered**

Map 60. Confirmed individuals. ▲ = before 1980 and ● = 1980 to 1989 (Michigan DNR and Michigan State University).

Status Rationale: The leadplant moth is known from only one small population that is under heavy collecting pressure. The larval food plant, leadplant (*Amorpha canescens*), is rare and very localized in southwestern Michigan. Most of the known leadplant populations have been surveyed for the leadplant moth without success.

Field Identification: The leadplant moth (Lepidoptera: Noctuidae) is a small (1.0–1.1 in. [25–28 mm] wingspan) (Covell 1984) moth that is readily identifiable. The dorsal surface of the forewings is a mottled dark and light purple with a whitish cast and dark streaks at the outer margin; the dorsal surface of the hindwing is yellowish with dark brown blotches (Covell 1984; Forbes 1954) (refer to plate 29 no. 6 in Covell 1984 or plate 27 no. 47 in Holland 1968). The ventral surfaces are a mosaic of white, purple, and yellowish brown, with a few scattered black spots. The larva has a buff head and seven pairs of lateral stripes (Forbes 1954). During the day, camouflaged adults rest on the flowers of leadplant. Flight dates are late June through mid- to late July, with peak numbers the first two weeks of July.

Total Range: The range of the leadplant moth is from South Dakota to Illinois and Texas, then east to North Carolina and Florida (Covell 1984). Covell (1984) reports that it is uncommon to rare throughout its range.

370

State History and Distribution: Historically, this species was known from only St. Joseph and Newaygo counties, with only the St. Joseph County population known to still be extant. The larval food plant has not been found growing in the wild in Newaygo county and only a single specimen of the moth has been taken there.

Habitat: The leadplant moth is restricted to populations of leadplant, the larval food plant. Leadplant, which is typically found in sand prairies and silt-loam prairies, today is found primarily on roadside and railroad right-of-way remnants of prairie habitat. These sites are usually extremely degraded and many are threatened with mechanical disturbances and pesticides. The only known Michigan population of the moth occurs in a small area of dense leadplant between a road and a railroad track.

Ecology and Life History: Very little is known of the life history of this striking moth in Michigan. The adults spend most of their time resting, well camouflaged, on the flowers of the larval food plant, as do most *Schinia*. The larvae feed on the flowers and seeds of leadplant; as they approach maturity they may forage on the outside of the seedhead. The species overwinters as a pupa. The St. Joseph County site has apparently burned several times, which indicates that the moth is either fire tolerant during some stage of its life cycle or that a portion of the population escaped the fires.

Conservation/Management: Protection of both the habitat and the moths themselves at the only known location for this species is critical to its survival in Michigan. The habitat should be managed to enhance the leadplant and moth populations and to protect the site from physical and chemical disturbances. Prescribed burns, if necessary, should be done in subunits and the response of the moth to fire should be determined. Protection from collectors should be strictly enforced.

LITERATURE CITED

Covell, C. V., Jr. 1984. *A field guide to the moths of eastern North America*. Houghton Mifflin Co., Boston.

Forbes, W. T. M. 1954. Lepidoptera of New York and neighboring states. Part 3. Noctuidae. *Cornell Univ. Agric. Exp. Sta. Mem.* 329.

Holland, W. J. 1968. *The moth book: A guide to the moths of North America*. Dover, New York.

Phlox Moth

· *Schinia indiana* (Smith)
· **State Endangered**

Map 61. Confirmed individuals. No historical records; ▲ = before 1980 and ● = 1980 to 1989 (Michigan DNR and Michigan State University).

Status Rationale: Known in Michigan from two sites within a few miles of each other in Newaygo and Montcalm counties, these populations are especially vulnerable to collecting.

Field Identification: The phlox moth (Lepidoptera: Noctuidae) is a small (1.3 in. [33 mm] wingspan) (Hardwick 1958) moth that is readily identified. The forewing is deep, purplish red above and predominantly chocolate-brown beneath with a wine-red spot at the apex. The hind wing is dark chocolate-brown above; beneath it is wine-red on the outer and anterior half and chocolate-brown on the posterior and inner half. The fringe on both wings is whitish gray (Hardwick 1958). Early larvae have a dark head capsule and cream to pale greenish yellow body. Older instars have an orange-fawn head capsule and grayish green to bright green body suffused with reddish brown and marked with whitish gray lateral stripes and a reddish brown mid-dorsal band highlighted by a greenish gray median line (Hardwick 1958). The light period extends about three weeks from late May to early June.

Total Range: Originally described from Hessville (now Hammond), Indiana (Kwiat 1908), the phlox moth is now known from verified specimens from Minnesota, Wisconsin, Illinois, Indiana, Michigan, and Arkansas. Reports of records from Nebraska, North Carolina, and Texas have not been verified by Hardwick (Balogh 1987).

State History and Distribution: First discovered in Michigan in 1985 in Montcalm County, the phlox moth has since been found a short distance away in Newaygo County. Surveys of phlox populations in several other counties have been unsuccessful. Herbarium records and field observations indicate that *Phlox pilosa* occurs in many counties of the southern Lower Peninsula, from Berrien to Oakland and northwesterly to Montcalm and Newaygo.

Habitat: The phlox moth occurs in prairies and pine-oak barrens on sandy soils. In Wisconsin it occurs in pine barrens and is associated with the karner blue (*Lycaeides melissa samuelis*). The Michigan sites are in an area that formerly was pine-oak barrens with associated prairie openings, but now has succeeded to closed oak woods in which most of the prairielike openings are caused by disturbances such as utility rights-of-way, roads, trails, off-road vehicle activities, and home sites. The karner blue is found at both Michigan phlox moth sites, though the phlox moth has not been found at any other karner blue sites. In Minnesota, it occurs on virgin tallgrass prairie along with the dacota skipper, *Hesperia dacotae* (Balogh 1987).

Ecology and Life History: Adult phlox moths are well camouflaged on the flowers of *Phlox pilosa*, where they spend much of their time. Their wings are slightly darker than fresh blossoms, more closely resembling a partially dried corolla. Hardwick (1958) has established the life history for the phlox moth based on observations of larval activity from laboratory-reared individuals raised on *Phlox divaricata*. The phlox moth has a single generation per year. Eggs are laid on the inner surfaces of the sepals, and the newly hatched larvae feed on buds, if flowers are closed, or upon the seed capsules if the flowers have already dropped off. The larvae seal their entrance holes with silk and ultimately are released from the capsules when they dehisce. Occupied capsules become discolored and hard, which distinguishes them from healthy capsules. Older larvae feed from outside the seed capsule until it is empty, then cut the stem to drop the boll. This species is vulnerable to fires in May, June, and July, the presumed duration of the adult and larval stages; Kwiat (1908) reported observed larvae feeding on seed pods on July 4 of the year in which adults were observed on May 30. The rest of the year is spent underground as pupae (Schweitzer 1985).

Conservation/Management: Management of the barrens and prairie landscape complex for high-quality prairie openings would benefit this species as well as several others associated with it. The management unit should be large enough to accommodate multiple patches of each habitat type and designed so that species can migrate from one patch to another as the habitat at a given site becomes unsuitable. Appropriately timed, prescribed burns could be an effective management tool for phlox moth habitat, but the fire tolerance of associate species must be considered (refer to *Hesperia ottoe* for more details). Protection of existing populations from collection and surveys to locate additional populations are necessary. Research on the ecology of the species is also needed, particularly on its response to habitat management, its dispersal capabilities, and its population dynamics.

LITERATURE CITED

Balogh, G. J. 1987. New localities for *Schinia indiana* (Smith) (Noctuidae). *Ohio Lepid.* 9 (2): 15–16.

Hardwick, D. F. 1958. Taxonomy, life history, and habits of the elliptoid–eyed species of *Schinia* (Lepidoptera: Noctuidae), with notes on the Heliothidinae. *Can. Entom., Suppl.* 6.

Kwiat, A. 1908. One day's collecting, with a description of a new noctuid. *Entom. News* 19:420–24.

Schweitzer, D. F. 1985. The Nature Conservancy global rank form for *Schinia indiana*.

Silphium Borer Moth

· *Papaipema silphii* Bird
· **State Threatened**

Map 62. Confirmed individuals. ▲ = before 1980 and ● = 1980 to 1989 (Michigan DNR and Michigan State University).

Status Rationale: Known historically from nine populations, an extensive survey found only seven extant populations. Several of these small, isolated populations are threatened by collecting pressure and roadside maintenance activities.

Field Identification: The silphium borer moth (Lepidoptera: Noctuidae) is the largest (1.6–2.0 in. [40–50 mm] wingspan) (Bird 1915) species of *Papaipema* in Michigan. It could only be confused with the occasional large specimen of the sunflower borer, *Papaipema necopina* (refer to Holland 1968, plate XXVI, no. 12). The silphium borer moth is brownish black with a dusting of white scales on the dorsal forewings. When fresh, this species has a distinctive purplish cast and a large tuft of hairlike scales on the thorax. Larvae are pinkish in color with a large, brown head and may reach a length of 50 mm or more at maturity (Bird 1915). They bore in the root of their food plant, prairie dock (*Silphium terebinthenaceum*) or other *Silphium* species. Signs of feeding are a few brown or yellow leaves, a wilted flower stalk, and large amounts of brown frass around the base of the plant (Hessel 1954). Larvae are most easily located between mid-July and mid-August. Adult flight dates are mid-September through the third week of October.

Total Range: This species is restricted almost exclusively to the northeastern fringe of the tallgrass prairie region of North America.

375

State History and Distribution: Known historically from nine sites in six counties of southern Michigan, a recent survey for the silphium borer moth found it to be extant at only seven locations. The populations are very localized, and two have been lost in the last eight years, one apparently to fire and the other to bulldozing and herbicides. Most remaining populations are small, occur in habitat that requires management, and are threatened by roadside and railroad right-of-way maintenance activities.

Habitat: In Michigan, the silphium borer moth occurs on a variety of prairie habitats, including mesic prairie, prairie fen, and lake plain mesic prairie. In many cases, only a remnant of the former habitat remains and frequently it is along a roadside or railroad where past maintenance activities have kept the habitat open. Formerly, controlled burns were frequently used to maintain railroad rights-of-way, thus enhancing the remnant prairie habitat for the silphium borer moth and other prairie species. Now, however, common maintenance practices include spraying with herbicides and bulldozing, both of which can destroy host plant populations.

Ecology and Life History: The silphium borer moth is restricted to large colonies of the larval food plant, prairie dock (*Silphium terebinthenaceum*) or other *Silphium* species. Eggs are laid on or near the food plant in the fall and hatch in late spring. By early July, larvae have moved to their final feeding place by burrowing into the stem and moving down into the rootstock of the host plant. They create extensive tunnels while feeding, causing the plant to wilt slightly or to lose a few leaves. The final instar pupates in the soil under the root (Bird 1915) or near the plant. Adults are quite sedentary, though they will come to a blacklight.

Conservation/Management: Protection of known populations is essential for the persistence of this species in Michigan. Landowners and managers should be contacted at all sites and advised of protection and management concerns. Habitat management for prairies typically includes brush removal and prescribed burns. The eggs and young larvae of this species and many other *Papaipema* are sensitive to fires. Only the later instar larvae and pupae are protected from all but the hottest fires because they are underground. Prescribed burns should be conducted only in late summer, after the larvae are within a rootstock and before the adults emerge in early fall. Prudent management would include dividing the site into subunits and burning only part of the site each year. Adults are quite sedentary and would not be expected to quickly recolonize a site from which they had been extirpated (Hessel 1954). Additional surveys and monitoring are needed. Of particular importance is information about the minimum size of a *Silphium* population necessary to support the moth indefinitely and the effects of management on both the moth and host plant populations.

LITERATURE CITED

Bird, H. 1915. New species and histories in *Papaipema* Sm. (Lepidoptera). *Can. Entom.* 47:109–14.

Hessel, S. A. 1954. A guide to collecting the plant-boring larvae of the genus *Papaipema* (Noctuidae). *Lepid. News* 8:57–63.

Holland, W. J. 1968. *The moth book: A guide to the moths of North America.* Dover, New York.

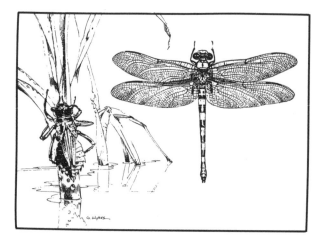

Greyback

- *Tachopteryx thoreyi* (Hagen)
- **State Threatened**

Map 63. Confirmed individuals. ▲ = before 1980 and ● = 1980 to 1989 (Michigan DNR).

Status Rationale: This species was recently rediscovered in the state, having last been reported 70 years ago. Recent surveys of several dozen wetland complexes in southern Michigan found only a single location, indicating this species is quite rare.

Field Identification: The greyback (Odonata: Anisoptera: Petaluridae), is a large (total length = 2.8–3.2 in. [71–80 mm]; hindwing length = 1.9–2.1 in. [48–53 mm]), gray, yellow, and black dragonfly (Needham and Westfall 1955; Walker 1958). The face is pale yellow with black bands across the front. The thorax is olivaceous to grey with two black, oblique stripes; the abdomen is black with dull orange markings forming saddle marks on segments 1 and 2, and interrupted half-rings on segments 3 to 7 (Needham and Westfall 1955; Walker 1958). The wings are clear with many veins. Mature nymphs can attain a length of at least 1.5 in. (38 mm) (Williamson 1901). They are characterized by thick, seven-segmented antennae, a stout spine at the base of the movable hook on the *labium* (a lower lip), a closed labial cleft, and lateral spines and dorsal tubercles with hairlike setae on abdominal segments 2 through 9 (Byers 1930, cited in Dunkle 1981; Needham and Westfall 1955). Adult flight dates from the two Michigan records are mid-June to late July. Records from Pennsylvania and Ohio extend that period to early June through early August (Hine 1925; Williamson 1901).

378

Total Range: Known only from the eastern United States, the greyback occurs from Massachusetts and New Hampshire west to Michigan, Indiana, and Missouri and south to Texas and Florida (Dunkle 1981; Needham and Westfall 1955). In Virginia it is locally distributed (Carle 1979).

State History and Distribution: The two state records are from Berrien (1919) and Cass (1989) counties. One literature report of a specimen from "near Detroit" (Walker 1958) is unverified, but not unlikely, especially if it were an older specimen that was taken while there was still suitable habitat in Wayne County.

Habitat: Adult males utilize low vegetation around seeps as perches when they await females. Nymphs inhabit the headwaters of wooded seeps and can be found among decaying leaves along the uphill edges (Dunkle 1981). Nymphs require vertical surfaces near seeps for transformation. Laboratory experiments (Dunkle 1981) determined that nymphs prefer the shelter of fallen leaves to bare soil, mud to open water, and shade to brightly lit areas.

Ecology and Life History: The only detailed study of the life history of this species was conducted in the vicinity of Gainesville, Florida, by Dunkle (1981). Adults typically flew in a smooth pattern for short distances before perching and were very unwary, making them easy to capture. Adults were on the wing for about two months, though the numbers of newly emerged adults declined rapidly two weeks into the flight season. Maturation took about two weeks from the time of emergence, and sexual maturity apparently took another week, based upon the behavior of the adults.

Males searched for females on the sunlit sides of tree trunks by flying up the trunk. They also waited for females near seeps, chasing away all other males that came within their range of view. Mated pairs tended to fly into the forest, where two females were observed ovipositing; one was on a low, nearly dry vertical surface in a seep, and the other on the ground at the edge of a seep. Away from the seeps, adults frequently basked on sunny tree trunks at an average height of 5.5 ft. (1.7 m). Adults preyed on large insects, primarily Lepidoptera, but also odonates, beetles, and other small insects. Nymphs emerge from their seep habitat when they transform into adults. Dunkle (1981) found transformation to occur on vertical stems ranging in size from herb to tree trunks, at an average height of 2.0 ft. (0.6 m), and not more than 3.3 ft. (1.0 m) away from a seep.

Conservation/Management: The recently discovered location for the greyback is contiguous with state land that should be carefully surveyed for nymphs and adults. Additional surveys are needed to identify new populations, including the nymphal habitat. Forest cover and hydrological patterns should be maintained at intact sites and restored at disturbed locations. Research into the ecological requirements (habitat type, size, and quality; food resources) and population dynamics of the greyback in Michigan is needed.

LITERATURE CITED

Carle, F. L. 1979. Environmental monitoring potential of the Odonata, with a list of rare and endangered Anisoptera of Virginia, United States. *Odonatologica* 8 (4): 319–23.

Dunkle, S. W. 1981. The ecology and behavior of *Tachopteryx thoreyi* (Hagen) (Anisoptera: Petaluridae). *Odonatologica* 10 (3): 189–99.

Hine, J. S. 1925. The dragonfly, *Tachopteryx thoreyi*, recorded for Ohio, with notes on its near relatives. *Ohio J. Sci.* 25:190–92.

Needham, J. G., and M. J. Westfall, Jr. 1955. *A manual of the dragonflies of North America,* (Anisoptera). Univ. Calif. Press, Berkeley.

Walker, E. M. 1958. *The Odonata of Canada and Alaska.* Vol. 2. Univ. Toronto Press, Toronto.

Williamson, E. B. 1901. On the manner of oviposition and on the nymph of *Tachopteryx thoreyi* (Order Odonata). *Entom. News* 12 (1): 1–3.

Lake Huron Locust

G. Wykes

Map 64. Confirmed individuals. ▲ = before 1980 and
● = 1980 to 1989 (Michigan DNR).

· *Trimerotropis huroniana* Walker
· **State Threatened**

Status Rationale: This species is a Great Lakes endemic that occurs only on high-quality, sparsely vegetated, coastal sand dunes. Most of the known occurrences worldwide occur in Michigan.

Field Identification: The Lake Huron locust (Orthoptera: Acrididae) is a small (length to end of folded forewings: males 1.00–1.24 in. [24–30 mm], females 1.1–1.6 in. [29–40 mm]) (Otte 1984), grasshopperlike insect. The Lake Huron locust is usually silvery to ash-gray, with darker brown and white markings. Brick red, burnt orange, and ocher color morphs occur occasionally, especially among females. The tegmina (toughened forewings) of the adults have darker bands that may be weakly or strongly expressed. The hindwings are light yellow near the body with a smoky patch near the tip. Sexes can be easily distinguished by the males' stronger mottling, their noisy (crepitating) flight, and, as in other Orthoptera, their significantly smaller size. The Lake Huron locust is one of four species

381

in the Great Lakes region with the pronotum (the saddlelike structure behind the head) cut across by two well-defined grooves called sulci; the other three species occur farther south than the Lake Huron locust.

The similar seaside locust, *T. maritima*, occurs in southern Michigan dunes and can be distinguished from the Lake Huron locust by the two narrow, blackish bands on the inner surface of the hind femora near the distal end. The Lake Huron locust has a broad band covering half of the inner surface of the hind femora near the body and a narrow band near the distal end. Other grasshoppers that occur with the Lake Huron locust have one or no sulcus cutting across the pronotum. Nymphs can be found before mid-July. Adults are present from early to mid-July through October until the time of frequent heavy frosts and snow.

Total Range: It is restricted to the Great Lakes sand dunes in northeastern Wisconsin (Ballard 1989), eastern Upper Peninsula and northern Lower Peninsula of Michigan, and on the central Lake Huron shoreline of Ontario (Otte 1984).

State History and Distribution: The Lake Huron locust is historically from the shorelines of Lakes Huron and Michigan from Huron to Manistee counties, including the offshore islands, and from Chippewa to Schoolcraft counties and on the Lake Superior shoreline from Chippewa to Alger counties. Otte (1984) has documented a southward range expansion on the eastern shore of Lake Michigan of about 50 miles to lakeshore areas between Muskegon and Pentwater. On the western shore of Lake Huron, the southern edge of the range of the Lake Huron locust has contracted to Iosco County. These range changes have been reflected in complementary changes in the range of *Trimerotropis maritima*.

Habitat: In Michigan, the Lake Huron locust is restricted to sparsely vegetated, high-quality coastal sand dunes. A similar habitat affinity has been reported from Wisconsin (Ballard 1989). In these areas, it occurs in high numbers and is always the dominant species. Where the open dunes grade into heavily vegetated or disturbed areas, their numbers quickly decline (Ballard 1989). Unfortunately, significant parts of the high-quality dune habitat have been degraded or destroyed by shoreline home and recreational development in Michigan and elsewhere.

Ecology and Life History: The life history of the Lake Huron locust has not been published. Its communication system and courtship behavior are similar to that of the pallid-winged locust, *Trimerotropis pallidipennis* (Otte 1970). Egg masses for the single generation per year are laid in the soft soil where they overwinter. Nymphs hatch in late spring and mature by mid-July. Adults may be found in large numbers through the fall, most likely succumbing to the first hard frosts (Ballard 1989).

Adults communicate through visual and auditory signals (Otte 1970). Only males crepitate in flight by flashing and snapping their wings, making a cracking noise with each snap. Crepitation occurs during a hovering courtship flight in which the males snap their wings two or three times while hovering; this display typically occurs on sunny days when temperatures reach 80°F. Crepitation also occurs during flight elicited by a disturbance. On the ground, courting males stridulate by rubbing the femora against the forewings, producing a trill in bursts of two to three pulses (Otte 1970). Females are cryptically colored against the light sand of the back dunes, whereas the males are virtually invisible on the gravel-dominated upper beaches of the foredunes.

The Lake Huron locust is strictly ground dwelling, essentially never climbing on foliage or supports (Ballard 1989). Its mandible structure restricts it to feeding on grass species, but which dune grasses it feeds upon has not yet been determined. In a mark-recapture study at a Cheboygan County site in 1989, Ballard (1989) estimated more than 6,000 individuals along a 1.25 mi. (2 km) shoreline of open dune habitat ranging from 20–70 ft. (6.10–21.5 m) wide. Individuals became active between 9:30 and 10:00 A.M., after the sun has risen far enough to warm the foredune shoreline. Earlier in the morning, some individuals, mostly females, were flushed from patches of sedges and rushes growing in saturated sand next to the water's edge. On sunny, windless days, the locusts were most common on sparsely vegetated sands, where they were evenly distributed with territories of several feet in diameter. In windy, overcast weather, individuals were densely distributed within the heavy dune grass cover, apparently seeking shelter.

Conservation/Management: Protection of its dune habitat is the most significant action that could be taken for the conservation of this species in Michigan. The Lake Huron locust apparently has a very limited tolerance of degraded habitat, so degradation of a site results in the loss of a local population, not just a reduction in population numbers. The extent of the dunes protected at a site should be large enough to allow natural processes to locally change the nature of the dunes through blowouts (creating more habitat) or stabilization by plants (reducing habitat). Additional surveys should be conducted to verify the current ranges of the Lake Huron locust and the seaside locust. Additional information on the ecology and life history of the Lake Huron locust is also needed to provide a stronger basis for a management plan.

LITERATURE CITED

Ballard, H. E., Jr. 1989. *Trimerotropis huroniana* (Orthoptera: Acrididae), a new record for Wisconsin. *Great Lakes Entom.* 22 (1): 45–46.

Otte, D. 1970. A comparative study of communicative behavior in grasshoppers. *Univ. Mich. Mus. Zool. Misc. Publ.* No. 141:1–168.

Otte, D. 1984. *The North American grasshoppers.* Vol. 2. Acrididae: *Oedipodinae.* Harvard Univ. Press, Cambridge, MA.

Great Plains Spittlebug

· *Lepyronia gibbosa* **Ball**
· **State Threatened**

Map 65. Confirmed individuals. ▲ = before 1980 and ● = 1980 to 1989 (Michigan DNR).

Status Rationale: Isolated populations of this species are restricted to narrow, moist zones of undisturbed sand prairie depressions. It is currently known from only three locations, two of which are being destroyed by off-road vehicles. Surveys of degraded prairie sites have not found any populations of this species and it has not been found in other community types.

Field Identification: The Great Plains spittlebug (Homoptera: Cercopidae) is the largest member of the Cercopidae in Michigan, ranging from 3–5 in. (8–12 cm) in length (Hanna and Moore 1966). It is a light, gray-brown with darker markings on the elytra (forewings) and the beak does not extend as far as the bases of the hind legs (Hamilton 1982). It is easily distinguished from other Michigan Cercopidae by its large body size and its very large head with an enlarged "sucking pump" or frons, which operates the mouthparts.

Total Range: Found in the Great Plains of the United States and Canada and in the northeastern United States in sandy and gravelly areas (Ball 1919).

State History and Distribution: Historically, known only from Newaygo, Lake, and Mecosta counties (Hanna 1970). Currently it is known from only three locations in Newaygo County, where it can be locally abundant.

Habitat: The Great Plains spittlebug is restricted to a narrow, moist zone within sand prairie depressions in sandy glacial outwash. This zone is dominated by *Andropogon scoparius*, *Carex* spp., *Sorghastrum nutans*, *Spartina pectinata*, and *Veronicastrum virginicum*.

Ecology and Life History: Little is known about the life history of most spittlebugs, except a few species of economic importance. Adult Great Plains spittlebugs have been found in June and July in Newaygo county and likely persist through the fall, when they would be expected to lay their eggs (Hamilton 1982). The nymphs of this species are unknown and may be subterranean, therefore not easily located by their protective mass of bubbly "spittle" that gives the group its name. This species is not expected to be restricted to a single food plant either in the adult or nymph stage, though it is clearly restricted to prairie habitat across its range. Doering (1942) reports that it feeds on prairie grasses.

Conservation/Management: Protection of known populations is the first priority for maintaining this species in Michigan. Destruction of its habitat by off-road vehicles should be halted immediately. Additional surveys should be conducted in Newaygo, Lake, Mecosta, Allegan, and other counties with dry sand prairie communities and strong protection given to newly identified populations. Until more is known of this species' life history, it should be considered sensitive to fire during all life stages. Management of the surrounding prairie and oak communities with prescribed burns should exclude known population sites. Each burn unit should remain unburned for a minimum of four or five years to allow the species to adequately recolonize burned areas.

LITERATURE CITED

Ball, E. D. 1919. Notes on the Cercopidae with descriptions of some new species. *Proc. Iowa Acad. Sci.* 26:143–50.

Doering, K. C. 1942. Host plant records of Cercopidae in North America, north of Mexico (Homoptera). *J. Kans. Entom. Soc.* 15 (2–3): 65–92.

Hamilton, K. G. A. 1982. The Spittlebugs of Canada. Homoptera: Cercopidae. The insects and arachnids of Canada, Part 10. *Biosyst. Res. Inst.*, Ottawa, Can.

Hanna, M. 1970. An annotated list of the spittlebugs of Michigan (Homoptera: Cercopidae). *Mich. Entom.* 3 (1): 2–16.

Hanna, M., and T. E. Moore. 1966. The spittlebugs of Michigan (Homoptera: Cercopidae). *Pap. Mich. Acad. Sci., Arts, Ltrs.* 51:39–73.

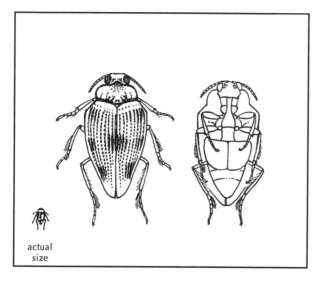

actual
size

Hungerford's Crawling Water Beetle

· *Brychius hungerfordi* Spangler
· **Federally and State Endangered**

Map 66. Confirmed individuals. ▲ = before 1980 and
● = 1980 to 1989 (Michigan DNR).

Status Rationale: This species is a postglacial relict, apparently endemic to the upper Great Lakes that has likely declined in abundance due to natural and human impacts on its habitat. Its rarity and the vulnerability of its shallow stream habitat to various disturbances call for strong protection.

Field Identification: Hungerford's crawling water beetle (Coleoptera: Haliplidae) is a small, 0.16 in. (approximately 4.0 mm), yellowish brown beetle with irregular dark markings and longitudinal stripes over much of the elytra (wing covers) and lighter areas along the margins. Each dark stripe is comprised of a series of fine, closely spaced and darkly pigmented punctures. Hungerford's crawling water beetle can be distinguished from other haliplids and from members of the similar aquatic family, Elmidae, by the shape of the pronotum on the front of the head, the sides of which are nearly parallel for the basal 2/3 and are widened midlaterally (Hilsenhoff and Brigham 1978). Larvae are light yellowish brown, elongate, rather stiff, and with the hooked tail (Wilsmann and Strand 1990) noted for the California *Brychius* (White 1986a).

Total Range: Hungerford's beetle is known from only three locations, two in the northern Lower Peninsula of Michigan and one in the Bruce Peninsula of Ontario, Canada. Extensive aquatic insect surveys in Wisconsin (Hilsenhoff and Brigham

386

1978) and Michigan (White 1986b), as well as Wallace's earlier work in Minnesota and southern Canada, have failed to locate additional populations. The genus *Brychius* is Holarctic in distribution with four North American species, of which Hungerford's beetle is the most eastern. The western species are typically northern in distribution, with southern range extensions occurring only in mountainous regions. These species are localized and disjunct in their distribution, a pattern that is emerging for Hungerford's beetle as well.

State History and Distribution: Hungerford's beetle is known from its type locality, the East Branch of the Maple River in Emmet County, where it was discovered by Spangler in 1952 (Spangler 1954) and from the East Branch of the Black River in Montmorency County (Wilsmann and Strand 1990). Extensive general aquatic insect surveys, as well as those focusing on this species (White 1986b), have been conducted in the vicinity of the type locality over the past several decades without finding new populations. The Maple River population is estimated at 200 to 500 individuals.

Habitat: Adult and larval Hungerford's beetle are found in cool (15–20°F), moderately swift, well-aerated streams over a sand, gravel, and cobble substrate in areas of open to partially open canopy. Water depths where they are found are typically a few inches to a few feet, though in late summer and fall some larvae have been found buried in the sand several inches above the waterline, where it is assumed they pupate (Wilsmann and Strand 1990). Larvae are frequently found in association with aquatic plants with a dense growth form, principally *Chara*, and adults have been observed crawling on the gravel substrate (Wilsmann and Strand 1990) and on the filamentous alga *Cladophora glomerata* (White 1986b). Historically, beaver dams may have been an important component of the environment of Hungerford's beetle. The impoundments would have moderated extremes in water levels and the dams would have provided riffle habitat on their downstream sides. Today, in the East Branch of the Maple River, the beetle is found in the well-aerated riffles downstream from beaver dams, wing dams, and culverts; at Scone, Ontario, in the North Saugeen River, it is found below a millrace.

Ecology and Life History: Little is known about Hungerford's beetle, though its basic biology is assumed to be similar to that of other Michigan haliplids (Hickman 1929, 1930, and 1931). Both larvae and adults are thought to be herbivorous, feeding on algae and periphyton. As do other Haliplidae, adults respire air from a bubble held beneath their elytra and the fused metacoxal (upper) plates of their hind legs; they must occasionally replenish this physical gill at the surface. The bubble gives them a glistening appearance when they are submerged. White (1986b) has suggested that their microhabitat preference among rocks in shallow, swiftly flowing water places them out of the range of midwater and benthic predators. The larva, with its stiff body and hooked tail, is adapted for crawling through vegetation where it, too, would avoid predation.

The life cycle of Hungerford's beetle is probably annual. The number of adults declines late in the summer as the number of larvae increases (Wilsmann and

Strand 1990). Whether the larval or pupal stage overwinters is not known, nor is the time of emergence, though adult numbers are high in early summer. Adults have not been taken at blacklights, nor have they been observed to fly at other times. It is possible that they are capable of flight only for a short time after emergence, if at all.

Conservation/Management: Originally, Hungerford's beetle probably inhabited the many streams that flowed off the glaciers as they retreated northward through Michigan. Today, the western *Brychius* species are often found in plunge pools below waterfalls in cool mountain streams where the conditions are similar to the periglacial streams of Michigan 10,000 years ago. Michigan has changed greatly since then and it is likely that Hungerford's beetle has become rarer due to natural changes in its habitat and in the species associated with it. Humans have also played an important role in changing its habitat through logging, stream channel modification, fisheries management, and beaver trapping.

Hungerford's beetle now remains only in isolated and disjunct remnant populations where the most important conservation tool is habitat protection. Habitat buffers both upstream and downstream, as well as the adjacent riparian corridor, are critical to the survival of Michigan's populations. Any activities that would modify the physical, chemical, or biotic environment, such as logging, channelization, rotenoning, or fish stocking or removal must be carefully reviewed and allowed to proceed only if their impact on the beetle is judged to be negligible.

Nondestructive field and laboratory life history and ecology studies would provide the information necessary to permit the successful establishment of new populations in suitable habitat if additional natural populations are not found.

LITERATURE CITED

Hickman, J. R. 1929. Life-histories of Michigan Haliplidae (Coleoptera). *Pap. Mich. Acad. Sci., Arts, Ltrs.* 11:399–424.

Hickman, J. R. 1930. Respiration of the Haliplidae (Coleoptera). *Pap. Mich. Acad. Sci., Arts, Ltrs.* 13:277–89.

Hickman, J. R. 1931. Contribution to the biology of the Haliplidae (Coleoptera). *Ann. Entom. Soc. Am.* 24:129–42.

Hilsenhoff, W. L., and W. U. Brigham. 1978. Crawling water beetles of Wisconsin (Coleoptera: Haliplidae). *Great Lakes Entom.* 11 (1): 11–22.

Spangler, P. J. 1954. A new species of water beetle from Michigan (Coleoptera: Haliplidae). *Entom. News* 65:113–17.

White, D. S. 1986a. The status of *Brychius hungerfordi* and *Stenelmis douglasensis* in Michigan. A proposal to The Nature Conservancy Michigan Field Office.

White, D. S. 1986b. The status of *Brychius hungerfordi* and *Stenelmis douglasensis* in Michigan. The Nature Conservancy Mich. Field Office, Unpubl. Rept.

Wilsmann, L. A., and R. M. Strand. 1990. A status survey of *Brychius hungerfordi* (Coleoptera: Haliplidae) in Michigan. *U.S. Fish Wildl. Serv.*, Unpubl. Rept.

American Burying Beetle

· *Nicrophorus americanus* Olivier
· **Federally and State Endangered**

Map 67. Confirmed individuals. ▲ = before 1980 (Michigan DNR). No record since 1980.

Status Rationale: Known historically from at least ten counties, this species has not been found in the state since 1961 and has declined dramatically throughout its range over the past 50 years.

Field Identification: The American burying beetle (Coleoptera: Silphidae) is the largest (1.0–1.4 in. [25–35 mm] total length) carrion beetle in North America (Anderson and Peck 1985). It is black with red-orange markings and can be easily distinguished from closely related species by the orange-red dorsal thorax. It also has orange-red markings on the elytra (wing covers), but this is a common trait among all *Nicrophorus* species. Males are distinguished from females by the size and shape of the clypeus (area above the mandibles), which is orange-red in both sexes. In males, it is large and rectangular; in females it is low and triangular (Schweitzer and Master 1986). The American burying beetle's large size and color pattern distinguish it from all other beetles. Flight dates are late May through early September in Michigan. Peak breeding season in Rhode Island is in June, with the greatest number of adults collected from June 10 to July 5 (Schweitzer and Master 1986).

Total Range: Historically known from 32 states, the District of Columbia, and three Canadian provinces, it ranged from Nova Scotia, Quebec, Ontario, Minnesota, and South Dakota south to Florida, Oklahoma, and Arkansas (Anderson and Peck 1985; Hecht 1989). It was frequently noted as common by collectors from the mid-1800s through the early 1900s, then showing a marked decline with few collection records by the 1940s (Davis 1980). A similar decline has been noted in the large European species *Nicrophorus germanicus* (Anderson 1982). The only extant population of the American burying beetle that has been precisely located and studied is on an island off Rhode Island. Several individuals have also been taken at a light trap in eastern Oklahoma over the past ten years (Hecht 1989). Single individuals have been taken at three other locations since the early 1980s, the most recent of which was found in Nebraska. Intensive trapping operations at the earlier two sites yielded regular captures of other *Nicrophorus* species but no additional *N. americanus*; the Nebraska site has not yet been thoroughly investigated (Hecht 1989 and 1990).

State History and Distribution: Known historically from ten counties from the eastern Upper Peninsula to the southern Lower Peninsula, the American burying beetle was last observed in northern Kalamazoo County in 1961. Between July, 1957, and June, 1961, specimens were taken at a blacklight that was run routinely in the area each summer.

Habitat: The American burying beetle was associated primarily with the eastern deciduous forest biome, except at the western edge of its range, where it occurred in the tallgrass prairie region and, farther west, in the shortgrass prairie region (Raithel 1990). Specific habitat information is not available for most collection records, including the type locality. Michigan habitat information is available only for the northern Kalamazoo County site. Since the beetle was captured at a light trap, the actual habitat it occupied is unknown. The site is on rolling moraine and the surrounding area contains an old field, a pond in what once may have been a fen before it was mined for marl, and a small mesic, deciduous woodlot. The general vicinity also is a matrix of woodlots, old fields, and active row crop and hay fields. The Rhode Island population is on a 23.25 sq. mi. (60.5 sq. km) island formed from glacial moraine. Originally it was forested, but it is now dominated by maritime shrub thickets, coastal moraine grasslands, and pastures (Kozol et al. 1987). The beetle requires humus and top soil suitable for burying carrion and is not expected on barren, sandy soils.

Ecology and Life History: The life history of the American burying beetle is similar to that of other *Nicrophorus* species; Kozol et al. (1987) and Schweitzer and Master (1987) provide the following information. Burying beetles require carrion for reproduction as the food resource for the developing larvae. Brood size, which varies with carrion size, has been reported to be between 8 and 23 emerging adults. During the night, both sexes are attracted to fresh vertebrate carrion between 50 and 200 gm. Intrasexual competition occurs until only one male and female, usually the largest, remain. The pair then buries the carrion, at times moving it to a more suitable location above ground first, or laterally up to a meter below ground. The carrion is then shaved, rolled into a ball, and treated with anal secretions by both sexes. Eggs are laid on the carrion and they hatch into altricial larvae requiring parental care for survival. The larvae are fed regurgitated food by both adults until they are able to feed on the carrion themselves. Typically, the female remains with the larvae until they disperse as adults, about 48–56 days after the carrion is buried. Young emerge in July and August, but probably do not reproduce until the following year. Adults feed on carrion and the live insects they capture. Adults overwinter, probably singly, in the soil.

Raithel (1990) suggests that the Rhode Island burying beetle population occurs under the unusual circumstances of an abundance of large, ground-nesting birds, especially the ring-necked pheasant (*Phasianus colchicus*) which has flourished on the island in the absence of large, vertebrate predators and with an abundance of open habitat. The pheasant's timing of reproduction, large clutches, tendency to replace lost nests, and chick size make it a suitable source of appropriately sized carrion during the beetle's peak reproductive period (June and July). He suggests that historically, fire-maintained habitats may have provided appropriate carrion biomass conditions of native birds and suggests looking for the burying beetle in large tracts of grassland or light agriculture, far from urban centers, that support high densities of galliformes other than turkey (*Meleagris gallopavo*) or ruffed grouse (*Bonasa umbellus*).

Conservation/Management: There currently are no known populations of this species in Michigan, so the most important priority must be to locate any existing populations. Surveying techniques should include bait trapping and black lighting during the peak activity periods of mid-June through August. Pitfall traps can be easily constructed from quart jars buried in the ground to their rims and baited with aged beef kidney (approx. 20 gm) placed in a 4 oz. glass jar with a screen mesh lid (Kozol et al. 1987). Putting the bait in a small jar with a screen lid prevents vertebrate scavengers from taking the bait and the beetles from developing blocked spiracles. A cover of 2 cm mesh screen placed over the trap and secured with tent stakes or large rocks is an additional deterrent to vertebrates. A raised cover of bark or shingles should be placed over the jar to deflect rain. Jars should be spaced 82 ft. (25 m) apart in a trap line, and traps should be cleared daily, as early as possible, to prevent mortality from heat and dehydration (Kozol et al. 1987). Small carrion (7 oz. [200 gm]) can also be left on the ground attached to 3 ft. (1 m) of dental floss to facilitate locating the buried carcass. The carcass may be fed upon by the adults but not buried, and it should be checked a few hours after dusk for feeding beetles (Master and Schweitzer 1986).

Until further information is available about the habitat and prey requirements of this beetle and the factors leading to its decline, newly discovered populations should be protected from habitat destruction. Although there is no known way to stop the decline of a population once it has begun (Hecht 1989), Raithel (1990) provides a basis for analyzing the situation and developing a cautious habitat and resource enhancement plan.

LITERATURE CITED

Anderson, R. S. 1982. On the decreasing abundance of *Nicrophorus americanus* Olivier (Coleoptera: Silphidae) in eastern North America. *Coleop. Bull.* 36 (2): 362–65.

Anderson, R. S., and S. B. Peck. 1985. The carrion beetles of Canada and Alaska (Coleoptera: Silphidae and Agyrtidae). The insects and arachnids of Canada, Part 13. *Biosyst. Res. Inst.*, Ottawa, Canada.

Davis, L. R., Jr. 1980. Notes on beetle distributions, with a discussion of *Nicrophorus americanus* Olivier and its abundance in collections (Coleoptera: Scarabaeidae, Lampyridae, and Silphidae). *Coleop. Bull.* 34 (2): 245–52.

Hecht, A. 1989. Endangered and threatened wildlife and plants; determination of endangered status for the American burying beetle. *Fed. Reg.* 54 (133): 29652–55.

Hecht, A. 1990. American burying beetles in Nebraska. Memorandum dated 6 March 1990.

Kozol, A. J., M. P. Scott, and J. F. A. Traniello. 1987. Distribution and natural history of the American burying beetle (*Nicrophorus americanus*). *The Nature Conservancy*, Unpubl. Rept.

Master, L., and D. Schweitzer. 1986. Update on survey for *Nicrophorus americanus*, the American burying beetle. Memorandum dated 13 August 1986.

Raithel, C. J., Sr. 1990. Some non-random thoughts about *Nicrophorus americanus*. Ms.

Schweitzer, D. F., and L. L. Master. 1986. American burying beetle, *Nicrophorus americanus*. Memorandum dated 10 April 1986.

Schweitzer, D. F., and L. L. Master. 1987. *Nicrophorus americanus* (American burying beetle): Results of a global status survey. *U.S. Fish Wildl. Serv.*, Unpubl. Rept.

Mollusks

Endangered and Threatened

Clubshell (*Pleurobema clava*)
Lake floater (*Anodonta subgibbosa*)
Salamander mussel (*Simpsoniconcha ambigua*)
Purple lilliput (*Carunculina glans*)
Catspaw (*Dysnomia sulcata*)
Northern riffleshell (*Dysnomia torulosa*)
Snuffbox (*Dysnomia triquetra*)
Wavy-rayed lamp mussel (*Lampsilis fasciola*)
Round hickorynut (*Obovaria subrotunda*)
Bean villosa (*Villosa fabalis*)
Deepwater pondsnail (*Stagnicola contracta*)
Petoskey pondsnail (*Stagnicola petoskeyensis*)
Acorn rams-horn (*Planorbella multivolvis*)
Cherrystone drop (*Hendersonia occulta*)

An Introduction to Mollusks

John B. Burch
University of Michigan

Michigan is home to a diversified mollusk fauna, exceeding 400 species. The richest freshwater mussel fauna in the world occurs in North America and has been the subject of numerous systematic reviews. In-depth studies have been few, however, particularly regarding species' habitat and biological requirements. A total of 46 freshwater mussels (or bivalves) are recorded in Michigan. The gastropod fauna is larger and is separated into two groups: land snails and freshwater snails. The status of many species is still unknown, however the following four species are extremely limited in distribution and all are endemic to Michigan.

BIBLIOGRAPHY

Baker, F. C. 1978. The fresh-water Mollusca of Wisconsin (part 2): Pelecypoda. *Bull. Wisc. Geol. Nat. Hist. Surv.* 70.

Barnes, D. W. 1823. On the genus *Unio* and *Alasmodonta;* with introductory remarks. *Am. J. Sci. Art.* 6:107–27, 258–80.

Burch, J. B., and C. M. Patterson. 1976. Key to the genera of freshwater pelecypods (Mussels and Clams) of Michigan. *Univ. Mich. Univ. Zool. Cir.* No. 4.

Burch, J. B., and J. L. Tottenham. 1980. North American freshwater snails. *Walkerana* 1:217–365.

Burch, J. B., and Van Devender. 1980. Identification of eastern North American land snails: The Prosobranchia, Opisthobranchia and Pulmonata (Actophila). *Walkerana* 1:33–80.

Call, R. E. 1900. A descriptive illustrated catalogue of the Mollusca of Indiana. *Twenty-fourth Ann. Rept. Ind. Dept. Geol. Nat. Resour.* (1899), 335–535, 1013–17.

Goodrich, C. 1932. *The Mollusca of Michigan.* Univ. Mich., Mich. Handbook Ser. No. 5., Ann Arbor.

Gould, A. A. 1847. [Description of a new species of *Physa,* among the shells collected by Dr. C. T. Jackson on the shores of Lake Superior; together with two other species of North American shells.] *Proc. Boston Soc. Nat. Hist.* 2:262–63.

Johnson, R. I. 1970. The systematic and zoogeography of the Unionidae (Mollusca: Bivalvia) of the southern Atlantic Slope region. *Harvard Univ., Mus. Comp. Zool.* 140:263–449.

Johnson, R. I. 1978. Systematics and zoogeography of *Plagiola* (=*Dysnomia*=*Epioblasma*), an almost extinct genus of freshwater mussels (Bivalvia: Unionidae) from middle North America. *Bull. Mus. Comp. Zool.* 148:239–320.

Johnson, R. I., and H. B. Baker. 1973. The types of Unionacea (Mollusca:Bivalvia) in the Academy of Natural Sciences of Philadelphia. *Proc. Acad. Nat. Sci.,* Philadelphia 125:145–86.

La Rocque, A. 1967. Pleistocene Mollusca of Ohio. *Bull. Geol. Surv. Ohio* 62:113–356.

Simpson, C. T. 1914. A descriptive catalogue of the *Naiades* of Pearly Freshwater Mussels. Bryand Walker, Detroit.

Stansbery, D. H. 1970. Eastern freshwater mollusks (part 1): The Mississippi and St. Lawrence River systems. *Malacologia* 10:9–22.

van der Schalie, H. 1938. The naiad fauna of the Huron River, in southeastern Michigan. *Univ. Mich. Mus. Zool. Misc. Publ.* No. 40.

van der Schalie, H. 1975. An ecological approach to rare and endangered species in the Great Lakes Region. *Mich. Acad. Sci.* 8:7–22.

Walker, B. 1908a. New Michigan lymnaeas. *Nautilis* 22 (1): 4–9.

Walker, B. 1908b. New Michigan lymnaeas. *Nautilis* 22 (2): 16–19.

Clubshell

- *Pleurobema clava* (Lamarck)
- **Federally and State Endangered**

This species' shell is relatively thick and small; Michigan individuals range up to 2 inches (5 cm) in length. The shell is skewed (arched anteriorly) so that the beaks have a very noticably anterior displacement. The disk often has irregularly incised grooves in its ventral half. The periostracal color is tan; there is a patch of interrupted green rays on the disk. The hinge is heavy and the beak depression is moderately deep. The nacre is a dull white. The clubshell is most similar to the Ohio pigtoe (*Pleurobema cordatum*) but can be distinguished by its more arched shell with very anteriorly placed beaks.

The clubshell occurs in the Ohio, Cumberland, and Tennessee river systems. In Michigan, it is restricted to the St. Joseph River (of the Maumee River system) in Hillsdale County. This river inhabitant prefers waters 6 to 24 inches (5 to 61 cm) deep with sand and gravel substrates.

Lake Floater

· *Anodonta subgibbosa* (Anthony)

· **State Threatened**

The shell of the lake floater is ovate in shape, very inflated and thin. Its color is light or dark olive-green. It also has a double-looped beak sculpture with a hinge that lacks teeth; this characteristic is diagnostic for this genus and accounts for the name *Anodonta* ("without teeth").

This species can be confused with the giant floater (*Anodonta grandis grandis*) and the tall floater (*A. g. corpulenta*). The lake floater's shell is relatively higher than the giant floater; the length-to-height ratio for the lake floater is less than 1.6, while in the giant floater it is more than 1.6. Lake floaters average smaller than tall floaters: adults of six or more years in age are less than $3^1/_2$ inches (9 cm) long, while tall floaters of the same age reach $5^1/_2$ inches (14 cm) long. Furthermore, the nacre of the lake floater is white or bluish white, whereas it is usually tinged with salmon in the tall floater.

Confirmed records of the lake floater are known only from White Lake, Muskegon County. However, there are unconfirmed records from Black Lake, at the mouth of the Kalamazoo River, Allegan County. Specific habitats are not recorded, but it presumably occurs in natural river impoundments more than 3 feet deep with mud or mud-sand substrates.

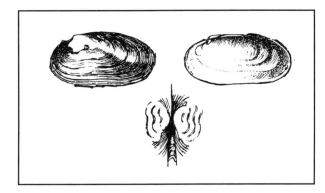

Salamander Mussel

· *Simpsoniconcha ambigua* (Say)
· **State Endangered**

The salamander mussel has an elongate shell, with the ventral margin slightly bowed dorsally. The largest individuals have nearly parallel dorsal and ventral margins; smaller ones are taller posteriorly. This small species, typically $1^3/_{16}$ inches (3 cm) or less in length, occasionally reaches $1^1/_2$ inches (4 cm). The beak sculpture is double looped. The periostracal color is a tannish olive-green. Hinge teeth are lacking, except for very small or nearly obsolete pseudocardinal teeth. The nacre is a grayish color, thicker anteriorly, but sometimes iridescent posteriorly and orange near the beak depression.

This species can be distinguished from other small, elongate Michigan mussels by its bowed-up ventral margin, grayish nacre, and very shallow beak depressions. These three features should separate it from the lilliput (*Carunculina parvus*), the purple lilliput (*Carunculina glans*), and the bean villosa (*Villosa fabalis*). The genus *Simpsoniconcha* is synonymous with the genus *Simpsonaias*.

This mussel is restricted to the Ohio River system: south to Arkansas, west to Iowa, north to Michigan, and east to Tennessee. In Michigan, specimens are recorded from Lake Erie, although it is also known from an Ontario river draining into Lake St. Clair. It inhabits rivers with mud and silt substrates and prefers to be under flat rocks, stones, and bedrock ledges. The mudpuppy (*Necturus maculus*) serves as the host during the mobile glochidal stage.

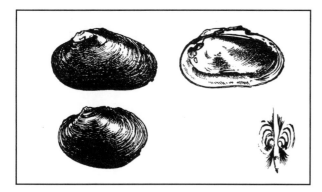

Purple Lilliput

· *Carunculina glans* (Lea)
· **State Endangered**

The purple lilliput's shell is evenly oval and moderately inflated. This is a small species; the largest individuals grow up to $1^1/_2$ inches (4 cm) but are usually smaller. The periostracal color is dark olive-green, nearly black. The beaks are moderately anteriorly placed. The beak sculpture is coarse and concentric and the beak depression is moderately deep. The hinge is moderate to relatively small and the teeth are well developed. The nacre is purple, sometimes iridescent brownish.

This mussel is separated from the lilliput (*Carunculina parvus*) by its purple or brownish (rather than white or bluish) nacre, more swollen shape, and coarser beak sculpture. The purple lilliput can be distinguished from the bean villosa (*Villosa fabalis*) by its nacre (not white), its swollen shape, less anteriorly placed beaks, and less tapering ends. The genus *Carunculina* is synonymous with the genus *Toxolasma*.

The distribution of the purple lilliput includes the Ohio River drainage and the St. Lawrence drainage in southern Michigan. It is known from Lake Erie and may occur in associated tributaries. Its historical range may have been larger, but it was formerly confused with the lilliput. Its habitat is rivers and headwater lakes.

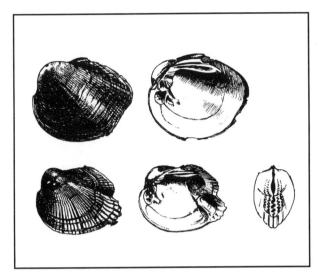

Catspaw

- *Dysnomia sulcata* (Lea)
- **State Endangered**

The catspaw is of moderate size, sometimes attaining a length of $2^1/3$ inches (6 cm). The shell is thick and has a very heavy hinge with well-developed teeth. The ground color of the periostracum is tan, and is marked with numerous, thin, closely spaced olive-green rays. The nacre is somewhat iridescent, and its color may be pink, bluish white, or white, depending on the locale. Michigan specimens in stream tributaries of western Lakes Erie and St. Clair have white nacre.

Species of this genus (synonymous with *Epioblasma*) are noticeably sexually dimorphic, with the females often having discriminating features. Catspaw females with eggs have a posterio-ventral bulge, which is sculptured with close-set radial ridgelets; this produces a serrated margin where they end at the shell edge. The shells of males lack this bulge, but instead have a groove. The overall shell shape of males is skewed, the beak is anteriorly placed.

The catspaw may be confused with the northern riffleshell (*Dysnomia torulosa*), but, in females of the catspaw, the bulge in the shell projects more laterally and is narrower. Also, in the catspaw, the sharpness of the rayed ridges makes a serrated edge to the shell, whereas, in the northern riffleshell, the shell margin at this extension is smooth or is more broadly and roundly scalloped.

This mussel occurs in the Tennessee River system in Tennessee and Alabama, the Cumberland River system in Kentucky and Tennessee, the Ohio River system in Ohio, and the Lake Erie drainages, including the Detroit River.

Northern Rifleshell

· *Dysnomia torulosa* (Rafinesque)
· Federally and State Endangered

The northern rifleshell is of moderate size, with large adults reaching or exceeding 2 inches (5 cm) in length. The shell is sturdy but not heavy; its light green to olive-green ground cover is overlaid with numerous narrow, closely spaced darker green or darker olive rays. The hinge is of moderate thickness, containing well-developed teeth. Male shells have a radiating groove running posterio-ventrally from just below the beaks. Females generally lack such a groove, but instead have a low bulge that accommodates the eggs. This bulge continues beyond the normal posterio-ventral edge of the shell as a well developed flange. The edge of this flange may be smooth or broadly and shallowly scalloped.

The northern rifleshell may be confused with the catspaw (*Dysnomia sulcata*), but the shells of males are not as skewed (i.e., the beaks are not as anteriorly placed) and the shell bulge is broader and flatter in the females. Shells also lack the rayed sculpture, which produces sharp serrations at the shell margin on the catspaw. The genus *Dysnomia* is synonymous with the genus *Epioblasma*.

Its range includes the Ohio River drainage and scattered locales in southwestern Michigan. Here, it occurs in a variety of lotic habitat types, from small headwater streams to large downstream rivers.

Snuffbox

- *Dysnomia triquetra* (Rafinesque)
- **State Endangered**

The shell of the snuffbox is generally of moderate size, although some individuals may reach $2^3/4$ inches (7 cm) in length. The shell is very inflated and has centrally placed beaks whose sculpture is double looped. The posterior ridge is angular and prominent (particularly in females). The posterior slope is truncated and sculptured with low, close-set radiating ridges, which give the posterior shell a scalloped margin. The periostracum is pale green, marked with dark green rays that generally are interrupted. The hinge is of moderate heaviness, and the hinge teeth are well developed, although the lateral teeth are short. The beak cavity depression is deep.

The snuffbox can be readily distinguished from the catspaw (*Dysnomia sulcata*) and the northern riffleshell (*Dysnomia torulosa*) by its extremely inflated shell with strong posterior truncation, the angular dorsal ridge, and the light green ground color. The genus *Dysnomia* is synonymous with the genus *Epioblasma*.

This mussel occurs in the following river systems and drainages: upper White River in Missouri, Missouri River in Kansas and Missouri, Mississippi River in Iowa and Wisconsin, Illinois River in Illinois, Tennessee and Cumberland rivers, Green River in Kentucky, and Ohio River from Indiana to Pennsylvania. Additionally, it inhabits parts of Lakes Michigan, Huron, Erie, and St. Clair as well as the Detroit River. It prefers habitats with stone, gravel, or sand bottoms in swift currents and usually is buried deeply in the substrate.

Wavy-rayed Lamp Mussel

- *Lampsilis fasciola* Rafinesque
- **State Threatened**

The wavy-rayed lamp mussel has a shell of moderate size measuring $2^7/8$ inches (7.3 cm) in length. The shell is evenly covered with color rays and the beaks are sculptured with small, double-looped ridges. The color rays are absent or restricted to the posterior slope in the pocketbook (*Lampsilis ovata*), and unlike the wavy-rayed lamp mussel has high protruding umbos. The other Michigan species in this genus, the fat mucket (*Lampsilis radiata*) differs by its elongate shell that has a length-to-height ratio for males of 1.6 or more. In the wavy-rayed lamp mussel, it is less than 1.6.

This species has a discontinuous distribution in the Great Lakes and their drainages. It occurs in the smaller rivers of southeastern Michigan, such as in the Raisin River.

Round Hickorynut

- *Obovaria subrotunda* (Rafinesque)
- **State Endangered**

The round hickorynut has a nearly round shell with centrally located beaks. This is a relatively small Michigan mussel; most individuals are 1³/₄ inches (4.5 cm) long or less. Rarely do they reach 2¹/₄ inches (6.0 cm) in length. The color of the periostracum ranges from tan to dark olive-green. Faint darker rays occur on some individuals. The posterior slope, and sometimes the anterior slope, has a lighter color. The beak sculpture is slightly double looped. The shell is moderately thick and the hinge is rather heavy. The nacre is white, with a tinge of iridescence; some individuals have a peach-colored hue, particularly in the disk.

This species can be distinguished from the hickorynut (*Obovaria olivaria*) by generally lacking color rays, in not being arched anteriorly, and by its round shape (versus the oval shape of the hickorynut). The round hickorynut differs from other Michigan mussels with a circular side view by lacking pustules on the shell surface. The Ohio pigtoe (*Pleurobema cordatum*) has a much heavier shell, with a slight medial groove and more inflated, forwardly arched beaks.

It occurs in the Ohio, Tennessee, and Cumberland river systems and in southeastern Louisiana and the Tombigbee drainage, north to Michigan and the St. Lawrence drainage. In Michigan, the round hickorynut occurs in the Grand, Huron, and Raisin rivers. It usually inhabits large rivers but occasionally uses smaller headwater streams.

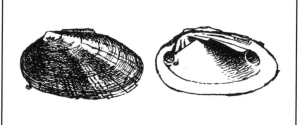

Bean Villosa

- *Villosa fabalis* (Lea)
- **State Endangered**

This freshwater mussel has an elongated shell that is relatively flattened laterally. The posterior end is tapered and the posterior ridge is high on the shell, bowed upward. This is a small species; individuals generally are less than $1^3/16$ inches (3 cm) in length but may reach $1^1/2$ inches (4 cm). The beaks are anteriorly placed and have a double-looped beak sculpture. Uneroded beaks are pointed. The periostracal color is green or dark olive-green to dark brown, often with narrow, light rays. The hinge is well developed, with strong pseudocardinal and lateral teeth. The nacre is white, often iridescent and occasionally with a pinkish cast. The beak depression is very shallow or lacking.

The bean villosa is separated from other Michigan mussels by its small size, generally white nacre, prominently anteriorly placed beaks, laterally compressed valves, tapering ends, and evenly rounded ventral margin. It is restricted to the Ohio River drainage, the Duck and Upper Tennessee rivers, and River Rouge in Michigan; it occurs in rivers and headwater lakes. Michigan records also are known for Lakes Erie and St. Clair and at the headwaters of the Clinton River. Here, it occurs in sand and fine mud substrates among roots of aquatic vegetation in approximately 4 inches (10 cm) of slowly flowing water.

Deepwater Pondsnail

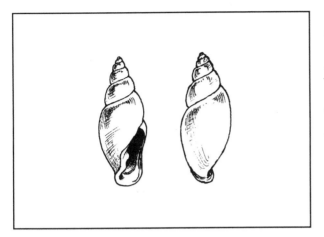

· *Stagnicola contracta* (Currier)
· State Threatened

The shell of this pondsnail is elongated, relatively solid, and white in color. Shells measure 7/8 inches (63 mm) in length. The last several whorls are shouldered, and the last whorl is noticeably elongated with parallel left and right margins. The columella is strongly twisted. This species' shell is diagnostic: its most characteristic feature is the long, cylindrical body (last) whorl.

The deepwater pondsnail is endemic to Michigan; live individuals have been found only in Higgins Lake, Roscommon County, in *Chara* at depths of approximately 33 feet (10 m).

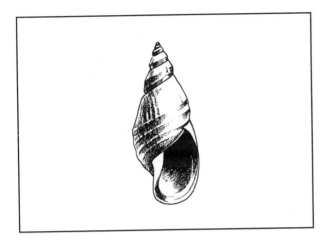

Petoskey Pondsnail

- *Stagnicola petoskeyensis* (Walker)
- **State Endangered**

The adult shell of this pondsnail grows up to one inch (2.5 cm) long, and has a translucent, whitish horn color. The shell is sculptured with fine growth and spiral lines. The spire is evenly tapered and its whorls are flattened, each with one or more spiral edges. This species has the general shape of *Stagnicola emerginata* (form *canadensis*), but it differs mainly by its flat-sided, spirally ridged spire whorls.

This Michigan endemic species is known only from a small, spring-brook flowing into Little Traverse Bay, near Petoskey.

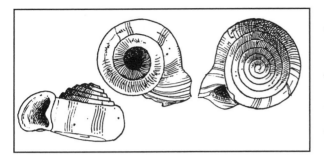

Acorn Rams-horn

· *Planorbella multivolvis* (Case)
· **State Endangered**

The acorn rams-horn's shell is dark brown and relatively large (up to 5/8 inches [15 mm]), with tightly coiled, numerous whorls, and a raised spire. All other members of the family Planorbidae in Michigan are discoidal. This endemic Michigan species is known only from Howe Lake, Marquette County. This lake is a small, oligotrophic lake with a red sandstone bottom along one side and washed-in organic debris along other shorelines. Surveys in 1988 found no individuals.

Cherrystone Drop

· *Hendersonia occulta* (Say)
· **State Threatened**

The cherrystone drop has a helicoid shell that is extremely small in diameter (1/4 inch [6 to 8 mm]) with $4^1/2$ to 5 whorls. Its color varies from cinnamon red to pale yellow. The periphery is slightly angular, sometimes weakly keeled (especially younger individuals). This land snail is distinguished by its helicoid shape, together with an operculum and a callous pad that covers the umbilical region. It occurs from Virginia west to eastern Oklahoma, north to Wisconsin and Michigan. In Michigan, it is known from wooded areas under and among fallen leaves, leaf mold, and rotting logs.